CMOS CELLULAR RECEIVER FRONT-ENDS

THE KLUWER INTERNATIONAL SERIES IN ENGINEERING AND COMPUTER SCIENCE

ANALOG CIRCUITS AND SIGNAL PROCESSING

Consulting Editor*: Mohammed Ismail. *Ohio State University

Related Titles:

WIRELESS BUILDING BLOCKS
J.Janssens, M. Steyaert
ISBN: 0-7923-7637-4

CODING APPROACHES TO FAULT TOLERANCE IN COMBINATION AND DYNAMIC SYSTEMS
C. Hadjicostis
ISBN: 0-7923-7624-2

RF CMOS POWER AMPLIFIERS: THEORY, DESIGN AND IMPLEMENTATION
Hella, Ismail
ISBN: 0-7923-7628-5

DATA CONVERTERS FOR WIRELESS STANDARDS
C. Shi, M. Ismail
ISBN: 0-7923-7623-4

STREAM PROCESSOR ARCHITECTURE
S. Rixner
ISBN: 0-7923-7545-9

LOGIC SYNTHESIS AND VERIFICATION
S. Hassoun, T. Sasao
ISBN: 0-7923-7606-4

VERILOG-2001-A GUIDE TO THE NEW FEATURES OF THE VERILOG HARDWARE DESCRIPTION LANGUAGE
S. Sutherland
ISBN: 0-7923-7568-8

IMAGE COMPRESSION FUNDAMENTALS, STANDARDS AND PRACTICE
D. Taubman, M. Marcellin
ISBN: 0-7923-7519-X

ERROR CODING FOR ENGINEERS
A. Houghton
ISBN: 0-7923-7522-X

MODELING AND SIMULATION ENVIRONMENT FOR SATELLITE AND TERRESTRIAL COMMUNICATION NETWORKS
A. Ince
ISBN: 0-7923-7547-5

MULT-FRAME MOTION-COMPENSATED PREDICTION FOR VIDEO TRANSMISSION
T. Wiegand, B. Girod

SUPER - RESOLUTION IMAGING
S. Chaudhuri
ISBN: 0-7923-7471-1

AUTOMATIC CALIBRATION OF MODULATED FREQUENCY SYNTHESIZERS

CMOS CELLULAR RECEIVER FRONT-ENDS

From Specification to Realization

by

Johan Janssens
KU Leuven, Belgium

and

Michiel Steyaert
KU Leuven, Belgium

KLUWER ACADEMIC PUBLISHERS
BOSTON / DORDRECHT / LONDON

A C.I.P. Catalogue record for this book is available from the Library of Congress.

ISBN 0-7923-7637-4

Published by Kluwer Academic Publishers,
P.O. Box 17, 3300 AA Dordrecht, The Netherlands.

Sold and distributed in North, Central and South America
by Kluwer Academic Publishers,
101 Philip Drive, Norwell, MA 02061, U.S.A.

In all other countries, sold and distributed
by Kluwer Academic Publishers,
P.O. Box 322, 3300 AH Dordrecht, The Netherlands.

Printed on acid-free paper

Printed in the Netherlands.

voor Ons Ann

Preface

During the last decade, the world of mobile communications has experienced an enormous growth. To a great extent, this growth has been made possible by the migration from the original all-analog mobile phones to handsets using digital technology. Another important enabler has undoubtedly been the rapid progress in silicon IC technology, which made it possible to squeeze ever more digital functions onto a single chip, reducing both the total terminal cost and the form factor.

Whereas today's mobile phones already feature a highly-integrated digital back-end, the architectures for the radio-frequency transceiver front-end generally still rely on external components to realize the most critical functions. Since the need for external components is tightly coupled to the architecture of the front-end, there is a strong drive towards more advanced, highly-integrated architectures that rely less on this class of external components, leading to significant cost savings. In addition, the high-frequency front-end is mostly still implemented in a relatively expensive technology instead of in a cheap CMOS process (as e.g. the digital part). By also integrating the high-frequency analog front-end in CMOS technology, the cost can be reduced further. In the end, this might lead to a mobile phone on a single CMOS IC.

The presented work deals with the design of the receive path of a highly-integrated CMOS transceiver front-end for mobile communications. It covers the whole design trajectory starting from documents describing the standard down to the systematic development of CMOS ICs which are measured to be compliant to the standard. In addition, it tackles CMOS receiver design at all abstraction levels: from architecture level, via circuit level, down to the device level, and the other way around. The DCS-1800 standard — better known as GSM-1800 — is used as the demonstration vehicle throughout this work. In parallel to the application-driven trajectory, special attention is given to the fundamentals and the fundamental limits of radio-frequency CMOS design. This material forms the basis of a systematic design methodology for high-performance low-noise amplifiers.

In this work, two highly-integrated low-IF receive paths have been described which are embedded in a single-chip CMOS DCS-1800 transceiver. These ICs demonstrate the feasibility of meeting the performance requirements of today's high-end cellular standards in a mainstream submicron CMOS technology. In addition, the limits of RF CMOS design have been explored by the realization of a high-performance low-noise amplifier (LNA) for the Global Positioning System. This LNA proves that CMOS can offer an extremely low noise figure and a large power gain, at the same power consumption as commercially available GaAs LNAs.

The outline of the work is illustrated in the figure below. After the introducing chapter, Chapter 2 introduces some general aspects of the radio-part of the DCS-1800 cellular system. Next, Chapter 3 compares different cellular receiver architectures with respect to integratability,

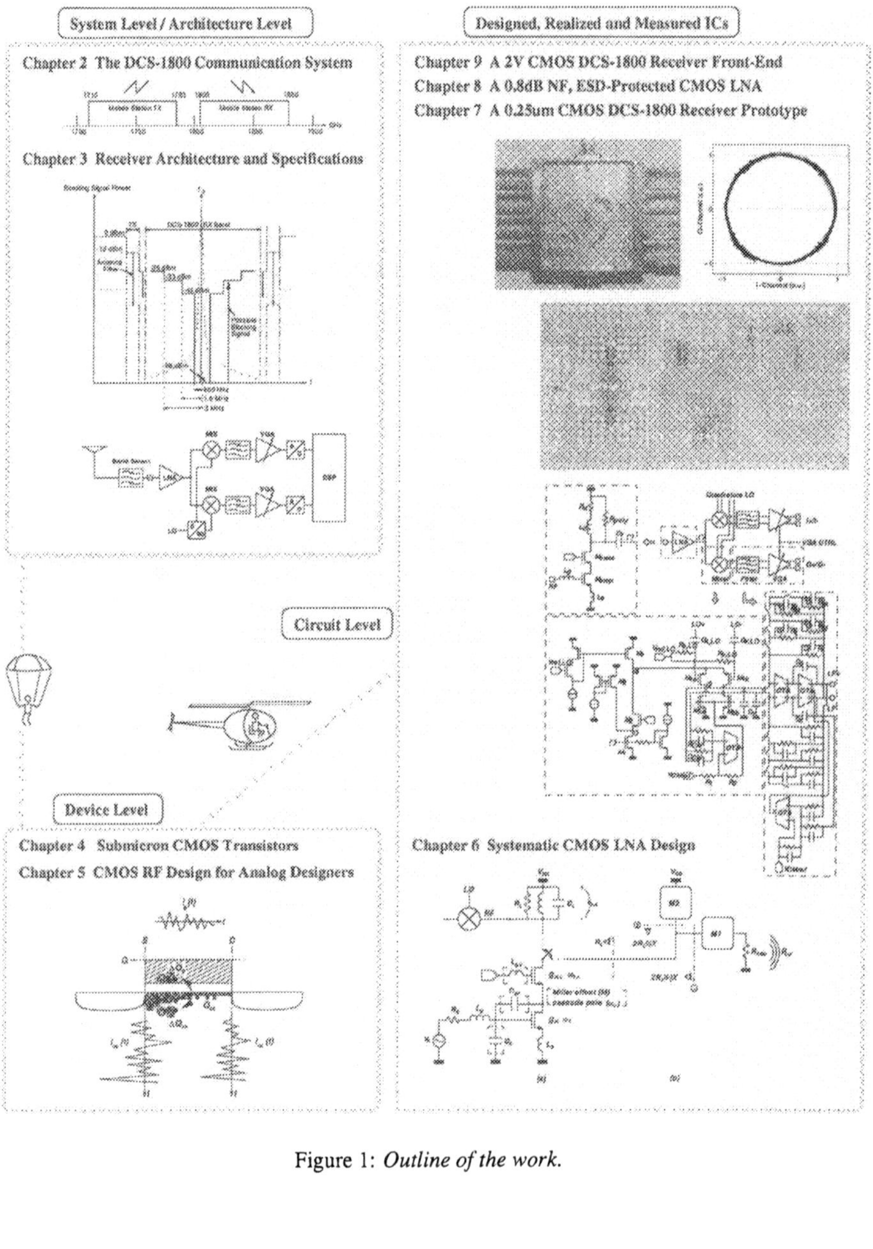

Figure 1: *Outline of the work.*

achievable performance and required building block specifications. After choosing the architecture which best suits our needs, the requirements of the DCS-1800 standard are translated into a set of specifications on the receiver as a whole. Eventually, the specifications are allocated to the different building blocks.

In Chapter 4 and Chapter 5, we make a steep descent from the architecture level down to the device level. In these chapters, the dynamics of elementary specifications are analyzed as a function of transistor parameters, boundary conditions and several non-idealities. In addition, these chapters cover the fundamentals, the peculiarities and the fundamental limits of RF CMOS design. In Chapter 6, the gathered knowledge is further refined and applied to the systematic design of CMOS LNAs.

Chapter 8, Chapter 7 and Chapter 9 present the design and measurement results of the most important ICs that have been realized as a consequence of this work: a high-performance CMOS LNA for GPS, and two highly-integrated CMOS receiver front-ends for DCS-1800. Special attention is given to the sizing procedures to design these topologies systematically for a set of specifications.

Acknowledgements

I am very grateful to my close colleagues Bram De Muer, Marc Borremans and Nobuyuki Itoh for the fruitful collaboration we had during the design of the different CMOS transceivers. Paul Leroux deserves special mention for the collaboration we had on CMOS low-noise amplifiers. In addition, I wish to thank Augusto Marques, Enzo Peluso and Koen Mertens for the many stimulating discussions we had on all aspects of analog design. Of course, I would also like to acknowledge Jan Crols, Jan Craninckx and Peter Kinget for bringing RF CMOS into the MICAS research group.

Finally, I would like to express my gratitude to Dr. Shimizu and Dr. Iwai from Toshiba Corporation for providing us with their CMOS technology and for their confidence in our design team.

Johan Janssens,
Heverlee, September 2001.

Contents

List of Symbols and Abbreviations

Abbreviations

A/D	Analog to Digital Converter
AGC	Automatic Gain Control
BER	Bit Error Rate
BiCMOS	Bipolar Complementary Metal Oxide Semiconductor
CAD	Computer Aided Design
CDMA	Code Division Multiple Access
CMFB	Common Mode Feedback
CMOS	Complementary Metal Oxide Semiconductor
DCS	Digital Cellular System
DIV	Frequency Divider
DSB	Double Side-Band
DSP	Digital Signal Processor
EDGE	Enhanced Data Rates for GSM Evolution
ESD	Electro-Static Discharge
ETSI	European Telecommunications Standard Institute
FDMA	Frequency Division Multiple Access
GMSK	Gaussian Minimum Shift Keying
GPRS	General Packet Radio Service
GPS	Global Positioning System
GSM	Global System for Mobile Communications
HBM	Human Body Model
HBT	Hetero-junction Bipolar Transistor
IF	Intermediate Frequency
IMRR	Image Rejection Ratio
LDD	Lightly Doped Drain
LF	Low Frequency
LNA	Low Noise Amplifier
LO	Local Oscillator
MCM	Multi-Chip Module
MSK	Minimum Shift Keying
nMOS	n-channel MOSFET
NQS	Non-Quasi Static
OTA	Operational Transconductance Amplifier

PA	Power Amplifier
PCN	Personal Communication Network
PCS	Personal Communications Service
PDC	Personal Digital Cellular
PLL	Phase Locked Loop
pMOS	p-channel MOSFET
PFD	Phase-Frequency Detector
PSD	Power Spectral Density
QAM	Quadrature Amplitude Modulation
RF	Radio Frequency
RX	Receiver
SAW	Surface Acoustic Wave
SMS	Short Message Service
SNR	Signal-to-Noise Ratio
SSB	Single Side-Band
TDMA	Time Division Multiple Access
TX	Transmitter
UMTS	Universal Mobile Communications System
VCO	Voltage-Controlled Oscillator
VGA	Variable-Gain Amplifier
VLSI	Very Large Scale Integration
WAP	Wireless Access Protocol

Symbols

$\triangleq$	Defined as
$\cong$	Approximately equals (in case of an expression)
$\approx$	Approximately equals (in case of a value)
$\propto$	Proportional to
✓	Specification is met
✗	Specification is not met
①, ②, ...	Node numbers
b_i	ith bit
BW	Bandwidth
CVG	Conversion gain of a mixer
c	Correlation coefficient between drain noise and gate noise
C_p	Parasitic parallel input capacitance
C_{gs}, C_{gd}, etc.	Device capacitances
e_n	Equivalent input noise voltage
F	Noise factor
F_E	Eigen noise factor (noise factor without the noise of the source)
F_d	Noise contribution of the classical drain noise current
$F_{g,c}$	Noise contribution of the correlated part of the gate noise current
$F_{g,u}$	Noise contribution of the uncorrelated part of the gate noise current

F_{min}	Minimum noise factor, corresponding to a noise match
F_{opt}	Optimum noise factor when constraints are imposed
f_T, ω_T	Unity current gain frequency and pulsation
GBW	Gain-bandwidth product
g_m	Transconductance of a MOS transistor
$g_{m,b}$	Bulk transconductance
$g_{m,eff}$	Effective transconductance
g_o, g_{ds}	Output conductance of a MOS transistor
G_t	Transducer power gain
G_u	Conductance associated with the uncorrelated part of i_n
$h(t)$	Impulse response of a filter
HD_2, HD_3	Second-order and third-order harmonic distortion ratio
i_c	Part of i_n that is correlated to e_n
i_u	Part of i_n that is uncorrelated to e_n
I_{DS}	Drain current
IM_2, IM_3	Second-order and third-order intermodulation ratio
$IMRR$	Image rejection ratio
i_n	Equivalent input noise current
i_{nd}	Drain noise current
i_{ng}	Non-quasi static gate noise current
i_{nb}	Non-quasi static bulk noise current
IP_2, IP_3	Second-order and third-order intercept point
K_n, K_p	Transconductance parameter of a nMOS, pMOS transistor
k	Boltzmann's constant (1.38×10^{-23} J/K)
L, W	Channel length and width of MOS transistor
L_g, L_s	Gate and source inductance
m	FM modulation index
M	Miller amplification factor
n	Factor modeling the bulk effect
NF	Noise figure
NF_{sys}	System noise figure
q	Elementary charge (1.60×10^{-19} C)
Q	Quality factor of a network
P_{av}	Available power
PCC	Power-to-current conversion
P_D	Dissipated signal power
P_{DC}	DC power consumption
P_{in}	Input power
$P_{N,LSB}$	Noise power in the lower side-band
$P_{N,USB}$	Noise power in the upper side-band
PVC	Power-to-voltage conversion
R_b	Resistance of the back-gate
R_g	Resistance of the gate fingers
$r_{g,NQS}$	Non-quasi static gate resistance
R_{in}	Input resistance

R_n	Resistance associated with the equivalent input noise voltage e_n
R_p	Equivalent parallel resistance of a passive element
R_S	Source resistance
R'_S	Source resistance, as seen by the inductive matching network
$R_{S,eq}$	Equivalent source resistance, as seen by the transistor
$R_{S,local}$	Characteristic impedance of a local reference plane
r_{vs}	Relative amount of velocity saturation
S_{21}, S_{11}	Forward gain and input reflection
$S_{\Delta f}$	Selectivity of a filter at a frequency offset Δf
$S(f,t)$	Time-varying power spectral density
T	Absolute temperature
T_b	Bit period (inverse of the bit rate)
V_{GST}	DC Gate-source over-drive voltage, i.e. $V_{GS} - V_T$
v_{DS}, v_{GS}	Drain to source and gate to source voltages
V_{DSsat}	Drain to source saturation voltage
v_{sat}	Saturation speed of an electron in a high electric field
V_{si-vs}	Transit voltage between strong inversion and velocity saturation
V_T	Threshold voltage of a MOS transistor
Y_c	Correlation admittance (ratio between i_c and e_n)
Y_{opt}	Source admittance corresponding to a noise match
Y_s	Source admittance
Z_t, Z_c, Z_s	Terminating, characteristic and source impedance
Z_{in}	Input impedance
α_{gd}, α_{db}, etc.	Ratio between the device capacitances and C_{gs}
α	Inverse of n
β	Current factor of a MOS transistor
δ	Parameter modeling the gate noise current
ϵ	Small phase difference
γ	Excess noise factor
Γ	Reflection coefficient
κ	Elmore constant of the channel
λ, Λ	Parameter modeling the channel length modulation
μ_0, μ_{eff}	Mobility and effective mobility
$\phi(t)$	Instantaneous phase
θ, Θ	Parameters modeling the mobility degradation
Υ	Reflection coefficient when the matching network is omitted
ω_0	Operating frequency (or in fact the corresponding pulsation)
ω_b	Bit pulsation
ω_c	Center frequency (or in fact the corresponding pulsation)
$\omega(t)$	Instantaneous frequency
ω_{max}, ω_{min}	Maximum and minimum instantaneous frequency

Chapter 1

Introduction

1.1 The Explosive Growth of Mobile Communications

During the last decade, the world of mobile communications has experienced an enormous growth (Fig. 1.1(a)). Whereas in 1992 only 7 million cellular handsets were sold all over the world, sales amounted to 405 million pieces in 2000. In less than a decade, more than 1.1 billion handsets have been sold in total. Along with the rising handset sales, also the number of subscribers increased at a very fast pace (Fig. 1.1(b)). In December 2000, the total number of subscribers already exceeded 700 million!

By now, in most developed countries mobile phones have become so popular that the decision not to own one is becoming a minority life-style choice [Geor01]. This is indicated in Fig. 1.1(c), which shows the market penetration of the mobile phone in different parts of the world. In Japan and Scandinavia, market penetration is so high that only the very young and very old tend to go without. In the countries of the European Union, mobile phones are widespread as well, especially among adults and adolescents. In the developing countries and the US on the other hand, handsets are comparatively rare, which implies that there is still an enormous growth potential.

To a great extent, the growth of the mobile communications market has been made possible by the migration from the original all-analog mobile phones to handsets using digital technology; Through the use of more efficient modulation schemes, digital signal processing and advanced communication protocols, the use of digital technology made it possible to accommodate a larger number of users in the limited radio band available. Now that we are in the Internet Age, the advantages of using a digital technology have never been more clear. Through the combination of digital error correction algorithms and encryption algorithms, handsets not only support speech transmission but also reliable and secure digital data communication. Today's digital handsets seamlessly connect to portable computers and digital organizers. In addition, a lot of services are available which were previously unthinkable, like SMS, WAP, mobile e-mail, Internet access, electronic payments etc.

Another important enabler has undoubtedly been the rapid progress in silicon IC technology, which made it possible to squeeze ever more digital functions onto a single chip, decreasing the cost per function — and hence the total terminal cost. Integration also enabled smaller form factors: from the bulky and expensive mobile phones which where almost literally indistinguishable

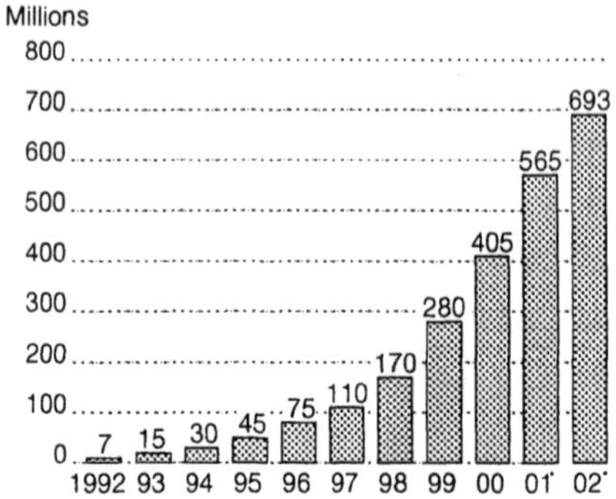

(a) Cellular handset sales in the last decade.

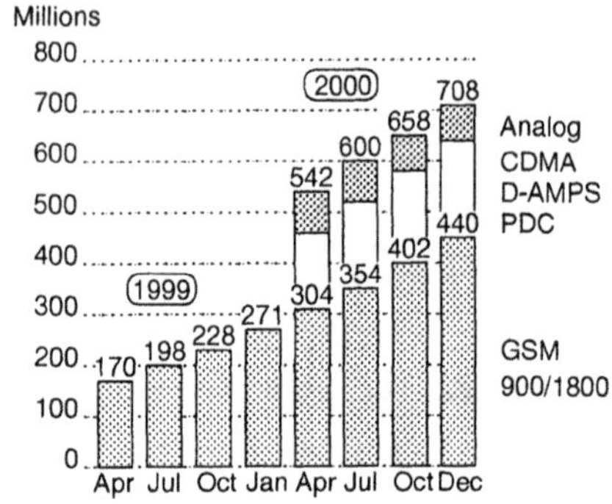

(b) World subscriber growth.

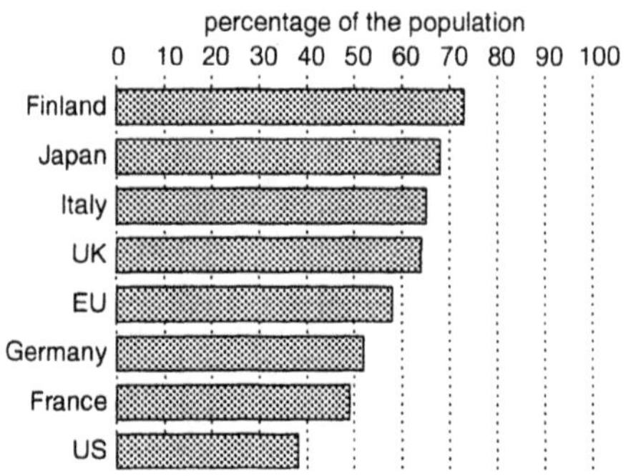

(c) Market penetration of cellular phones.

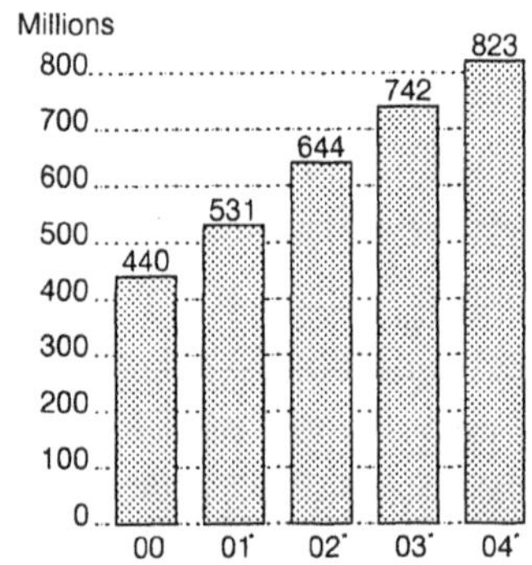

(d) World subscriber forecast (GSM).

Figure 1.1: *The dynamics of the mobile communications market (source: [Geor01] and [GSMW]). Predictions have been indicated by an asterisk.*

from a brick, we have evolved towards handy[1], affordable digital handsets.

At this moment, the most prevalent digital cellular standard is the GSM system (Fig. 1.1(b)), a standard which exists in three flavors: GSM-900, GSM-1800 (also known as DCS-1800 or PCN) and GSM-1900 (also known as PCS-1900). With 440 million subscribers, GSM[2] accounts for 71 percent of the world's digital market, and 64 percent of the world cellular market (i.e. including the analog systems). The remaining 36 percent consists of the other digital standards (i.e. CDMA and D-AMPS in the US and PDC in Japan) and some surviving analog standards.

The coming years, the so-called third generation systems (3G) will be introduced, which will enable high speed data communication and multimedia, high-speed mobile Internet, video-on-demand, etc. In the mean time, GPRS, a packet-switched version of the originally circuit-switched GSM standard, already significantly boosts the data capability of GSM by a factor two to ten (i.e. 172 kbit/s). With the introduction of EDGE — an enhanced version of GPRS

[1] sometimes so small that it becomes clumsy.

[2] An overview of the birth of the GSM system can be found at [GSMW].

— the data rate may even reach peak values of 384 kbit/s! Eventually, the IMT-2000 family of 3G systems is expected to take over, consisting of three standards — UMTS, CDMA2000 and TDMA —, each standard being tailored to the existing infrastructure of mobile operators in different parts of the world.

1.2 The Need for Cost Reduction

In industry, everything revolves around the following basic equation:

$$\mathit{Profit} = \mathit{Sales} \cdot \mathit{Margin}. \tag{1.1}$$

Consequently, it is not difficult to imagine that the fast growth of the mobile market has attracted a lot of companies, each of them wanting to get a piece of the pie. However, as could have been expected, the market soon turned into a highly competitive, turbulent market, leading to genuine price wars. The fact that in the US — one of the most important expansion markets — other cellular technologies like CDMA directly compete with GSM, doesn't make things easier for the manufacturers.

A second important issue is that, according to some [Geor01], the expansion may be coming to an end in the foreseeable future; When the market penetration reaches the saturation point in Europe, this will inevitably mean a slowdown in sales: the growth market will eventually become a replacement market[3], a much less dynamic market which is highly vulnerable to economic fluctuations.

Needless to say that the above issues put severe pressure on the margin and on the profit growth of today's handset manufacturers. More than ever it is critical for manufacturers to offer the functionality required by the customer at the lowest possible manufacturing cost in order to safeguard the profit. Basically, there exist two ways to tackle this problem: reducing the cost of labor by moving the production to the Far East[4] and reducing the cost of the electronic system by decreasing the number and cost of the components, or in other words, by working towards a higher degree of integration. As a result, the pressure more or less shifts from the mobile phone manufacturers to the semiconductor industry. Nowadays, a lot of semiconductor companies are vigorously trying to bring on more cost-effective solutions. This work will focus precisely on this issue: the cost reduction of the internal electronics of the mobile phone.

1.3 Cost Reduction through Integration

The electronics of a mobile phone consists roughly of two parts (see Fig. 1.2): the radio-frequency (RF) and intermediate-frequency (IF) analog front-end — responsible for the reception and the transmission —, and the digital back-end along with the baseband analog circuits — responsible for the (de)modulation and the intelligence of the mobile station.

Whereas today's mobile phones already feature a highly-integrated back-end — more than is the case in Fig. 1.2 —, the architectures for the analog (RF/IF) front-end generally still rely on high-quality external components to realize certain critical functions. Typical examples of this

[3]Already, replacements for existing handsets represent 40 to 50 percent of Nokia's global sales.

[4]If it isn't a fact already.

Figure 1.2: *Exploded view of a mobile phone.*

are the highly selective discrete filters that are used for interstage image filtering and IF channel selection, and high quality passives. While the use of discrete components and SAW filters enables a very performant solution, they are generally very expensive; Accuracy, selectivity and quality factor do come at a price! In addition, these components take up significant board space, complicate the assembly process and decrease the manufacturing yield. Since the need for external components is tightly coupled to the architecture of the front-end, there is a strong drive towards more advanced, highly-integrated architectures that rely less on this class of external components, leading to significant cost savings. And, one could go even further...

Today's high-end RF front-ends are still implemented in BiCMOS, SiGe HBT, Bipolar or GaAs technology, performant technologies which are very suitable for building RF circuits. However, as a result of the high complexity of these processes and because these technologies do not benefit from a high-volume market as CMOS does (with e.g. DRAM memory and digital VLSI), the above-mentioned technologies are generally more expensive than CMOS. Whereas in the past CMOS was anything but an option for implementing RF circuits, CMOS technology is evolving so rapidly that today's submicron CMOS technologies have actually become a viable alternative for building applications in the GHz range, a range that traditionally required bipolar or GaAs technologies; As indicated in Fig. 1.3, the f_T of today's 0.25 μm CMOS devices is already comparable to the f_T of the traditional bipolars (20-25 GHz) and will soon reach the 40 GHz region, challenging the SiGe heterojunction bipolar transistors. Since CMOS is already the technology of choice for the baseband analog part of a mobile phone, and since the large digital back-end almost exclusively employs CMOS, an attractive low-cost solution would be to integrate the RF front-end in CMOS as well. In the future, this may eventually lead to a single-chip CMOS mobile phone. [5]

Unfortunately, there is no such thing as a free lunch; First of all, omitting external components

[5] Single-chip CMOS is by no means a dogma; Another alternative which is gaining acceptance in industry is the *single-package* solution: an MCM-D module with several flip-chip bonded dies in a single package. This solution enables one to build a highly-integrated system using the most optimal combination of IC technologies (e.g. an old BiCMOS technology for the RF analog, and a deep submicron CMOS technology for the highly-scalable digital part) and offers high-quality MCM-D passives as a replacement for external components. Another advantage of the single-package approach is that it offers a path to further integration without requiring analog and digital to peacefully share the same substrate, an issue which unfortunately has remained unsolved until now. The future will show whether the larger production and assembly costs of the MCM-D module can effectively be offset by using cheaper (i.e. older) technologies than deep submicron CMOS for some of the chips on the module. Depending on each company's specific situation (like e.g. available technologies, tradition, in-house skills) and the product, either the single-chip CMOS or the single-package approach will be considered the most cost-effective solution.

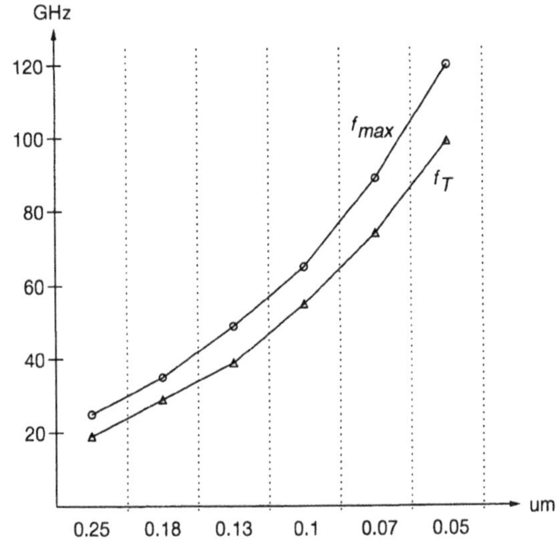

Figure 1.3: *MOS f_T versus technology generation (source: International Technology Roadmap for Semiconductors, 1999 edition).*

and going to more integrated architectures inevitably means tougher specifications for the analog building blocks. Probably the best example is that, because of the absence of interstage filtering, the circuits need to be able to handle a very large dynamic range: low noise and high linearity are mandatory in order for small wanted signals to peacefully coexist with large adjacent channels and blocking signals. And, as is always the case in analog design, the tougher the specifications the larger the power consumption tends to be.

Secondly, despite the fact that CMOS is capable of achieving the speed required for building wireless front-ends, many people question whether CMOS is capable of achieving the performance of high-end RF systems at a reasonable power consumption. Disbelievers of CMOS RF claim that if CMOS is used at all for RF applications, CMOS will only be used in low-end applications, like e.g. Bluetooth, which require less performance and hence consume less power.

Consequently, the main question which needs to be answered is: "Can CMOS offer the performance required for today's high-end cellular standards... and can it offer this performance at a reasonable power consumption?".

1.4 Research in the World: Transceivers from Past to Present

In the early years of mobile communication, virtually all ICs for the radio-frequency front-end of both cellular phones and cordless telephones were implemented in Si bipolar technology. The transceiver front-ends in [Koul93, Seve94] were for instance implemented in 12 GHz and 16 GHz f_T Si bipolar technologies, respectively. At that time, the path towards further integration evidently pointed in the direction of a BiCMOS solution where the bipolars executed the (high-frequency) analog functions and CMOS did the digital processing. Nevertheless, a single-chip

BiCMOS radio was still considered "a strategic utopia because few foundries provided BiCMOS technologies with sufficient bipolar f_T" [Seve94]. However, that did not mean that one was not thinking about other possibilities for further integration, like e.g. going towards more integrated receiver architectures; The front-end in [Seve94] already used the zero-IF (direct-conversion) architecture.

As could have been expected, it didn't take long before BiCMOS actually became an option. Whereas the elementary BiCMOS receiver in [McDo93] still featured a relatively poor performance, [Meye94, Taka95, Long95] demonstrated that the bipolar transistor in BiCMOS had indeed become powerful enough to build the critical building blocks of a receiver front-end: an LNA and a mixer. From the above, only [Taka95] used a highly-integrated direct-conversion architecture. 1995 was also marked by the publication of the first GSM transceiver ICs: [Mars95] in BiCMOS technology and [Stet95] in Si bipolar technology. Again, both transceivers still used a double-conversion heterodyne topology for the receive path, so that the performance came at the cost of external components like SAW filters, loop filters, varactors, tank circuits, etc..

In the same year CMOS suddenly made its entry in GHz range with a 900 MHz CMOS mixer designed for insertion in a highly-integrated low-IF receiver [Crol95a, Crol95b] and a high-performance 1.8 GHz VCO [Cran95]. From that moment on, research on integration of wireless transceivers in CMOS took off very rapidly; Stimulated by the sceptical attitude of the industrial world regarding the feasibility of high-performance CMOS transceivers, universities all over the world were vigorously trying to earn some credits in this field. This is precisely when this research work started.

From the beginning, CMOS research was mainly focused on receivers. First, some basic building blocks like CMOS LNAs, mixers and VCOs were explored, in an attempt to squeeze some performance out of it [Rofo96, Kara96, Shae97, Cran97]. One of the first publications that exceeded the building block level was [Rude97], with a 1.9 GHz wide-band IF double-conversion CMOS receive path for cordless telephone applications (DECT). However, this IC only contained the receive strip and still required an external LO. [Raza97, Wu98] presented some more elementary receive strips. In the mean time, industry went on using BiCMOS and bipolar technologies [Banu97, Yama97].

In 1998, the first single-chip CMOS transceiver demonstrating the feasibility of achieving cellular specifications in CMOS was presented [Stey98]. This IC — which was targeted at the DCS-1800 application — combined a highly-integrated low-IF receive strip and a direct-conversion transmit strip with an on-chip VCO. In the same conference, [Cran98] demonstrated the first single-chip CMOS PLL frequency synthesizer, also for the DCS-1800 application. Not much later, [Rofo98a, Rofo98b] presented a single-chip transceiver for the 900 MHz ISM band, consisting of a direct conversion receiver and a direct conversion transmitter. Even though ISM-band applications are much less demanding than GSM or DCS-1800, this IC was a nice example of a highly-integrated CMOS transceiver. Still in 1998, [Piaz98] presented a CMOS transceiver front-end for GSM-900, yet, at that moment the prototype IC merely contained a heterodyne receive path and a transmit pre-amplifier. The next version of this IC — presented in [Orsa99] — did include a direct-conversion transmit path, which finally made it a true CMOS transceiver.

And, CMOS integration went even further... The first breakthrough came with the integration of full-CMOS cordless telephone transceivers. In 1999, [Cho99] demonstrated an impressive single-chip CMOS transceiver for 900MHz spread-spectrum cordless telephony. The IC integrated a direct conversion receiver and a direct VCO-modulation transmitter together with a PLL (with off-chip loop filter). The rest of the industrial world would continue their efforts to obtain integrated solutions either in BiCMOS [OptE00] or in bipolar technologies [LiPu00].

It took until 2000 before complete transceiver front-ends for cellular phones appeared; In 2000, [Stey00a] presented a 2V CMOS cellular transceiver front-end for DCS-1800. This IC was the first single-chip CMOS transceiver which combined a highly-integrated receive and transmit path with an on-chip PLL, including VCO and loop filter. In addition, this IC achieved the major requirements of the DCS-1800 cellular system while operating from a 2V supply. Industry, on the other hand, had always been working towards more integrated BiCMOS or bipolar solutions [Yama98, Taki99, Mago00]. Nevertheless, in 2001, the first commercial CMOS transceiver chip-set [Aero01, Fren01] for GSM finally came onto the market.

1.5 Research Work

The presented work deals with the design of the receive path of a highly-integrated CMOS cellular transceiver[6] front-end. It covers the whole design trajectory starting from documents describing the standard to the systematic development of CMOS ICs which are measured to be compliant to the standard. In addition, it tackles CMOS receiver design at all levels: from architecture level, via circuit level, down to the device level... and the other way around. The DCS-1800 standard — better known as GSM-1800 — is used as the demonstration vehicle throughout this work.

- Different receiver architectures are compared with respect to integratability, achievable performance and required building block specifications.
- The requirements of a cellular standard are mapped onto a set of measurable specifications for a highly-integrated receiver; It is shown how a minimum required reception quality can be translated into specifications on the receiver's noise figure, image rejection, intermodulation, etc. and how these specifications can be distributed among the different blocks.
- Several circuit topologies are presented that realize the main functions of the receive path. The dynamics of the elementary specifications of these circuits are explained in terms of the operating point of the involved devices i.e. in terms of the performance at the transistor level. Whenever possible, this is done using closed analytical expressions. Based on these insights, detailed sizing procedures are developed to systematically design these RF circuits for a set of specifications.
- The feasibility of meeting the tough requirements of today's high-end cellular standards at a reasonable power consumption is demonstrated in a mainstream submicron CMOS technology by the development of two highly integrated DCS-1800 receivers.

In parallel to the application-driven design trajectory, special attention is given to the fundamentals and the fundamental limits of CMOS RF design. This material forms the theoretical core of this thesis, and provides the link between the device level and the circuit level during the design of the receiver. The insights developed in this theoretical part mainly find their application in the design of high-performance low noise amplifiers.

[6]The issues involved in the design of the PLL, the transmit path and the power amplifier in submicron CMOS technology will be covered in [Demu02], [Borr01] and [Mert03].

- Traditional CMOS analog designers are provided with insight in the fundamentals of RF CMOS design. Answers are provided to questions like: "Is input matching a prerequisite for good performance?", "Does a matched MOS device always efficiently use the power it is provided with?" "Is a 50 Ω environment a nuisance or not?", etc.— questions that are often a subject of controversy between analog designers and RF designers. Attempts are made to reconcile the analog designer's and the RF designer's point of view on how look at CMOS transistors.

- Analytical expressions are derived for the gain, the noise figure and the linearity of an input-matched MOS device; The dynamics of these specifications are analyzed and explained as a function of transistor parameters, source impedance, reflection coefficient and several non-idealities. Some important differences between submicron transistors and long channel devices are covered as well. The fundamental limitations of CMOS at low-GHz radio frequencies are examined and tracked down to the non-quasi static gate resistance and its associated noise.

- A systematic design methodology for the design of power-to-voltage and power gain CMOS LNAs is developed. The method is based on a bank of equations describing the performance of the LNA in the design space — including all non-idealities — and allows an easy identification of un-realizable, inactive and binding constraints. In particular, the method reveals the precise impact of a constraint on the optimization goal: power minimization.

- The design of a GPS LNA demonstrates that even in a standard submicron CMOS technology, an extremely low noise figure can be combined with a high power gain at the same power consumption as commercially available GaAs LNA solutions.

As mentioned above, as a consequence of this work, two highly-integrated receive paths have been fabricated which are embedded in a DCS-1800 CMOS transceiver. In addition, a high performance LNA for the Global Positioning System has been fabricated. The detailed design procedures along with the measurement results are also reported in this text. A brief summary of their main characteristics is presented next.

- A 0.25 μm CMOS DCS-1800 receiver prototype, the receive path of the single-chip transceiver presented at ISSCC 1998. This transceiver was — to the author's knowledge — the first transceiver demonstrating the feasibility of achieving cellular specifications in CMOS.

  The receive path integrates the low noise amplifier, I/Q mixers with active coils and IF trans-impedance amplifiers. The front-end achieves a conversion gain of 32.5 dB, a total noise figure of 5 dB, an IP_3 of -9 dBm and an image rejection ratio of 33 dB. The total receiver (including active coils) draws 46.5 mA from a 2.5V supply.

- A 2V, 0.25 μm CMOS DCS-1800 receiver front-end, the receive path of the transceiver front-end presented at ISSCC 2000, the first CMOS transceiver with a receive path, a transmit path and a complete PLL integrated on a single die.

  The low-IF receiver integrates a low noise amplifier together with two (I&Q) down-conversion mixers and two (I&Q) variable gain amplifiers/filters. The receiver features a conversion gain of 54.5 dB, a total noise figure of 6.2 dB, an IP_3 of -6 dBm, an image rejection of 32.2 dB and draws 56 mA from a 2V supply.

- A 0.8 dB NF, ESD-protected, 9 mW CMOS LNA for GPS applications, presented at ISSCC 2001.

  The LNA is implemented in an advanced 0.25 μm CMOS process and offers the lowest noise figure published to date in CMOS (0.8 dB), the second highest power gain published to date in CMOS (20 dB), and features a power consumption which is among the lowest in CMOS (9 mW).

The earlier work on broadband CMOS low noise amplifiers and the associated IC realizations presented in [Jans97](VLSI 1997) and [Jans98b](CICC 1998) have been omitted in this work. The interested reader is referred to these publications.

Chapter 2

The DCS-1800 Communication System

2.1 Introduction

DCS-1800 — also known as PCN or GSM-1800 — is a digital cellular communication system. Handsets (mobile stations) roaming throughout a blanket of cells constantly keep in radio contact with the base station in the nucleus of the current cell.

Because discussing the whole DCS communication protocol would lead us too far, we confine ourselves to giving some relevant background information on the air-interface between the mobile station and the base transceiver station. In this introductory chapter on DCS-1800, we will consecutively cover the frequency plan, the employed modulation method, and its associated spectrum and constellation diagram.

2.2 Frequency Plan

The DCS-1800 system is basically a frequency-translated version of the GSM-900 digital cellular system. The digital part of the standard (i.e. the communication protocol, the modulation method, the algorithms, etc.) is essentially identical. The only difference lies in the requirements for the radio interface[1].

Although DCS-1800 is considered a *digital* system — using digital information coding and a digital control protocol —, its physical layer still consists of an *analog* radio interface; The allocated frequency bands are 1710-1785 GHz and 1805-1880 GHz, of which the lower band is used for transmitting and the higher band is used for receiving (Fig. 2.1).

[1]The difference between the GSM-900 and the DCS-1800 standard is discussed in Section 3.7.

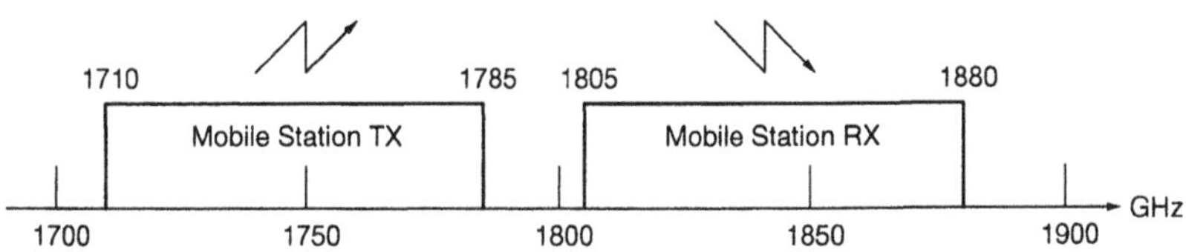

Figure 2.1: *The DCS-1800 frequency plan.*

Frequency-division multiple access (FDMA) and time-division multiple access (TDMA) techniques are adopted in order to allow a maximum number of users in the cell accessing the network. First of all, each 75 MHz band is subdivided into 372 carrier frequencies spaced 200 kHz apart (FDMA). To each base station a few carrier frequencies are assigned. These frequencies are further subdivided into elementary time slots or *burst periods*, containing the data (TDMA). Eight consecutive bursts are grouped together in a so-called TDMA frame, each burst being assigned to one of eight 'virtual' communication channels. The data stream in the ith channel is then formed by the ith burst in each incoming frame. Each mobile station is dynamically assigned a carrier frequency and a channel (time slot) through which the communication takes place. During this time slot, the mobile station is the only transmitter/receiver using that particular carrier frequency. In DCS-1800 and GSM, 13 frames occupy exactly 60 ms, so that the TDMA structure looks like:

$$60 \text{ ms} = \overbrace{\underbrace{\frac{15}{26} \text{ ms/burst}}_{\approx 577\ \mu s} \cdot 8 \text{ bursts/frame}}^{\approx 4.615 \text{ ms}} \cdot 13 \text{ frames}. \tag{2.1}$$

Each of these 15/26 6ms bursts contains 156.25 bits — a 148 bit frame, containing 2×57 information bits, and 8.25 additional bits —, leading to a gross bit rate of 270.833 kbps.

2.3 Modulation Method

DCS-1800 uses Gaussian Minimum Shift Keying (GMSK) to modulate the data bursts onto a carrier. GMSK modulation is a smoothed form of MSK (Minimum Shift Keying) which is essentially binary digital FM. In both schemes the data is coded in the instantaneous frequency (i.e. the derivative of the phase) rather than directly in the phase itself (like e.g. in QPSK).

One of the most important properties of both GMSK and MSK is that they are constant envelope modulation techniques. As a consequence, a high power-added efficiency, non-linear power amplifier (e.g. a Class-E PA) can be used in the transmitter part without generating too much interference in the adjacent channels. In practice there remains a trade-off between out-of-channel radiation (so-called spectral regrowth) and the steepness of the PA ramp-up/down. To alleviate this trade-off, adjacent carrier frequencies are often not used at the same time, giving rise to a chess-board pattern in the cell PSD.

Since GMSK modulation is a variant of MSK, it is useful to review the basic MSK modulation scheme first. It will become clear why GMSK has been selected in the DCS-1800 system, and not MSK.

2.3.1 MSK Modulation

MSK can be seen as binary digital FM with a modulation index h of 0.5 around a center frequency equal to the bit frequency, ω_b, given by

$$\omega_b = \frac{2\pi}{T_b}. \tag{2.2}$$

phase turn due to incoming symbol $\lvert\measuredangle\rvert = \pi/2$	nett $\Delta\phi$ in T_b
1 ↶	$\pi/2$ ①
0 ↷	$-\pi/2$ ②

Table 2.1: *MSK phase transition table.*

where T_b is the bit period (i.e. the inverse of the bit rate). The modulation index m is in this context defined as

$$m = \frac{\omega_{max} - \omega_{min}}{\omega_b}, \tag{2.3}$$

where ω_{max} and ω_{min} denote the maximum and the minimum instantaneous frequency, respectively.

In the MSK scheme, a 'zero' is coded by a slightly lower frequency,

$$\Delta\omega_{zero} = \omega_b - \frac{\pi}{2T_b} = \frac{3\omega_b}{4}, \tag{2.4}$$

whereas a 'one' is coded by a slightly higher frequency,

$$\Delta\omega_{one} = \omega_b + \frac{\pi}{2T_b} = \frac{5\omega_b}{4}, \tag{2.5}$$

which is indeed consistent with a modulation index of 0.5 with respect to ω_b.

In order to generate the MSK symbols, the bit stream is first mapped onto a NRZ (Non-Return to Zero) signal by applying the conversion rule

$$b_{i,NRZI}(t) = 2 \cdot b_i(t) - 1. \tag{2.6}$$

The resulting NRZ signal (shown in Fig. 2.2) consists of a stream of pulses (+1 or −1), each of unity height and of T_b width. Multiplying the NRZ signal by a reference frequency of $\pi/2T_b$ yields the instantaneous frequency deviation $\Delta\omega(t)$:

$$\Delta\omega(t) = b_{i,NRZI}(t) \cdot \frac{\pi}{2T_b}. \tag{2.7}$$

This frequency deviation $\Delta\omega(t)$ is subsequently converted into a running phase deviation $\Delta\phi(t)$ by a simple integration operation:

$$\Delta\phi(t) = \int \Delta\omega(t)\, dt. \tag{2.8}$$

Evidently, the resulting phase deviation (Fig. 2.2) is piece-wise linear. During each bit period, the phase covers $\pi/2$ radians in the direction determined by the incoming bit: the direction (sign) of the phase turn directly relates to the information bits. The net MSK phase transitions shown are summarized in Table 2.1. Adding this phase deviation to a linearly increasing phase representing the un-modulated frequency ω_b, and using the result as the argument of a sine, leads to the MSK signal shown at the bottom of Fig. 2.2:

$$MSK_{LF}(t) = \sin(\omega_b t + \Delta\phi(t)). \tag{2.9}$$

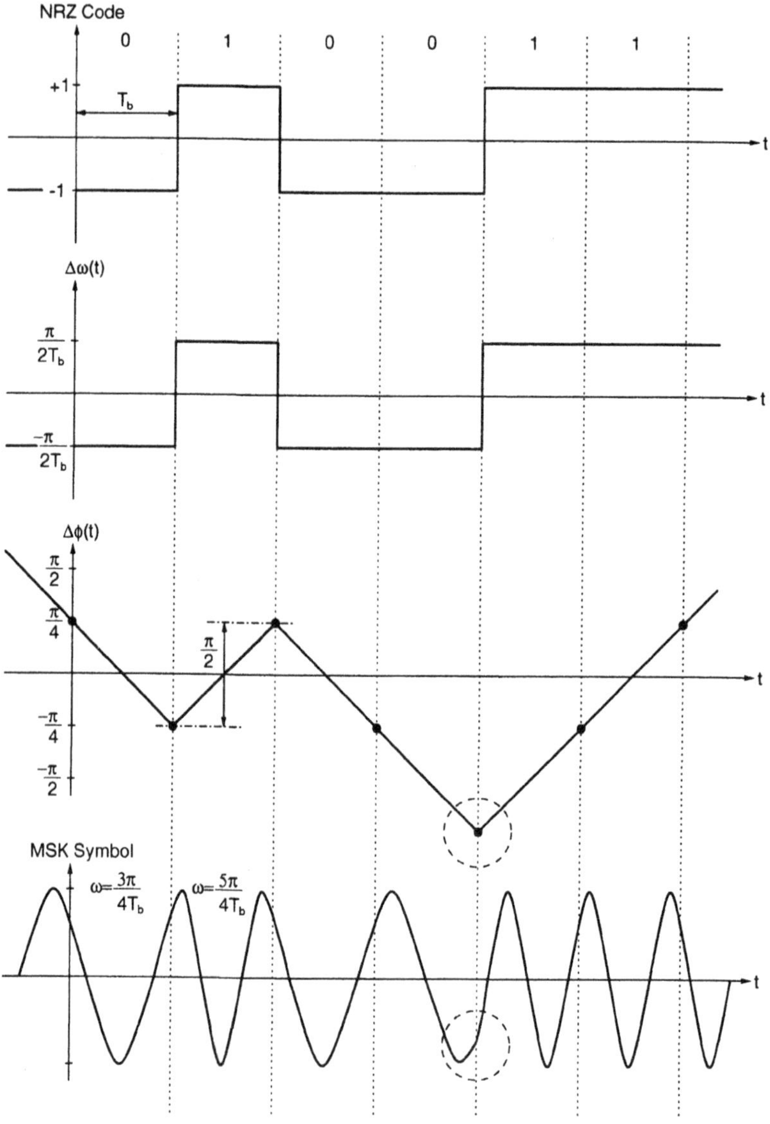

Figure 2.2: *Generating a MSK signal out of a bit-stream. All possible phase transitions are described in detail in Table 2.1.*

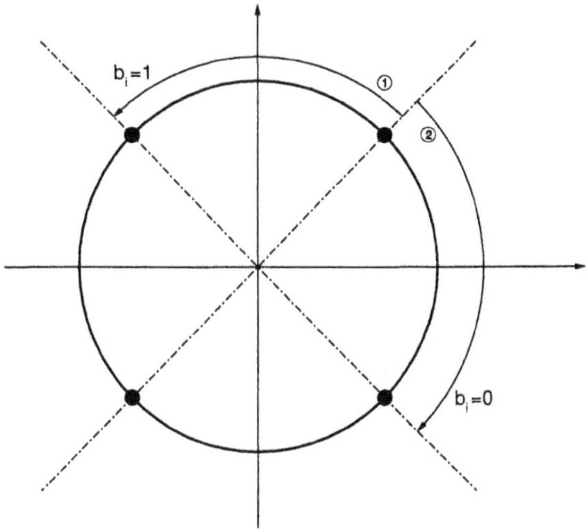

Figure 2.3: *MSK constellation with its phase trajectories (described in Table 2.1).*

In theory, the low-frequency signal in (2.9) can be up-converted and transmitted. In practice however, the MSK signal is not present in the one-dimensional form of (2.9) but rather as a two-dimensional quadrature signal: $\sin(\Delta\phi(t))$ and $\cos(\Delta\phi(t))$. Then, based on the quadrature representation of the MSK signal, one can directly construct the high-frequency SSB signal to be transmitted:

$$MSK_{RF}(t) = \sin(\omega_{center}t + \Delta\phi(t)) \tag{2.10}$$

$$= \sin(\Delta\phi(t))\cos(\omega_{center}t) + \cos(\Delta\phi(t))\sin(\omega_{center}t). \tag{2.11}$$

Hence, the phase trajectory $\Delta\phi(t)$ of the MSK signal can equally be represented on a circle. This is shown in the so-called *constellation diagram* of Fig. 2.3. Depending on the incoming bits, the phase ends up $\pi/2$ radians counter clock-wise (①) or clock-wise (②). The black dots indicate all possible phase trajectory endpoints. The constellation diagram is essentially equivalent to the $\Delta\phi(t)$ plot in Fig. 2.2.

Even though the phase varies continuously in time, it abruptly changes direction at the bit transitions. The corresponding abrupt change in momentous frequency[2] leads to sharp corners in the MSK waveform, as is indicated by the dashed circle at the bottom of Fig. 2.2. In his turn, this produces extra spectral components next to the MSK signal peak, leading to a broader spectrum and hence extra inter-channel interference. Luckily, there is GMSK to solve the problem.

[2]The derivative of the frequency (the so-called 'jerk' in classical mechanics) in principle approaches infinity

$$\lim_{b_i \to b_{i+1}} \left|\frac{d\,(\Delta\omega(t))}{dt}\right| = \infty. \tag{2.12}$$

2.3.2 GMSK Modulation

In contrast to MSK, where the frequency changes in a very short instant around the bit transitions, the frequency transitions in GMSK [Muro81, Muro85] are much smoother; As a result, the PSD of GMSK is much more compact. This is a very welcome property since it lowers the out-of-channel radiation and therefore relaxes the channel spacing requirements. The smooth frequency transitions are obtained by filtering the bit sequence before modulation by a Gaussian filter whose bandwidth B is a fraction of the bit rate $1/T_b$. Of course, as a result of the low filter bandwidth when compared to the bit rate, a controlled amount of inter-symbol interference (ISI) is created. For instance, the GMSK variant used in the GSM standard features a BT_b product of 0.3, which makes that the phase trajectory is determined by a running set of four consecutive bits in the bit-stream. Ultimately, this leads to a constellation where each quadrant contains three points instead of one. The details of the GMSK modulation method will be clarified further in the following paragraphs and figures.

Let's first look at Fig. 2.4. Every NRZ sequence can be viewed as a stream of pulses with a height of +1 or -1, coding for a one or a zero, respectively. In case GMSK modulation with a BT_b product of 0.3 is to be used, these pulses are pre-shaped by a $0.3/T_b$ Gaussian filter, or in other words, convolved with a $3T_b$ Gaussian pulse:

$$\Delta\omega_{\text{GMSK}}(t) = \Delta\omega_{\text{MSK}} \otimes h(t), \tag{2.13}$$

where $h(t)$ is given by

$$h(t) = \frac{k_1 B}{\sqrt{\pi}} \exp(-k_1^2 B^2 t^2) \tag{2.14}$$

with

$$k_1 = \frac{2\pi}{\sqrt{2\ln(2)}} \tag{2.15}$$

and B the -3 dB bandwidth of the Gaussian filter.

As a consequence, each frequency pulse is spread out over a $4T_b$ time span and contributes to the net instantaneous frequency in the same $4T_b$ window. The corresponding phase turn of $\pi/2$ is also spread out over $4T_b$, and occurs in steps of ϵ, $\pi/4-\epsilon$, $\pi/4-\epsilon$, ϵ[3]. The net phase turn in a T_b time span thus depends on 4 symbols. The thin solid lines show the frequency and phase deviations when the Gaussian filter is left out, corresponding to the 'original' MSK modulation technique.

How a bit-stream is transformed into a GMSK phase trajectory is shown in Fig. 2.5 . Exactly the same bit-stream is used as in Fig. 2.2. Basically, the instantaneous frequency and phase deviations can be obtained by superimposing the appropriately shifted 'templates' of Fig. 2.4 onto each-other. The GMSK trajectories are represented by the bold lines while the 'old' MSK trajectories are drawn using thin lines. Evidently, the frequency and phase trajectories of GMSK are much softer than those of MSK. The derivative of the instantaneous frequency now also exists in all time points, leading to the smooth GMSK signal at the bottom of Fig. 2.5.

Of course, as mentioned before, the instantaneous phase (frequency) is considerably influenced by the neighboring bits. As a consequence, the phase transition table (Table 2.2) becomes quite complicated. Luckily, the phase turn is largely determined by the previous and the current

[3]the symbol 'ϵ' does stand for a specific value, but it is irrelevant to quantify it in this context.

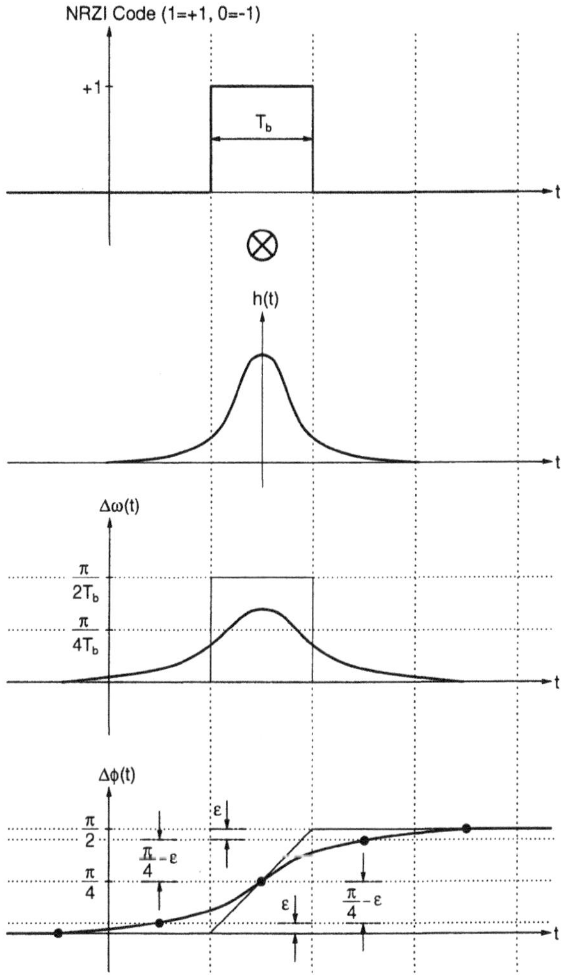

Figure 2.4: *In GMSK, the frequency pulses representing the bits are pre-modulated by a Gaussian filter (convolved with a $3T_b$ Gaussian pulse). As a consequence, each frequency pulse is spread out over a time span of $4T_b$ and contributes to the net instantaneous frequency in the same $4T_b$ window. The corresponding phase turn of $\pi/2$ is also spread out over $4T_b$, and occurs in steps of ϵ, $\pi/4-\epsilon$, $\pi/4-\epsilon$, ϵ. The net phase turn in a T_b time span thus depends on 4 symbols and is given in Table 2.2.*

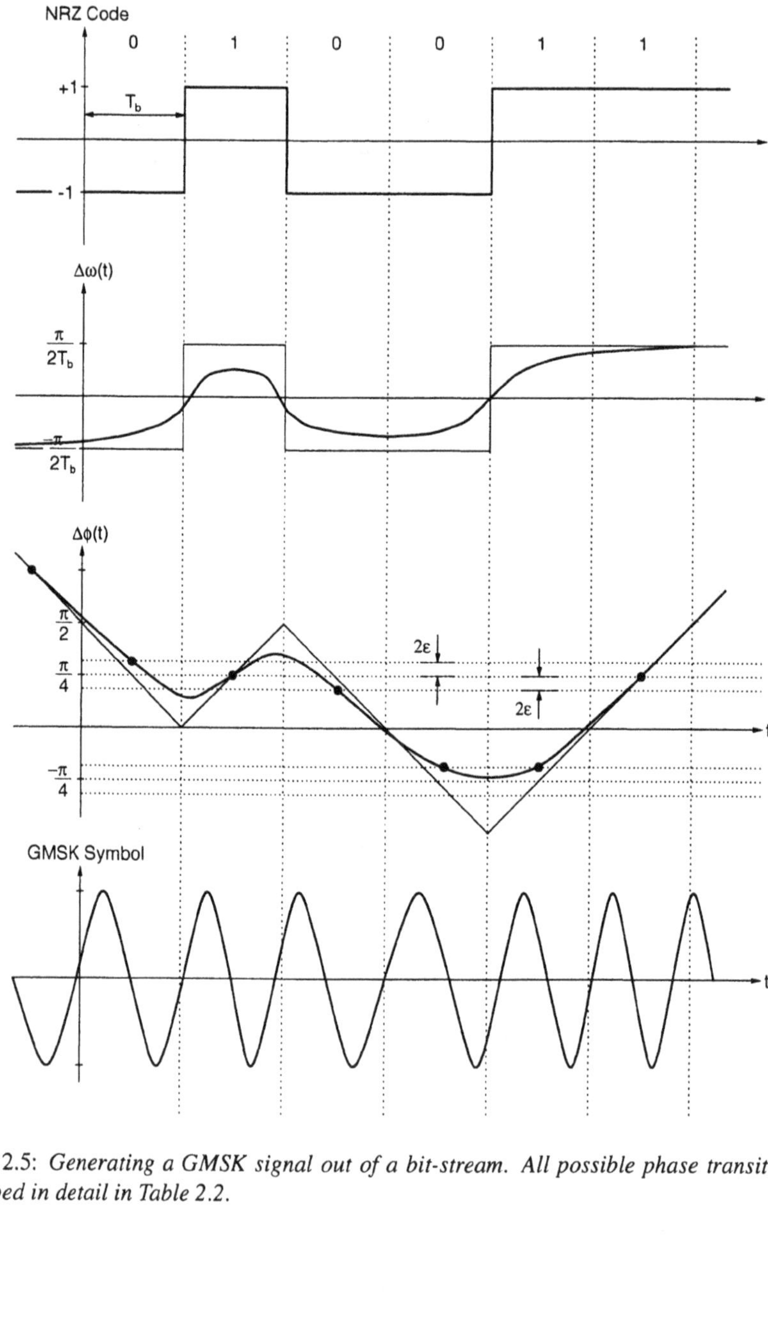

Figure 2.5: *Generating a GMSK signal out of a bit-stream. All possible phase transitions are described in detail in Table 2.2.*

Residual pulling of past symbol $\|\epsilon\|$	Rotation due to previous symbol $\|\frac{\pi}{4}-\epsilon\|$	Rotation due to current symbol $\|\frac{\pi}{4}-\epsilon\|$	Early pulling of future symbol $\|\epsilon\|$	nett $\Delta\phi$ in T_b
1 ↶	1 ↶	1 ↶	1 ↶	$\pi/2$ ①
1 ↶	1 ↶	1 ↶	0 ↷	$\pi/2-2\epsilon$ ②
0 ↷	1 ↶	1 ↶	1 ↶	$\pi/2-2\epsilon$
0 ↷	1 ↶	1 ↶	0 ↷	$\pi/2-4\epsilon$
1 ↶	1 ↶	0 ↷	1 ↶	2ϵ ③
1 ↶	1 ↶	0 ↷	0 ↷	0 ④
0 ↷	1 ↶	0 ↷	1 ↶	0
0 ↷	1 ↶	0 ↷	0 ↷	-2ϵ
1 ↶	0 ↷	1 ↶	1 ↶	2ϵ
1 ↶	0 ↷	1 ↶	0 ↷	0
0 ↷	0 ↷	1 ↶	1 ↶	0
0 ↷	0 ↷	1 ↶	0 ↷	-2ϵ
1 ↶	0 ↷	0 ↷	1 ↶	$-\pi/2+4\epsilon$
1 ↶	0 ↷	0 ↷	0 ↷	$-\pi/2+2\epsilon$
0 ↷	0 ↷	0 ↷	1 ↶	$-\pi/2+2\epsilon$
0 ↷	0 ↷	0 ↷	0 ↷	$-\pi/2$

Table 2.2: *GMSK phase transition table.*

symbol; If these symbols are different, the net phase turn in a time span of T_b is virtually zero. In the other case the phase turn is $+\pi/2$ or $-\pi/2$ when receiver a one or a zero, respectively. The influence of past and future symbols on the net phase turn is relatively limited.

Fig. 2.6 shows the constellation diagram of the GMSK modulation scheme. In contrast to MSK, where each quadrant contains a single phase trajectory endpoint, GMSK modulation features three clustered endpoints in each quadrant. The angular distance between the points of the cluster equals 2ϵ. Phase transitions ① to ④ are explained in Table 2.2. The table clearly shows that it is not important to be able to discern the points belonging to a cluster. To demodulate the GMSK symbols, it is sufficient to correctly detect the quadrant.

Fig. 2.7 shows the typical power spectrum of a DCS-1800 GMSK signal using an integration bandwidth of 30 kHz. Obviously, the GMSK spectrum considerably leaks outside of the allocated, 200 kHz wide channel (Section 2.2). Hence, even though the spectrum is much more compact than MSK, the leakage into the adjacent channels still requires a channel spacing of 400 kHz in a given cell.

2.4 Conclusion

In this chapter, some background information has been provided on the DCS-1800 system. The most important aspects of the frequency plan and the employed GMSK modulation method have been covered. In the next chapter we will actually look into how the DCS-1800 standard can be mapped onto a highly integrated receiver.

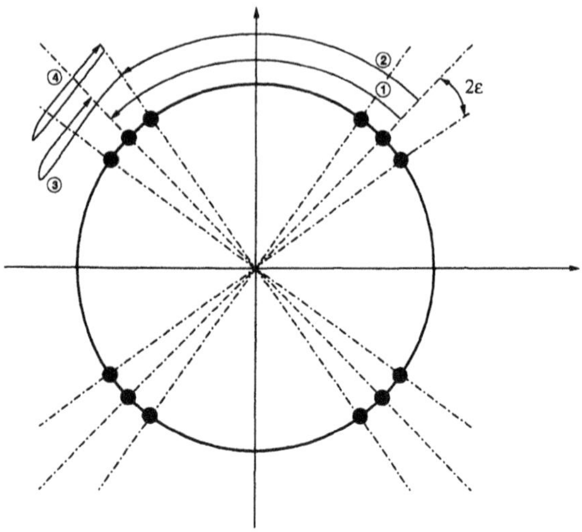

Figure 2.6: *GMSK constellation with a set of phase trajectories (described in Table 2.2).*

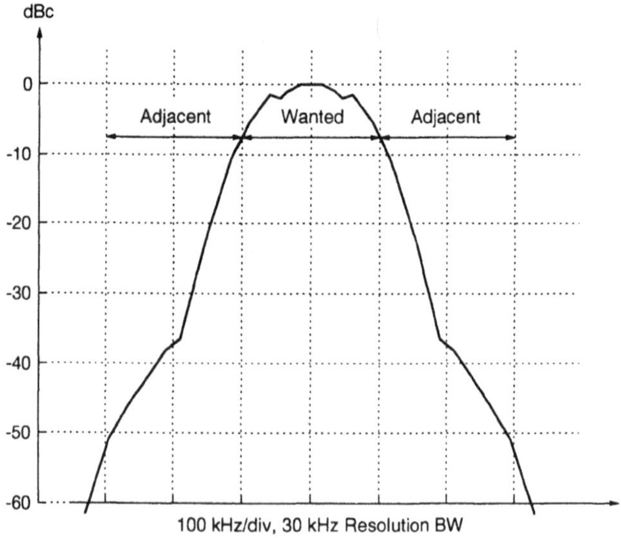

Figure 2.7: *The spectrum of a DCS-1800 GMSK signal.*

Chapter 3

Receiver Architecture and Specifications

3.1 Introduction

In the previous chapter, DCS-1800 has been introduced as a bidirectional cellular communication system at 1.8 GHz. Yet, the realization of a DCS-1800 receiver is not merely the implementation of a 1.8 GHz receiver; In fact, an important part of the DCS-1800 standard [ETSI] is dedicated to specifying the minimum reception quality that must be guaranteed under certain 'loading' conditions, like e.g. adjacent interference, blocking signals, low input powers etc. A look to these standard documents reveals that a DCS-1800 compliant receiver must be able to detect wanted signals with a power as low as 64 fW (1 fW = 10^{-15}W!), in the presence of unwanted signals which are ten million to one hundred million times larger! The implementation of a DCS-1800 receiver front-end thus involves the design of a very sensitive and at the same time very linear radio-frequency front-end.

In this chapter we will take the first step towards a silicon implementation. We will do this by exploring a few architectures that are suitable for realizing cellular receivers and determining what are the system level requirements for a receiver front-end (and its building blocks) to be DCS-1800 compliant.

The outline of this chapter is as follows. First, different cellular receiver architectures are compared with respect to integratability, achievable performance and required building block specifications. After choosing an architecture which suits our needs — basically, a high degree of integration —, the requirements of the DCS-1800 cellular standard are mapped onto a set of measurable specifications for the receiver as a whole; It is shown how a minimum required reception quality can be translated into specifications on the receiver's noise figure, image rejection, intermodulation, etc. and how these specifications can be distributed among the different blocks.

3.2 Cellular Receiver Architectures

In this section, three different receiver architectures will be contrasted against one another. We confine ourselves to discussing the main properties of the heterodyne receiver, the zero-IF or direct-conversion receiver, and the low-IF receiver. Many combinations of these architectures exist (like e.g. the wide-band double-IF receiver), but their properties can equally be understood from looking at the extremes covered in this section.

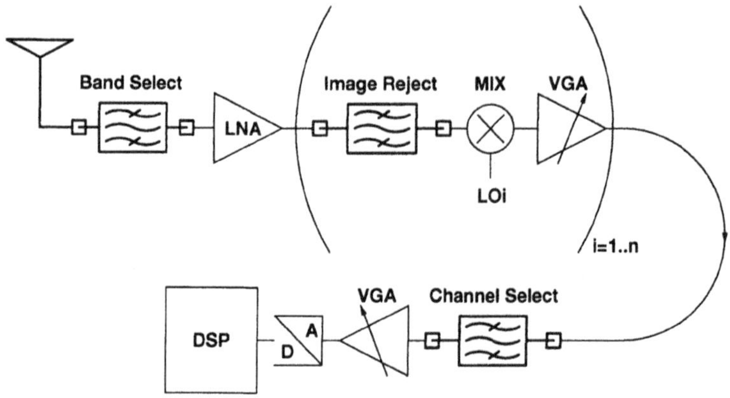

Figure 3.1: *A generic heterodyne receiver.*

3.2.1 The Super-Heterodyne Architecture

3.2.1.1 Structure

The most straightforward architecture for implementing a cellular receiver front-end is unquestionably the super-heterodyne receiver. Its generic scheme is shown in Fig. 3.1. External components have been accentuated by drawing them in bold.

The broadband antenna signal is fed into a highly selective bandpass filter (the pre-selector filter) which suppresses all interferers residing outside of the application band. By removing these out-of-band blocking signals the dynamic range requirements of the receiver can be relaxed considerably. A low noise amplifier (LNA) boosts the weakest channels — along with the more powerful channels — above the noise floor of the first mixing stage.

The receiver core consists of a cascade of n image-suppressing down-mixing stages that are converting the signal from the current IF, $IF_{current}$, to a lower IF_{new}. Before each mixing operation, an image filter (interstage filter, IF-filter) suppresses the mirror signal at

$$f_{image} = IF_{current} + 2IF_{new}. \tag{3.1}$$

In this way only a strongly attenuated version of the mirror signal is folded onto the wanted signal, preventing irreparable corruption of the information contents. The filter also strips off the noise on the image frequency, originating from the preceding building blocks. Note that the pre-selector filter behind the antenna also contributes to the mirror suppression.

Ultimately, the wanted channel is positioned at the center frequency of the channel select filter from where it can be (sub)sampled by an A/D.

3.2.1.2 Properties

In this architecture, the mirror signal is situated at a distance of $2IF_{new}$ from the wanted channel. Because the level of the mirror signal is in general much larger than the level of the wanted signal — it can even be the signal of a completely different application — the mirror signal needs to be immensely attenuated by the image filter. Since this attenuation must already be provided at a distance of roughly $2IF_{new}$ from the current channel, this makes very high demands upon the quality factor and the selectivity of the image filter. The required quality factor of the filter can be defined as

$$Q = \frac{f_{center}}{BW}, \tag{3.2}$$

where f_{center} is the center frequency of the IF-filter and BW is the IF-filter bandwidth. The filter selectivity on the other hand, can be defined as

$$S_{\Delta f} = \frac{S_{21} \text{ at } f_{center} + 0.5BW}{S_{21} \text{ at } f_{center} + 0.5BW + \Delta f}, \tag{3.3}$$

where Δf is a certain frequency offset, typically taken a few times the filter bandwidth. It is clear that the required quality factor is related to the ratio between the current IF and the next IF (which *must* be a few times larger than the filter bandwidth). The required shape factor on the other hand, is strongly related to the difference between the IF-filter bandwidth and the next IF.

In a cellular receiver, the ratio between the center frequency of the application band (basically the first *IF*) and the lowest *IF* typically ranges between 80 to 200. The quality factor required for this *IF* ratio is simply unattainable. Hence, this conversion ratio *must* be realized in two, three or even more steps, requiring as many image filters. We will come back to this shortly.

A super-heterodyne receiver features a single path topology; Path mismatch is simply not an issue here because the image rejection (mirror signal suppression) does not rely on any matching between two signal paths — as is the case in the next receiver topologies — but is done by the interstage image filters. Also LO feed-through and DC offset do not affect the signal quality since the wanted channel is never close to these frequencies. The same applies to self-mixing of either LO or RF signal. The accuracy-speed-dynamic range (mismatch-bandwidth-dynamic range) trade-off is thus much less pronounced in this topology.

Another important property is that the channel selection process occurs before the VGA-A/D cascade. Hence, the IF VGA and the A/D only need to handle a minimum dynamic range. Due to the bandpass nature of the channel, even a subsampling A/D can be used. Additionally, the number of bits can be kept low since both the out-of-band and the in-band blocking levels have already been removed.

Precisely because critical functions are being realized by passives and need therefore not be implemented in silicon, the power consumption of these receivers can be kept low. This, along with its high performance, is *the* reason why this architecture is so popular. Additionally, one must not forget that because the filters only pass the band of interest at each depth in the receiver chain, the specifications of the involved circuits are maximally relaxed.

However, although the filtering functions themselves do not consume any power — and come essentially only at the cost of area... and money — there is a shadow cost involved. The insertion loss in the high frequency filters must be compensated by a corresponding increase in gain to keep the *SNR*. Since the filters need to be driven at a low impedance level (e.g. 50 Ω) they generally require a specially designed interface and high quality passives (on-chip, M-CM or external ones) to perform an impedance transformation from the relatively high circuit impedance to the characteristic impedance of the filter. If a low output impedance buffer stage is used instead of impedance matching, gain comes at the cost of an extra amount of power: extra power for the buffer, and possibly extra power to overcome the loss that is associated with the resistive division in the buffer. While the last is a viable option for the lowest IF frequencies — because the filter impedance is allowed to be higher and passive matching would require large value, external inductors anyway —, this is to be avoided in the RF part. This is especially true for a CMOS implementation, since CMOS devices are all but suitable for driving low impedances.

Additionally, a heterodyne receiver heavily relies on (expensive) external, high-Q passive filters to perform both the mirror signal suppression and the channel selection. Integrating these high-Q filters on silicon is not workable because the power consumption of an active filter is proportional to the square of its quality-factor. An alternative solution could consist in integrating the passive filters on an MCM-like substrate in combination with flip-chip bonding. From a marketing perspective this is very attractive, since one obtains a very compact solution while keeping all degrees of freedom,e.g. the use of different technologies for the RF front-end (BiCMOS, bipolar or CMOS) and LF part (CMOS). Additionally, the impedance level of the interface between the different building blocks may be raised because the parasitics are now much smaller. However, it is unlikely that this solution can implement the selectivity that is required for heterodyne receivers.

From the above it is clear that the heterodyne receiver architecture is not a good option regarding integratability, because it simply does not offer a path to full integration. Hence, the sense of making a full CMOS implementation using this architecture can be questioned; A CMOS implementation is only useful if it can provide a more compact and cheaper solution than the existing BiCMOS and bipolar implementations, without consuming too much extra power. And in this topology one will probably never get rid of the external filters. Besides, the topology does not exploit the power of digital signal processing, which is after all the 'core competence' of CMOS.

3.2.2 The Zero-IF and Low-IF Architecture

Integratability can dramatically be improved if one could only find a way of getting rid of the external, high-Q filters. Why were the filters needed anyway? For removing the image frequency band, possibly containing high power interferers! Now, the channels in the direct neighborhood of the wanted channel — the so-called *adjacent channels* — are much less strong than the channels residing further away. Hence, if a single-stage down-conversion is done towards an IF that equals the frequency difference between a 'regulated' adjacent channel and the wanted channel, this is coupled with a lower image power, requiring less image rejection to keep the *SNR*. In the limit, the IF can even be made zero such that the signal becomes its own image. Of course, the image signal can never be suppressed by any filter before being folded onto the wanted signal so another image rejection method must be used: dual path, quadrature down-conversion.

Implementing this down-conversion scheme leads to two related receiver topologies: the zero-IF (or direct conversion) receiver [Abid95, Raza97b] and the low–IF receiver [Crol98]. For the sake of compactness, both topologies will be discussed together and contrasted against each other.

3.2.2.1 Structure

Fig. 3.2 shows the generic scheme of both a zero-IF and a low-IF receiver. As is the case in the heterodyne receiver topology, the antenna signal is first passed through a band select filter, suppressing all out-of-band blocking signals and relaxing the required dynamic range. After amplification by the LNA, the signal is demultiplexed and fed to two independent signal paths.

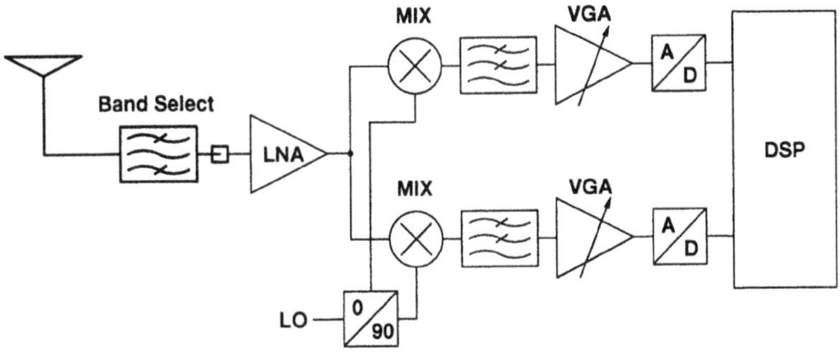

Figure 3.2: *A generic zero/low-IF receiver.*

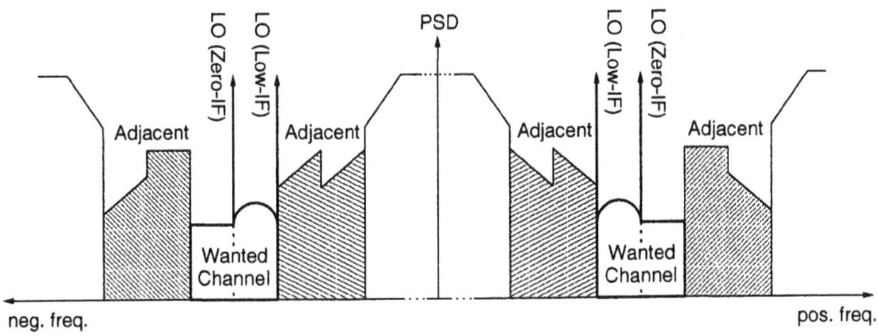

Figure 3.3: *PSD of the wanted signal and the adjacent channels before down-conversion by a zero-IF and a low-IF receiver. The local oscillator signal is also indicated for both receivers.*

It has to be noted that the (external) interstage filter is not required any longer, since its function has become obsolete; both the mirror signal *and* the noise will eventually be 'neutralized' by the recombination of the two signal paths after the quadrature down-conversion.

Each path performs a single stage down-conversion of the wanted signal towards an IF. Because each path's local oscillator consists of a cosine and a sine, respectively, the down-conversion gives rise to an in-phase component (I-path) and a quadrature component (Q-path).

Precisely in this down-conversion step lies the (subtle) difference between a zero-IF and a low-IF receiver; In a zero-IF receiver, the wanted channel is converted to DC and a mirrored version of the channel is superimposed onto the 'clean' version. In a low-IF receiver, the wanted signal is mixed down to a low, non-zero IF which is for instance set to half the channel bandwidth such that the mirror signal is the adjacent channel. Fig. 3.3 shows the starting point of the down-conversion process: a wanted channel flanked by adjacent channels. The local oscillator signals in both the zero-IF and the low-IF receiver are indicated as well. Fig. 3.4(a) and Fig. 3.4(b) show the signal behind the mixer in the I-path of a zero-IF and a low-IF receiver, respectively. It has to be noted that the image has not yet been canceled at this position in the receive chain; This can only be done by appropriate recombination of the I- and the Q-path.

Each path contains a coarse low-pass channel-select filter to reduce the in-band out-of-channel blocking levels. The headroom that becomes available is maximally exploited by imple-

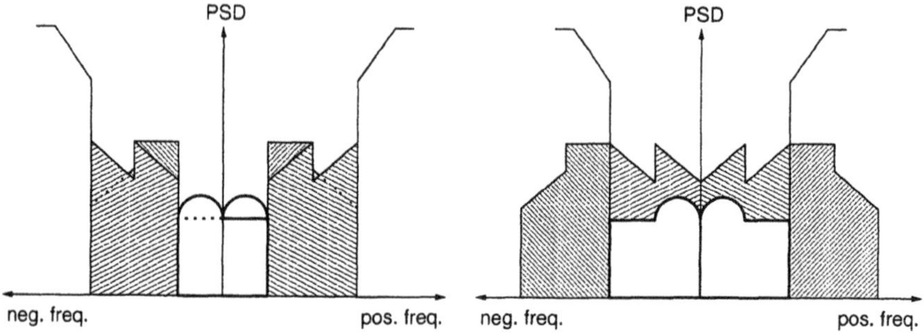

(a) Zero-IF receiver. (b) Low-IF receiver.

Figure 3.4: *Image folding in the zero-IF and the low-IF receiver.*

menting some gain in the VGA in order to relax the dynamic range spec for the A/D converters. The wanted signal, along with the filtered (but still considerably high) blocking signals, is sampled by low bandwidth, high accuracy A/D converters.

Ultimately, the image signal and the wanted channel are separated in the DSP by recombining the information that is present in both 'corrupted' channels. So, the real channel selection and image rejection is done in the digital back-end.

3.2.2.2 Properties

It is clear that both the zero-IF and the low-IF architecture feature an excellent integratability: Since the image rejection and the channel selection no longer rely on high-Q filtering, in principle no external filters are required. The inevitable loss associated with the image filters is no longer present, leading to a lower power consumption. The only external filter that still remains is the pre-selector filter, which is directly positioned behind the antenna. This is however no real limitation since today's cellular applications always require a switch multiplexer or duplex filter. Additionally, owing to the absence of the external filters, the number of external nodes can be kept to a minimum. This means that the (external) passives that were previously transforming the local on-chip impedance down to the characteristic impedance of the IF or interstage filters are no longer needed. Also the robustness of the architecture is improved because the interaction with the package parasitics is minimized. Sensitive nodes are not brought outside, decreasing the sensitivity to interference and crosstalk.

The down-conversion is done in a single step in two signal paths. This makes that, instead of a cascade of mixing stages, only one mixer is needed in each path, resulting in a low power consumption. Apart from the coarse channel select filter — which also functions as anti-aliasing filter for the oversampled data converters —, and some signal conditioning, the rest of the operations in the receive path — like image rejection, channel selection and demodulation — are carried out in the digital domain by a digital signal processor (DSP). This is an attractive property, since the digital domain is after all the natural biotope of CMOS. Additionally, in contrast to the analog part, the digital part features a much better scalability (i.e. portability towards next process generations) and a great flexibility (i.e. software reconfigurable chip). These properties make a zero-IF and low-IF receiver a very attractive solution.

However, the reception is distributed over two independent paths. This puts a lot of requirements on symmetry because the image rejection critically relies on that symmetry. Indeed, all amplitude and phase errors result in a limited image suppression so that the image is partly leaking through and affects the wanted signal. Since path mismatch becomes extremely important at all frequencies, the quadrature local oscillator, the down-conversion mixers as well as the low-frequency filters/VGAs need to be designed very carefully. At first sight the required image rejection for a zero-IF system appears to be much lower than the required rejection in a low-IF system (Fig. 3.4). After all, in a zero-IF system the power of the mirror signal only is a folded version of the wanted channel and hence has exactly the same power. whereas in a low-IF system, the — possibly 18 dB *stronger* — adjacent channel is mirrored onto the wanted channel. However, one may not jump to conclusions too fast. In a low-IF receiver, the image merely acts as colored *noise* because the adjacent channel is completely *uncorrelated* to the wanted signal. In the case of a zero-IF receiver, the folding operation results in *distortion* of the wanted signal, because the image is strongly *correlated* to the wanted signal. In fact, due to the distortion, the original, circular constellation becomes an ellipse, leading to a systematically decreased distance between the points of the constellation. This is especially a problem in QAM-based systems. Therefore, the image rejection specifications are generally quite comparable: 20 to 25 dB in a zero-IF receiver and 32 dB in a low-IF receiver.

Yet, other pitfalls lie in wait... Since the wanted signal of both zero- and low-IF receiver is situated at low frequencies, the signal is exposed to $1/f$ noise and mismatch-related DC-offset; Because the gain of receiver front-end is always limited by the (first-order filtered) blocking levels, the wanted signal at the receiver output is in general much smaller than the output range of the last building block in the receiver chain. This is illustrated in Fig. 3.5 and Fig. 3.6, showing the power spectral density (PSD) of the complex[1] GMSK output signal of a quadrature DCS-1800 receiver in zero-IF and low-IF mode, respectively. The DC-offset and the LF noise densities are indicated by an arrow and solid lines, respectively. It is clear that a zero-IF receiver is more susceptible to $1/f$ noise and DC-offset than a low-IF receiver. The total integrated noise is higher (lower *SNR* at a given signal power because of the $1/f$ noise) *and* the noise and the DC-offset obscure the signal contents around its peak PSD. Very sophisticated feedback mechanisms from the digital modem back to the analog front-end, can compensate for the DC-offset. However, since this feedback loop has a finite time-constant, also part of the signal contents is canceled by the loop, further degrading the signal quality. The low-IF receiver on the other hand is much less vulnerable because both the DC-offset and the $1/f$ noise lie at a much more comfortable position. As long as the DC-offset does not saturate the A/D converters, there is no signal degradation. In fact,the *SNR* can still be improved by filtering the DC component along with part of the low-frequency noise in the digital domain.

The low-frequency building blocks also need to handle high dynamic range signals, because of the absence of filtering in the RF part. Especially the A/D converters need to have a high accuracy. On the other hand, the wanted signal is situated at low frequencies (e.g. 100 kHz) instead of 10 MHz which is common in heterodyne receivers. It is therefore no problem to use oversampling converters, as long as the filter sufficiently attenuates the blocking signals to prevent aliasing.

Clearly, the accuracy-bandwidth-dynamic range trade-off is much more pronounced here

[1]Quadrature signals can efficiently be represented using complex notation.

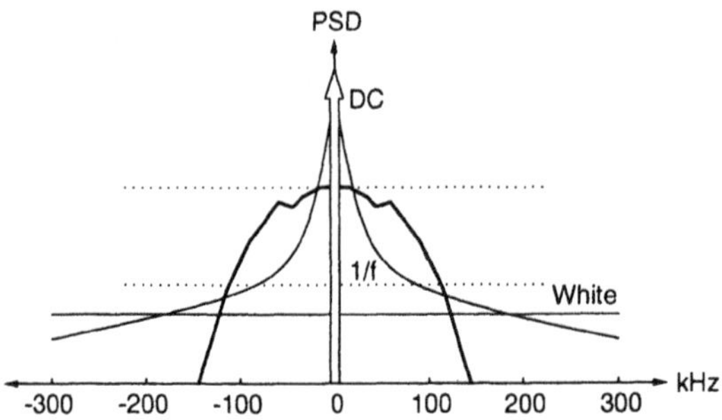

Figure 3.5: *PSD of the output signal of a quadrature receiver ($I + jQ$) in zero-IF mode, including low-frequency noise and DC-offset.*

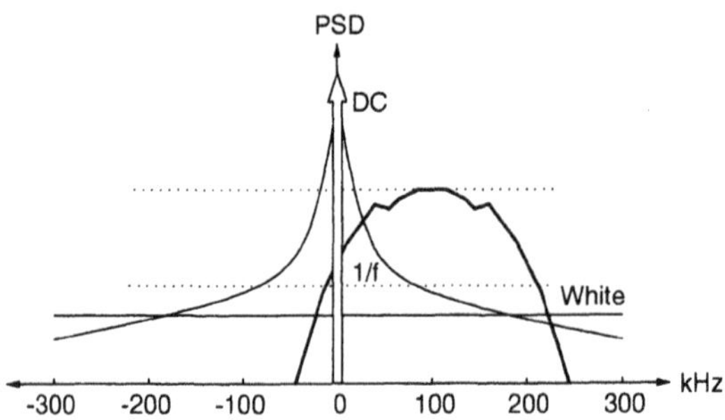

Figure 3.6: *PSD of the output signal of a quadrature receiver ($I + jQ$) in low-IF mode, including low-frequency noise and DC-offset .*

than it was in the heterodyne receiver. Non-idealities that previously were unimportant in heterodyne topologies (like e.g. mismatch, IP_2, offset and self-mixing) have now become real constraints. In fact, one must admit that it is the direct consequence of postponing both the image rejection and the channel selection to the digital part. Of course, all of this puts a lot of pressure on the power consumption of the analog part.

Hence, there is simply no substitute for cubic inches: combining a low power consumption with a maximum degree of integration is not straightforward. This is the major challenge when integrating these receivers.

3.2.3 Other Architectures

Other architecture variants exist which are suitable for building highly integrated receivers, like e.g. the wide-band IF architecture [Rude97], or the wide-band low-IF architecture [Crol97]. Basically, these receivers are two-stage low-IF receivers with an intermediate quadrature down-conversion to a fixed intermediate IF. The image rejection of this first mixing operation is ensured by the antenna filter and the quadrature nature of the down-conversion. Nevertheless, these architectures are not so suitable for mobile telephony (cellular) applications. The reason is that in these applications the signals residing at the mirror frequency during the first down-conversion can be extremely large with respect to the wanted signal. Even though two image-suppression mechanisms are active (a filter operation and a quadrature conversion), this will require both a very good quality of the quadrature LO and a large stop-band rejection in the antenna filter. For cordless telephony applications (like e.g. the DECT standard), the achievable performance is adequate however.

3.3 A Low-IF Receive Path for a DCS-1800 Transceiver

From the previous section, it is clear that the low-IF architecture offers both the most integrated solution and the best path to full-CMOS integration:

- No external filters are required except for the antenna filter. In this way, the number of external high-frequency nodes (read: low-impedance interfaces) and therefore also the number of external matching components is reduced to a minimum. The absence of low-impedance interfaces makes this architecture particularly CMOS-friendly.

- A low-IF receiver features the same high degree of integration as the zero-IF receiver but does not suffer as much from its non-idealities.

- The receive chain is very compact as it requires only a minimum of consecutive stages, resulting in a low power consumption (although the tough dynamic range requirements may somewhat damp this effect)

- The image rejection and the channel selection can be done in a highly scalable CMOS DSP along with the demodulation (see Fig. 3.8). Hence, if one manages to build the complete RF front-end in CMOS, a true path is offered to full-CMOS (single-chip?) integration.

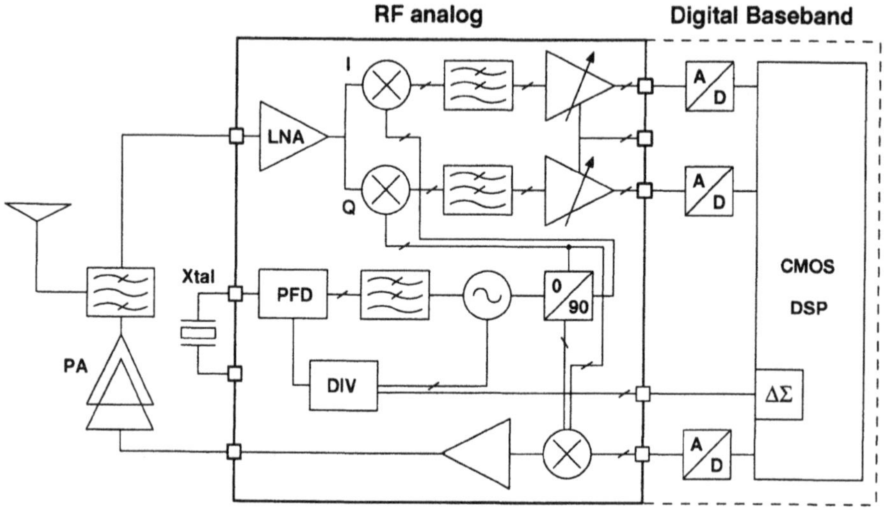

Figure 3.7: *A complete DCS-1800 cellular system can be built by flanking the transceiver IC by a minimum number of external components .*

Therefore, the remainder of this work focuses on the design of a low-IF receive path which fits in a full-CMOS DCS-1800 transceiver front-end.

Fig. 3.7 shows the goal of this work in a wider context: the development of a full-CMOS transceiver front-end. By combining the low-IF receiver with a direct up-conversion transmitter and an on-chip PLL frequency synthesizer, the receive and the transmit function can be realized on a single die. Just as the low-IF receiver, the direct up-conversion transmitter architecture does not require any filtering in between stages and hence requires no inter-stage components. A single synthesizer can serve both the receiver and the transmitter because the reception and the transmission of frames does not occur at the same time. In addition, the use of a high reference frequency enables th loop filter to be integrated on-chip. As a result, a complete DCS-1800 cellular system can be build by flanking the transceiver IC by a minimum number of external components: a duplexer, a power amplifier, a crystal and a baseband chip (containing the A/D, D/A and the all-digital delta-sigma fractional-N PLL steering).

If the digital coupling stands in the way of a true single-chip CMOS solution with the analog and the digital part on the same substrate — which is (still) the situation at the moment — this architecture still allows for a two-chip solution. In that case, the A/D converters and the D/A converters should be assigned to the analog front-end where they share the same, quiet ground. For all purposes they should be considered as analog building blocks. It will be shown in Subsection 3.5.3 that the specifications of these building blocks are relatively relaxed compared to converters that have been published in the past, like e.g. [Marq98, Geer98]. The interested reader is referred to these publications.

Fig. 3.8 shows a possible implementation of the digital modem [Crol97]. The quadrature outputs of the low-IF receiver are fed to a set of complex digital mixers. The quadrature mixing process implements the actual image rejection and translates the signal from 100 kHz IF down

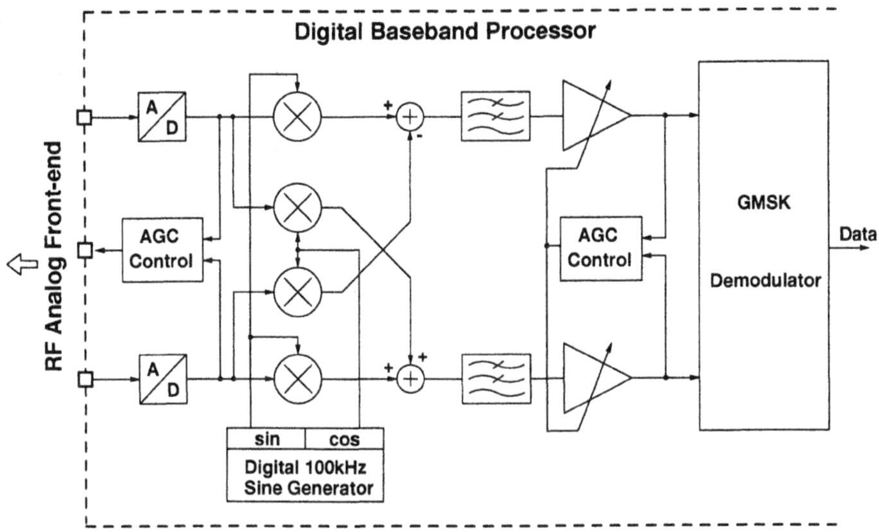

Figure 3.8: *The digital back-end.*

to a quadrature signal at DC. From this moment on, the digital back-end is the same as for a zero-IF receiver. Subsequently, the signals are fed to high-order digital low-pass filters that implement the actual channel selection. The low-pass filters also remove the DC-offset, which is now positioned around 100 kHz. After applying some digital AGC, the quadrature signals are fed to a GMSK demodulator that extracts the data bits from the constellation.

3.4 From DCS Standard to Receiver Specifications

The DCS-1800 standard [ETSI] specifies the minimum reception quality that must be guaranteed under certain 'loading' conditions, like e.g. adjacent interference, blocking signals, low input powers etc. In this section, the requirements of the DCS-1800 standard are systematically mapped onto a set of measurable specifications for a receiver front-end employing the low-IF architecture. The requirements for the local oscillator section will be derived as well since its phase noise critically determines the blocking performance of the receiver. In the end, this results in the overall system requirements summarized in Table 3.2 on page 39.

3.4.1 From Bit-error Rate to Signal-to-noise Ratio

In Section A.1, the receiver sensitivity is defined as the minimum signal level that is still detectable with a required, application-specific signal-to-noise ratio (*SNR*). In reality however, the sensitivity and the reception quality of a complete digital communication system are not defined in terms of signal-to-noise ratios but are defined in terms of the bit-error rate (*BER*) — the percentage of bits that have errors relative to the total number of bits received in a transmission. Hence, in order to be able to derive the specifications for the RF analog part of a digital re-

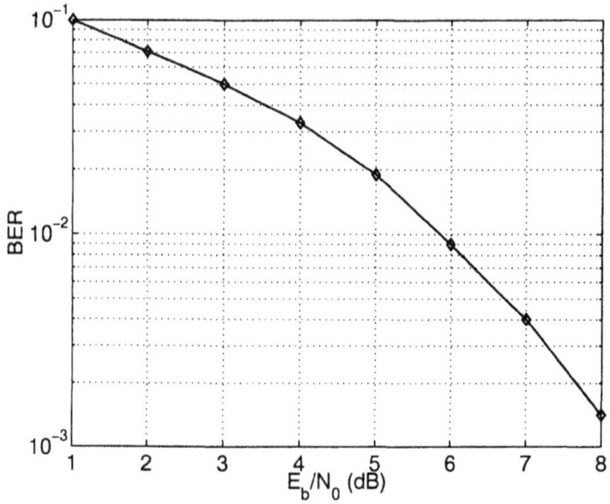

Figure 3.9: *The bit-error rate of a GMSK signal as a function of E_b/N_o, the SNR for a noise bandwidth equal to the bit rate r_b.*

	Reference sensitivity	Maximum *BER*
Class 1&2	-100 dBm	2%
Class 3	-102 dBm	2%

Table 3.1: *Reference sensitivity performance for a DCS-1800 mobile station.*

ceiver, the required *BER* needs to be mapped onto an equivalent minimum *SNR*. For a given (de)modulation method, the *BER* values can directly be translated into a corresponding *SNR*. As an example, Fig. 3.9 shows the theoretical bit-error rate of a GMSK modulated signal as a function of E_b/N_o, the *SNR* of the signal when taking into account the noise in a bandwidth equal to the bit rate r_b.

Table 3.1 states the reference sensitivity performance for a DCS-1800 mobile station [ETSI, p. 25, 6.2 and p. 27], consisting of a sensitivity level and a corresponding maximum *BER*. Clearly, of the three different DCS-1800 flavors that exist the so-called class 3 flavor is the most sensitive; In a class 3 mobile station, one must be able to achieve a *BER* of 2% at signal powers as low as -102 dBm — which is less than 10^{-13} W! We will therefore focus on achieving the sensitivity of this class. Using Fig. 3.9, this *BER* translates into a required E_b/N_o of about 4.9 dB. Knowing that the bit-rate during each DCS-1800 burst is 270.83 kbps, and taking into account the channel spacing of 200 kHz, this requires a minimum *SNR* of 6.2 dB in a 200 kHz bandwidth. In the next subsections, we will work towards an *SNR* of 7 dB for -102 dBm signals — corresponding to a *BER* of 1.2% in a Class 3 mobile station — or equivalently, towards an *SNR* of 9 dB for -100 dBm signals — corresponding to a *BER* of 0.3% for a Class 1/2 mobile station.

3.4.2 Noise Figure

The relative degradation in *SNR* due to system noise, and hence also the degradation in sensitivity, is quantified by the noise figure *NF*, which is defined as

$$NF = 10\log_{10}\left(\frac{SNR_{IN}}{SNR_{OUT}}\right). \tag{3.4}$$

The noise figure requirement can be derived from the required receiver sensitivity; In the previous subsection, it has been shown that to meet the DCS-1800 reference sensitivity performance in Table 3.1, an *SNR* of about 7 dB must be guaranteed at the respective power levels. The equivalent input noise power at the antenna input must therefore be lower than:

$$-102\text{ dBm} - 7\text{ dB} = -109\text{ dBm}. \tag{3.5}$$

Knowing that the noise power of a noise source at an absolute temperature T is given by

$$10\log_{10}\left(\frac{kTB}{1\text{ mW}}\right)\text{ dBm}, \tag{3.6}$$

the noise power of a source at 290 K in a bandwidth of 200 kHz can be calculated as

$$-173.87\text{ dBm} + 53\text{ dB(integration over the 200 kHz channel)} = -120.87\text{ dBm}. \tag{3.7}$$

The noise figure required at the level of the antenna is thus

$$-109\text{ dBm} - (-120.87\text{ dBm}) = 11.87\text{ dB}. \tag{3.8}$$

Taking into account the 3 dB attenuation of the antenna filter, the noise figure requirement for the receive path becomes:

$$11.87\text{ dB} - 3\text{ dB} = 8.87\text{ dB}. \tag{3.9}$$

3.4.3 Image Rejection Ratio

The image rejection ratio quantifies the factor by which the mirror signal — i.e. the parasitic signal which is folded onto the wanted signal and ideally is removed after down-conversion by combining the quadrature paths — is suppressed. Evidently, this factor must be large enough in order not to distort the wanted signal.

The image rejection specification can be derived from the DCS-1800 reference interference specification in [ETSI, p. 26, 6.3]. This specification requires that the so-called reference interference performance — basically a minimum *BER* of 4 % [ETSI, p. 29, 6.3] — must be guaranteed in the presence of co-channel and adjacent channel interference. A co-channel interferer is a strongly attenuated signal from a distant cell that is being transmitted at the same frequency as the wanted signal. The adjacent channels are predominantly caused by users in the same cell (like e.g. the one at 400 kHz) and adjacent cells (like e.g. 200 kHz).

Fig. 3.10 shows the co-channel and adjacent channel interference levels a DCS-1800 receiver must be able to cope with: a co-channel interferer (9 dB lower than the wanted signal), a first adjacent channel (9 dB larger than the wanted signal), and a second adjacent channel (41 dB larger than the wanted signal). All GMSK power spectra are drawn on scale, as if they were

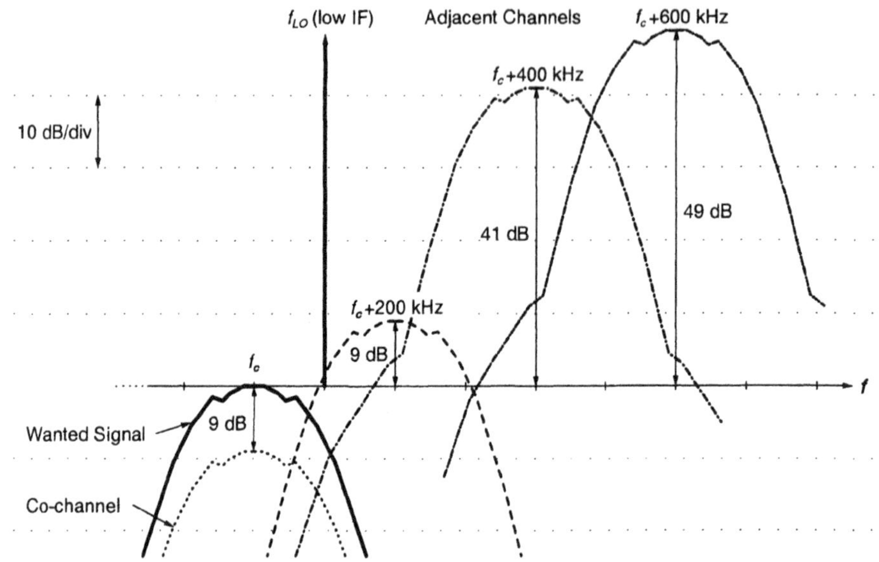

Figure 3.10: *Co-channel and adjacent interference in DCS-1800. The required BER must still be achieved after down-conversion when these interferers are present (one at a time).*

measured on a spectrum analyzer with a 30 kHz integration bandwidth. Before moving on to the actual calculation, one must be aware of a few things:

- The interference specification applies for a wanted signal input level of 20 dB above the reference sensitivity — so that noise is not important — and for a random, continuous GSM modulated interfering signal.
- The reference interference performance must be met for each radio-frequency interferer separately i.e. only one at a time.
- The relative power of the interferers are the same for all DCS-1800 classes.
- The adjacent channel at 600 kHz does not require testing [ETSI, p. 26, 6.3].
- In the subsequent calculation, we will not use the *SNR* values from Fig. 3.9 since the conversion from *BER* to *SNR* in this figure assumes that the noise is completely white, whereas here the 'noise' is an uncorrelated GMSK channel or its side-lobe. The *IMRR* specification will be derived by an indirect reasoning [Crol97].

The worst case interference occurs when the interferers and the wanted signal are at opposite sides of the local oscillator: in this case the suppressed interferers are directly folded onto the wanted signal. In a high-injection receiver this happens when the interferers lie above the LO frequency while in a low-injection receiver, when the interferers lie below the LO frequency. Fig. 3.11 shows the PSD of the $I + jQ$ output signal of a low-IF DCS-1800 receiver in the presence of co-channel and adjacent interference. The example depicted in the figure is a

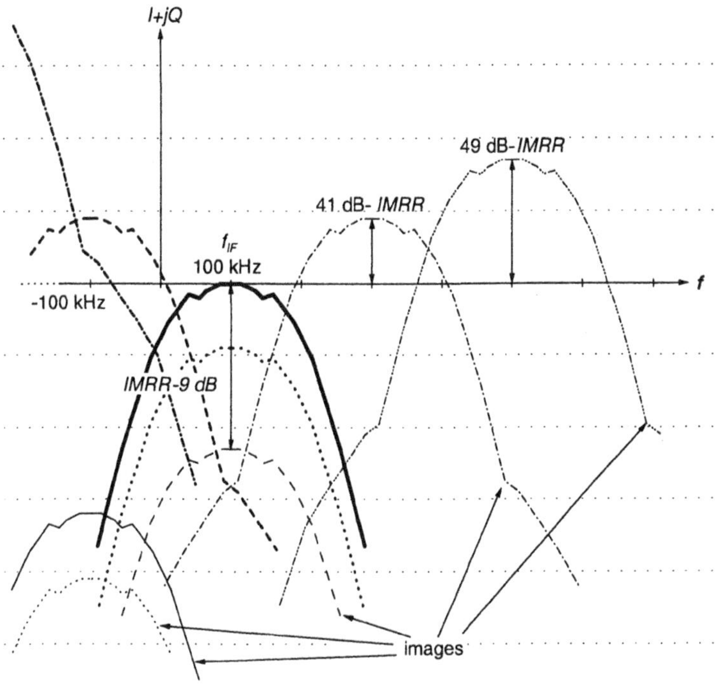

Figure 3.11: *$I + jQ$ output signal of a low-IF DCS-1800 receiver in the presence of co-channel and adjacent interference. In this example, the IMRR is chosen 32* dB. *Again, when judging the interference, only one interferer (and its image) must be accounted for at a time.*

high-injection receiver which mixes the signal down with a positive frequency. In the following paragraphs we will look at the effect of a finite *IMRR*.

If there is a co-channel interferer, and if we have a receiver with an infinite *IMRR*, the *SNR* of the wanted signal by definition equals 9 dB: the co-channel can not be removed because it can not be discerned from the wanted signal. Since the *BER* in the presence of a co-channel must — by construction — be achievable, the *SNR* required to meet the *BER* associated with the reference interference performance is necessarily less than 9 dB. This insight allows one to derive the minimum *IMRR* to sufficiently suppress the 200 kHz adjacent channel; Because the image of the adjacent channel at 200 kHz is directly superimposed onto the wanted signal, at exactly the same position where the co-channel can interfere with the wanted signal, it suffices to suppress this interferer to the level of the co-channel interference, requiring a mirror signal suppression of

$$9\ \text{dB} - (-9\ \text{dB}) = 18\ \text{dB}. \tag{3.10}$$

Probably, this is even an overkill...at least for the adjacent interference at 200 kHz; In case the interference is white — and a GMSK signal has a relatively flat PSD — even an *SNR* as low as 7 dB already suffices to achieve a *BER* of 2%, two times better than the 4% *BER* required by the reference interference spec. The image rejection spec required by the adjacent channel at 400 kHz offset (the dash-dotted line in Fig. 3.10) is much more stringent, however. After down-conversion, this channel will become centered around 300 kHz, 200 kHz above the 100 kHz IF. This is indicated by the thin dash-dotted line in Fig. 3.11. At 200 kHz, the tail of this adjacent channel directly interferes with the wanted signal and must be suppressed by ensuring a high enough image suppression. As shown in Fig. 3.10, the DCS-1800 standard requires that the reference interference performance is met when the adjacent channel is 41 dB larger than the wanted signal. Since the *BER* is by construction preserved when the un-removable tail of the adjacent channel at 200 kHz offset is interfering with the wanted signal (the dashed line in Fig. 3.10), and the image of the 400 kHz adjacent channel appears at exactly the same frequency, it suffices to suppress the image of the 400 kHz adjacent channel to the level of the adjacent channel at 200 kHz. Since the 200 kHz adjacent signal is 9 dB higher than the wanted signal, the required image rejection for the adjacent channel at 400 kHz offset from the wanted signal is:

$$41\ \text{dB} - 9\ \text{dB} = 32\ \text{dB}. \tag{3.11}$$

It is precisely this situation that is depicted in Fig. 3.11. Observe that the un-removable part of the interference caused by the 400 kHz adjacent signal — the dash-dotted tail around DC in Fig. 3.11 — is another 10 to 15 dB lower in power than the interference of the image.

3.4.4 LO Leakage

According to the DCS-1800 standard [ETSI, p. 19, 4.3.3], the spurious power emissions can not be higher than -47 dBm (20 nW) between 1 GHz and 12.75 GHz. In case of a receiver, this refers to the parasitic emission of the local oscillator signal (LO leakage) due to the finite reverse isolation of the LNA and the mismatch in the differential LO phases of the mixer. Since the attenuation of the antenna filter is 3 dB, at the level of the receiver input the spurious emissions are allowed to be 3 dB higher. This results in a -44 dBm spec for the IC itself.

3.4.5 Intermodulation Performance

As was already the case in the noise figure specification, the class 3 standard turns out to be the most stringent; A stated in [ETSI, p. 24, 5.3], the class 3 standard requires that a wanted signal 3 dB above the reference sensitivity (-102 dBm + 3 dB = -99 dBm) is still detectable with a *BER* of 2% in the presence of a -45 dBm static sine (at 800 kHz offset from the wanted signal) and a -45 dBm GMSK signal (at 1.6 MHz offset from the wanted signal). In the class 1/2 system, the wanted signal and the interferers are -97 dBm (2 dB more) and -49 dBm (4 dB less), respectively.

Since the nominal loss of the RF filter is 3 dB, the two interferers that reach the input of the receiver are both -48 dBm. The GMSK interferer that is generated due to third order intermodulation can under these conditions not be larger than an equivalent input signal with the following power:

$$-99 \text{ dBm} - 3 \text{ dB} - 7 \text{ dB} = -109 \text{ dBm}. \tag{3.12}$$

Knowing that

$$IP_3 = P_{in} - IM_3/2, \tag{3.13}$$

the required IP_3 to meet this specification is thus:

$$-45 \text{ dBm} - 3 \text{ dB} + (-45 \text{ dBm} - 3 \text{ dB} - (-109 \text{ dBm}))/2 = -17.5 \text{ dBm}. \tag{3.14}$$

The above calculation assumes that the effect of an uncorrelated GMSK signal is approximately equivalent to the effect of white noise of the same power. This is a reasonable assumption since the PSD of a GMSK signal is quite flat in the center.

When the out-of-channel IP_3 value is used to prove that the DCS-1800 spec in [ETSI, p. 24, 5.3] is met, it must always be checked whether the out-of-channel IP_3 is indeed relevant for the 'almost in-channel' 800 kHz and 1.6 MHz interferers. This is done by showing that amplification of the interferers does not result in saturation of the output amplifier.

3.4.6 Phase Noise

According to [ETSI, p. 22, 5.1 and p. 25, 6.2], a wanted signal 3 dB above the reference sensitivity level of -102 dBm must still be detectable with a *BER* of 2% in the presence of large, *sinusoidal* blocking signals. the maximum power of these sinusoidal blocking signals is shown in Fig. 3.12 as a function of the frequency offset from the wanted channel. As the out-of-band blockers are strongly attenuated by the antenna filter-duplexer, the largest remaining blocking signals are the in-band blockers.

If the local oscillator signal were ideal, the only requirement would be that the signal path does not saturate when a blocker propagates through the receiver and that the A/D converters have a large enough dynamic range to sample the wanted signal together with the blocker. However, in practice the carrier exhibits a phase noise skirt (see Fig. 3.12) which down-converts the blockers onto the wanted signal, degrading the signal's *BER*. In addition, the phase noise down-converts the adjacent channels, compromising the reference interference performance stated in [ETSI, p. 29, 6.3]. For these two reasons, the local oscillator needs to feature a very low phase noise.

The most critical unwanted signals that can be down-converted onto the desired signal are the adjacent channel at 400 kHz— which can be 41 dB larger than the wanted signal —, the

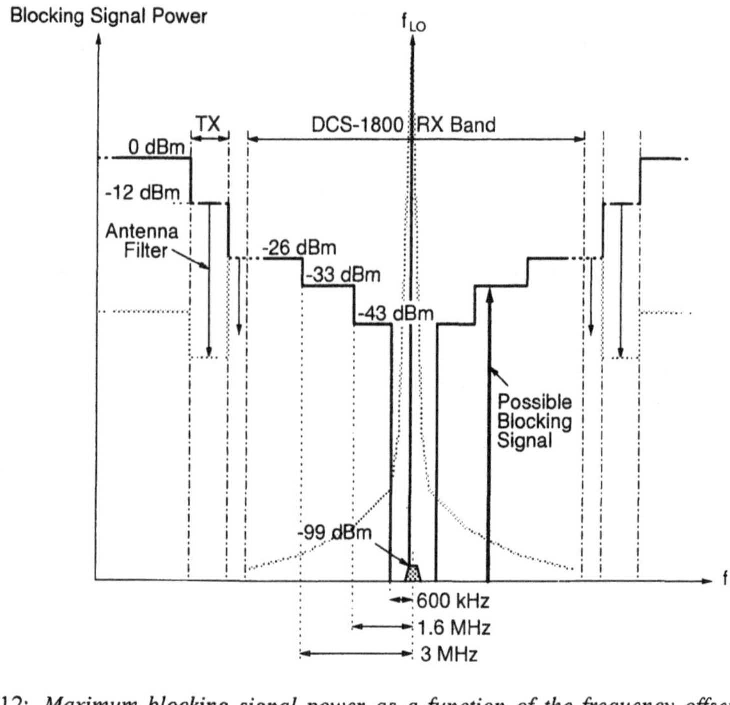

Figure 3.12: *Maximum blocking signal power as a function of the frequency offset from the wanted channel. (The 3* dB *in-band loss of the antenna filter is not drawn here.)*

Overall System Requirements	Noise figure (*NF*)		
	DCS-1800 spec	<8.87 dB	3.4.2
	Goal	5 dB	3.5.2
	Image rejection ratio (*IMRR*)	>32 dB	3.4.3
	Third-order intercept point (IP_3)		3.4.5
	DCS-1800 spec	>-11 dBm	3.5.2
	Goal	-7.5 dBm	3.5.2
	LO leakage	<-44 dBm	3.4.4
	Phase noise		3.4.6
	at 600 kHz offset	<-125.8 dBc/Hz	
	at 3 MHz offset	<-143.7 dBc/Hz	
	Spurious suppression	>80 dBc	3.4.7

Table 3.2: *Summary of the derived specifications for the low-IF receiver as a whole.*

-43 dBm blocking signal at 600 kHz, the -33 dBm blocking signal at 1.6 MHz and the -26 dBm blocking signal at 3 MHz. Hence, in order to obtain a *SNR* of 7 dB after down-conversion — which is sufficient for a BER of 2% —, the phase noise needs to be lower than:

$$\mathcal{L}\{400\text{ kHz}\} \leq -41\text{ dB} - 7\text{ dB} - 10\log(200\text{ kHz}) = -101\text{ dBc/Hz} \tag{3.15}$$

$$\mathcal{L}\{600\text{ kHz}\} \leq -102\text{ dBm} + 3\text{ dB} - (-43\text{ dBm}) - 7\text{ dB} - 10\log(200\text{ kHz}) = -116\text{ dBc/Hz} \tag{3.16}$$

$$\mathcal{L}\{1.6\text{ MHz}\} \leq -102\text{ dBm} + 3\text{ dB} - (-33\text{ dBm}) - 7\text{ dB} - 10\log(200\text{ kHz}) = -126\text{ dBc/Hz} \tag{3.17}$$

$$\mathcal{L}\{3\text{ MHz}\} \leq -102\text{ dBm} + 3\text{ dB} - (-26\text{ dBm}) - 7\text{ dB} - 10\log(200\text{ kHz}) = -133\text{ dBc/Hz}. \tag{3.18}$$

The above calculation makes use of the fact that the phase noise at the frequency offset of interest is quasi-white a 200 kHz scale. Assuming that the phase noise of a frequency synthesizer has a -20 dB per decade roll-off for the critical offset frequencies, the phase noise specification at 3 MHz turns out to be the most stringent.

3.4.7 Spurious Suppression

The specification in [ETSI, p. 22, 5.1 and p. 25, 6.2] not only determines the LO phase noise, but also the allowable spur levels; A spur at a certain frequency offset from the carrier can be seen as an LO signal for a possible blocking signal at that frequency offset. Since the maximum blocking signal is -26 dBm (see Fig. 3.12), and since under these conditions, a -99 dBm wanted signal (-102 dBm + 3 dB) must still be detectable with a *BER* of 2%, this results in a spurious specification of roughly -80 dBc:

$$-99\text{ dBm} - 7\text{ dB} - (-26\text{ dBm}) = -80\text{ dBc}. \tag{3.19}$$

3.5 From Receiver Specifications to Circuit Specifications

In this section, it is shown how the receiver specifications are assigned to the different building blocks. In the end, this results in the building block specifications summarized in Table 3.3 on page 45 and the level diagram in Fig. 3.14 on page 46.

Note that the mapping of the specifications is often subject to change; Only during the actual design phase, when the building block topologies are known, the true impact of each specification can be determined and specifications can be traded against one another.

3.5.1 The Low Noise Amplifier

The LNA gain is set by two considerations:

1. The gain needs to be be large enough in order to bring the smallest possible wanted signal sufficiently above the noise floor of the mixers. Since it is quite a challenge to realize mixer noise figures below 18 dB, it has little sense designing an LNA with a gain less than 18 dB.

2. The gain must be low enough in order to make sure that the down-converters are not saturated by the amplified blocking signal. The maximum gain is determined by the largest in-band blocking signal (-26 dBm according to [ETSI, p. 23, 5.1]) and the input capability as well as the linearity of the mixer. This limits the gain to

$$\textit{gain} \leq \textit{input capability mixer} - (-26\ \text{dBm} - 3\ \text{dB}). \tag{3.20}$$

Assuming a realistic value of -10 dBm (i.e. 0.2 V_{ptp}) for the mixer's input capability, the allowable gain is 19 dB.The design window is thus quite small.

In this work, the target LNA gain (including AC coupling) has been set to 18 dB. Even when the input capability of the mixer is larger, it is not recommended to further increase the LNA gain; Achieving higher gains requires either a lot of current in the LNA (that is, in case of a resistor loaded LNA) or a high-Q LC tank, which makes the LNA sensitive to process variations. In addition, a higher gain inevitably toughens the linearity requirements of the mixer.

It also follows from the above discussion that the input range of the LNA needs to be larger than

$$-26\ \text{dBm} - 3\ \text{dB} = -29\ \text{dBm}. \tag{3.21}$$

After all, wanted signals 3 dB above the reference sensitivity — signals that *need* the LNA — must be able to be received in the presence of these blocking signals. At large wanted signal powers, there comes a moment when the LNA needs to be bypassed, as the input capability of the mixers would be exceeded. According to [ETSI, p. 25, 6.1]) a wanted signal can be maximally -23 dBm. In case the LNA would still be in the receive chain, the mixer gets an input signal of

$$-23\ \text{dBm} - 3\ \text{dB} + 18\ \text{dB} = -8\ \text{dBm}. \tag{3.22}$$

Hence, bypassing is indeed required in practice.[2]

The LNA noise figure must be low, yet, it has little sense to design for a noise figure less than 3 dB as the mixers will most probably dominate the total noise figure anyway. Therefore, we

[2] In this work, the bypass circuit is not implemented.

have chosen 3 dB as target noise figure for the LNA. The IP_3 spec of the LNA (set to −4 dBm) is clarified in the next subsection, along with the mixers.

3.5.2 Quadrature Mixers

The required noise figure of the complete receive path is 8.87 dB, of which 3 dB has already been assigned to the LNA. Taking into account the 18 dB gain of the LNA, the required noise figure for the quadrature down-converter becomes 25.6 dB. However, in order to get some margin, the noise figure specification of the receive path has been set to 5 dB, requiring a noise figure less than 18.6 dB from the quadrature down-converter. For a single down-converter this boils down to a required noise figure of 21.6 dB, provided that all the output noise between DC and 200 kHz is taken into account. The detailed proof is given in Subsection A.3.2.

According to Subsection 3.4.5, the required IP_3 of the receive path is -17.5 dBm. However, one must also guarantee that when receiving the maximum wanted signal (i.e. -23 dBm-3 dB) the signal is never distorted by its own intermodulation products. In first order (and in worst case) this requires that

$$IM_3 = -2(IP_3 - (-23 \text{ dBm} - 3 \text{ dB})) < -30 \text{ dB}, \tag{3.23}$$

which yields

$$IP_3 < -11 \text{ dBm}. \tag{3.24}$$

Since the third-order harmonic distortion products are minimally 10 dB lower than the intermodulation products, and since they experience some extra attenuation, folding of harmonics though multiplication with the higher order harmonics of the local oscillator is not a problem. In this work, -7.5 dBm is considered the target IP_3, which corresponds to a margin of 10 dB with respect to [ETSI, p. 24, 5.3] and 3.5 dB with respect to (3.24). Assuming a worst case combination of the different intermodulation products leads to a required IP_3 of -4 dBm for the LNA and +14 dBm for the down-conversion mixer.

3.5.3 VGA – Filter

The main task of the low-pass filter and the variable-gain amplifier consists in reducing the dynamic range requirements of the subsequent data converters by providing some filtering of the blocking signals and the adjacent channels and adjusting the signal to the input range of the A/D converter. The following calculations assume a simple first-order, un-tuned (±25%) 300 kHz filter. Such a filter can easily be implemented using a trans-impedance configuration. The order of this filter can be increased to relax the specifications for the A/D converters. Let us first calculate the required dynamic range of the A/D converters.

The 400 kHz adjacent channel (maximally 41 dB larger than the wanted signal), the 600 kHz adjacent channel (maximally 49 dB larger than the wanted signal) the 600 kHz blocker (maximally -43 dBm), and the 3 MHz blocker (maximally -26 dBm) experience an minimum attenuation of 2 dB, 4.5 dB, 5.5 dB, and 23.5 dB respectively. When these signals are properly scaled to the input range of the A/D converters, the converters need to feature a minimum dynamic range of respectively

$$41\ \text{dB} - 2\ \text{dB} + 7\ \text{dB} = 46\ \text{dB}, \tag{3.25}$$

$$49\ \text{dB} - 4.5\ \text{dB} + 7\ \text{dB} = 51.5\ \text{dB}, \tag{3.26}$$

$$-43\ \text{dBm} - 5.5\ \text{dB} - (-102\ \text{dBm} + 3\ \text{dB}) + 7\ \text{dB} = 57.5\ \text{dB}. \tag{3.27}$$

$$-26\ \text{dBm} - 23.5\ \text{dB} - (-102\ \text{dBm} + 3\ \text{dB}) + 7\ \text{dB} = 56.5\ \text{dB}. \tag{3.28}$$

The 3 dB in the above equation stems from the fact that blocking performance is specified for a wanted signal 3 dB above the reference sensitivity. As a result, these requirements put a minimum of 60 dB on the dynamic range of the data converters, corresponding to 10 bit converters. In the next paragraphs it will be shown that the required number of bits is somewhat larger: some margin (two to three bits) must be foreseen 1) to implement the channel equalizer and the GMSK modem in the digital back-end. 2) to accommodate the DC offset and 3) to meet the reference sensitivity performance given the limited set of VGA gain levels.

Let us first look at the VGA. The VGA function must ensure that the wanted signal is maximally amplified to optimally exploit the dynamic range of the converter and to ensure that the above conditions are actually realized. Assuming a $2V_{ptp}$ (+11 dBm) output capability, a gain of

$$+11\ \text{dBm} - (-102\ \text{dBm} + 20\ \text{dB} - 3\ \text{dB} + 49\ \text{dB} - 4.5\ \text{dB}) = 51.5\ \text{dB} \tag{3.29}$$

must be implemented to realize (3.26) (which is more stringent than (3.25)). The 20 dB stems from the fact that the adjacent channels are specified for signals starting from 20 dB above the reference level. The 3 dB stands for the in-band loss in the blocking filter. To implement (3.27), a total receiver gain of

$$+11\ \text{dBm} - (-43\ \text{dBm} - 5.5\ \text{dB} - 3\ \text{dB}) = 62.5\ \text{dB} \tag{3.30}$$

must be generated. Apparently this requirement is the most stringent one. This gain can be realized as 18 dB gain in the LNA along with 44.5 dB gain in the VGA.

The minimum gain must be realized when the maximum wanted signal enters the receiver. If one wants to amplify this signal to the full input capability of the data converters, this would require a receive path gain as large as

$$+11\ \text{dBm} - (-23\ \text{dBm} - 3\ \text{dB}) = 37\ \text{dB}. \tag{3.31}$$

However, this is not really necessary since the dynamic range of the A/D converters is large enough anyway. Moreover, the 18 dB gain of the LNA can safely be bypassed when these signals enter the receiver. In this work, the VGA (including the mixer) offers a gain between 19 dB and 43 dB, in five 6 dB steps.

3.5.4 A/D Converters

In the previous subsection we have seen that the dynamic range of the A/D converters must at least be 57.5 dB. In this section we will derive how much extra dynamic range must be foreseen.

Let us again look at the VGA. Whereas the 6 dB steps in the VGA ease the implementation of the VGA, it must be checked whether this doesn't come at the cost of one extra bit (6 dB *SNR*) in the A/D converters. After all, if an adjacent channel or a blocking signal would be amplified a tiny bit above +11 dBm, the VGA would have to lower the gain by 6 dB to prevent saturation of the converters. If there would be no extra bit foreseen, this would in that case result in the inability to detect signals with a power below 9 dB above the reference sensitivity. Fortunately, this is not the case in this receiver, as the blocking signals (−43 dBm at 600 kHz and −26 dBm at 3 MHz) in combination with the largest VGA gain are still lower than +11 dBm:

$$-43\text{ dBm} - 3\text{ dB} + 18\text{ dB} + 43\text{ dB} - 5.5\text{ dB(attenuation of filter)} = 9.5\text{ dBm} \tag{3.32}$$

$$-26\text{ dBm} - 3\text{ dB} + 18\text{ dB} + 43\text{ dB} - 23.5\text{ dB(attenuation of filter)} = 8.5\text{ dBm}. \tag{3.33}$$

The presence of DC offset does require some extra margin in the A/D converters. The reason is that the offset at the output of the VGA can be quite large (in the order of 500 mV) because of the large amplification factor...and the fact that we are working in CMOS. The margin can be calculated as

$$20\log_{10}\left(\frac{2\text{ V}_{\text{ptp}} + 1\text{ V}}{2\text{ V}_{\text{ptp}}}\right) = 3.5\text{ dB}, \tag{3.34}$$

resulting in the requirement for an additional 0.5 bit.

Since the maximum gain of the VGA has been set to 43 dB, the -43 dBm blocking signal at 600 kHz is not amplified to +11 dBm, but to 9.5 dBm. This means that an extra 1.5 dB must be foreseen in the A/D. Note that in order to cope with the overloading behavior of the delta-sigma A/D converters this margin is necessary anyway.

One last correction stems from the fact that the A/D must be able to quantize the minimum signal level, which is 3 dB less that the signal assumed in (3.27).

As a result, the theoretically required dynamic range of the A/D converter is:

$$57.5\text{ dB(eqn. (3.27))} + 3.5\text{ dB} + 1.5\text{ dB} + 3\text{ dB} = 65.5\text{ dB} \quad (11\text{ bit}), \tag{3.35}$$

or alternatively,

$$11\text{ dBm} + 3.5\text{ dB} - (-102\text{ dBm} - 3\text{ dB} + 18\text{ dB} + 43\text{ dB} - 7\text{ dB}) = 65.5\text{ dB} \quad (11\text{ bit}). \tag{3.36}$$

Along with a 3-bit margin for the equalizer and the GMSK demodulation, this requires a 14 bit ADC (84 dB *DR*) with a 200 kHz bandwidth. [Marq98, Geer98] show that such converters (actually much better ones) can definitely be implemented in CMOS. The 14-bit delta-sigma converter in [Anse01] — which is specifically tailored for this purpose — consumes even less than 10 mW!

3.5.5 Overall Quadrature Accuracy

From Subsection 3.4.3 it follows that an image rejection ratio of 32 dB is required to meet the DCS-1800 specification. This property can be translated in a specification on the amplitude and phase mismatch between the in-phase and the quadrature path of the receiver [Rude97]:

$$IMRR = 10\log_{10}\left(\frac{A_{IQ} + \dfrac{1}{A_{IQ}} + 2\cos(\Delta\Phi)}{A_{IQ} + \dfrac{1}{A_{IQ}} - 2\cos(\Delta\Phi)}\right), \tag{3.37}$$

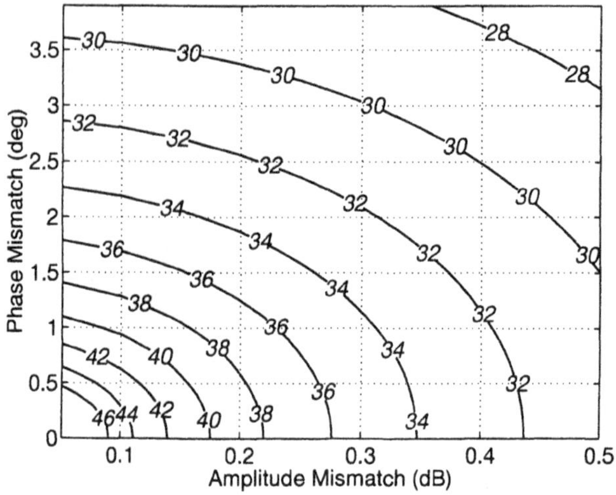

Figure 3.13: *Image rejection ratio versus amplitude and phase mismatch.*

where A_{IQ} and $\Delta\Phi$ are given by

$$A_{IQ} = \frac{A_I}{A_Q} \tag{3.38}$$

$$\Delta\Phi = \frac{\pi}{180}(\phi_I - \phi_Q), \tag{3.39}$$

with A_I and A_Q the gains of the I and Q paths, respectively, and ϕ_I and ϕ_Q the phases of the I and Q signals. This equation is plotted in Fig. 3.13. Obviously, an amplitude accuracy of 0.3 dB (0.4 dB) combined with a phase accuracy of 2 degrees (1 degree) does the trick.

3.6 Specification Summary

Table 3.3 summarizes the specifications for the low-IF receiver's building blocks. Along with the values, also the subsection is indicated where this value has been determined.

Finally, Fig. 3.14 shows the level diagram of the resulting low-IF receiver. In particular, the propagation of the following set of signals is shown:

① The minimum detectable signal (the reference sensitivity of -102 dBm).

② A signal 3 dB above the reference sensitivity level.

③ The largest blocking signal (measured at the output of the receiver) that can occur in the presence of ②, i.e. a -43 dBm signal at 600 kHz offset from the wanted signal.

④ The largest blocking signal (measured at the input of the receiver) that can occur in the presence of ②, i.e. a -26 dBm signal at 3 MHz offset from the wanted signal.

Block	Parameter	Value	Section
LNA	Reflection coefficient (S_{11})	<-10 dB	—
	Gain	18 dB	3.5.1
	Noise figure (NF)	<3 dB	3.5.1
	Third-order intercept point (IP_3)	>-4 dBm	3.5.2
	Input capability	-29 dBm	3.5.1
Mixers	IF frequency	100 kHz	3.2.2
	Conversion Gain	0 dB	—
	SSB Noise figure (NF)		3.5.2
	Quadrature mixer ($I + jQ$)	<18.6 dB	
	I path	<21.6 dB	
	Third-order intercept point (IP_3)	>+14 dBm	3.5.2
	Image rejection ratio ($IMRR$)	>32 dB	3.4.3
	Amplitude accuracy	<0.3 dB	3.5.5
	Phase accuracy	<2 deg	3.5.5
	Input capability	-10 dBm	3.5.1
VGA/Filter	Gain levels	19 dB	3.5.3
		25 dB	
		31 dB	
		37 dB	
		43 dB	
	Nominal cutoff frequency	300 kHz	3.5.3
	Output capability	+14.5 dBm	3.5.3
A/D	Dynamic range	14 bit	3.5.4

Table 3.3: *Summary of the derived specifications for the low-IF receiver's building blocks.*

⑤ The minimum signal for which adjacent channels must be taken into account.

⑥ The largest adjacent channel in the presence of ⑤.

⑦ The largest wanted signal.

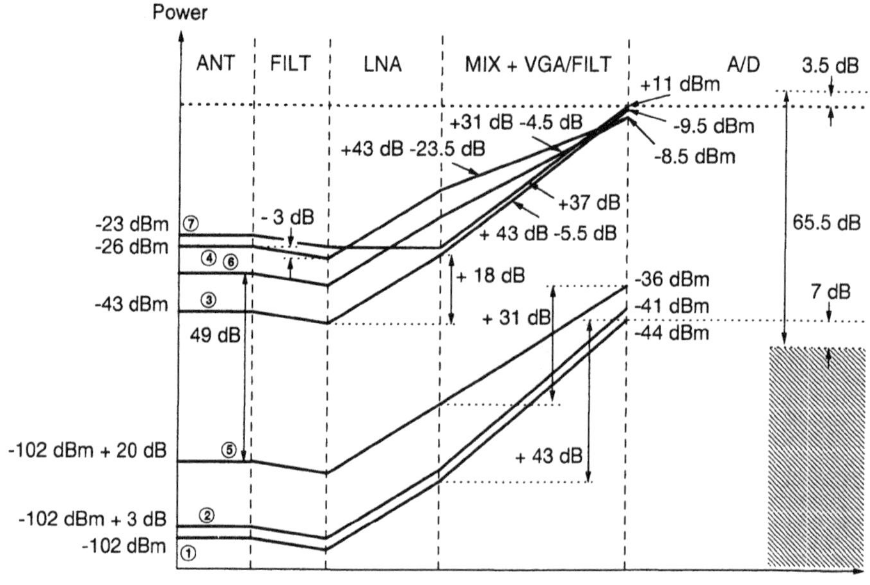

Figure 3.14: *Level diagram of the low-IF receiver.*

3.7 DCS-1800 versus GSM-900

In this chapter a suitable architecture has been derived for a DCS-1800 cellular receiver and the DCS-1800 specifications have been mapped onto the different building blocks. The question may arise whether implementing the GSM-900 standard would have lead to the same numbers, or whether a GSM-900 receiver is more difficult to implement or not. This section should give a clear understanding of the differences between DCS-1800 and GSM-900.

The DCS-1800 system operates at a frequency which is two times larger than GSM-900. Due to the higher operating frequency, it is intrinsically more difficult to achieve a low receiver noise figure:

- As will be shown later on, the noise factor of the low noise amplifiers contains a quadratic term in ω_0, the operating frequency.
- The higher operating frequency makes it more difficult to achieve sufficient gain in the low noise amplifiers. The noise contribution of the down-conversion mixers is therefore increased.
- Because of the larger operating frequency, internal poles in the signal path of the down-conversion mixers tend to degrade the signal strength, causing the $1/f$ noise — which is one of the main noise sources in CMOS mixers — to become dominant. In addition, the $1/f$ noise sources themselves tend to be larger since less capacitance can be tolerated in the mixer core. Otherwise, the local oscillator signal would be too much attenuated. This would also have lead to a larger noise figure, both due to $1/f$ noise as well as due to white noise.

Therefore, even though the sensitivity requirements of DCS-1800 are slightly more relaxed than GSM-900 — GSM-900 must be able to detect -104 dBm signals whereas the reference sensitivity of DCS-1800 is 2 dB higher (-102 dBm) — both systems are quite comparable if it comes to noise... at least at first sight.

If one looks at the intermodulation requirements of both standards, it turns out that GSM-900 needs to cope with interferers that are 2 dB larger, corresponding to a 3 dB tougher IP_3 requirement when compared to DCS-1800. However, in this work the target IP_3 already has a margin of 10 dB with respect to the DCS-1800 intermodulation spec in [ETSI, p. 24, 5.3], and therefore a margin of 7 dB with respect to the GSM-1900 intermodulation spec (which is also specified in [ETSI, p. 24, 5.3]).

The essential difference between GSM-900 and DCS-1800 lies in the magnitude of the blocking signals that can occur when a wanted signal 3 dB above the reference sensitivity is received; In GSM-900 these blocking signals can be considerably larger than in DCS-1800, lowering the allowable gain of the low noise amplifiers and therefore toughening the noise figure specification of the mixers. The origin of this specification has to do with the fact that the 'old' 8 W cellular phones must be able to coexist with the current 900 MHz cellular phones. The difference between GSM-900 and DCS-1800 is therefore more an 'historical' burden, than an essential property of a high-end mobile system; GSM-900 is tougher than DCS-1800,... but for the wrong reason.

3.8 Conclusion

This chapter has treated the high-level design of a DCS-1800 receiver front-end. First, a few receiver architectures have been explored with respect to integratability, performance and required building block specifications. The low-IF architecture has been selected as the best candidate for integration in a CMOS transceiver. Next, the system level requirements (noise figure, image rejection, intermodulation, etc.) have been derived in order for the front-end to be compliant to the DCS-1800 standard. In the end, it is shown how these specifications can be allocated to the different blocks.

If there is one thing worth remembering from this chapter, it is as follows:

- Traditional heterodyne receivers are capable of offering a high performance at a low power consumption; As the interstage filters pass only the band of interest at each depth in the receiver chain, the specifications of the RF circuits are maximally relaxed. Yet, its strength is at the same time a weakness: the heterodyne receiver features an extremely poor integratability since it heavily relies on external, high-Q passive filters to perform both the mirror signal suppression and the channel selection. Integrating these high-Q filters on silicon is simply not workable.

- The zero-IF and the low-IF architecture feature a much better — if not excellent — integratability since the image rejection and the channel selection no longer rely on high-Q filtering. Image rejection, channel selection and demodulation are all carried out in the digital domain by a digital signal processor. This is an attractive property, since the digital domain is the natural biotope of CMOS — the process of interest. Nevertheless, integratability does come at a price: the low-frequency building blocks need to be able to handle a high dynamic range because of the absence of filtering in the RF part. The accuracy-bandwidth-dynamic range trade-off is therefore much more pronounced than in the heterodyne receiver. In fact, one must admit that it is the direct consequence of postponing both the image rejection and the channel selection to the digital part. Of course, this puts extra pressure on the power consumption of the RF analog front-end.

Hence, it is clear that there is no substitute for cubic inches: combining a low power consumption with a high degree of integration is not straightforward and requires careful, power-aware design of all the involved building blocks.

Chapter 4

Deep Submicron CMOS Transistors

4.1 Introduction

In the previous chapter, the requirements of the DCS-1800 cellular standard have been mapped onto a set of specifications for a highly integrated low-IF receiver. Subsequently, these specifications have been distributed among the different building blocks of the receive path. However, since all these circuits need to be implemented in terms of elementary transistors — in our case submicron CMOS devices — it should come to no surprise that the transistor performance has a major impact on the overall receiver performance. It is therefore important to be aware of the performance limitations that are already imposed at the transistor level and to understand the dynamics of elementary specifications as a function of the device's operating point. In this chapter, and also in the next chapter, we will actually descend to the device level, and focus on all these aspects in detail. Later on, the obtained insights will allow us to systematically size the LNA's and mixer circuits in a receiver front-end for a set of specifications.

In this first chapter on transistor level issues, we will try to get a more realistic idea of the properties of deep submicron CMOS transistors when compared to long channel devices. In particular, we will look at the intrinsic linearity of a submicron device, expressed by its IP_3 and its IP_2, and its transconductance efficiency (g_m/I_{DS}). Since both properties relate to the DC current equation of the MOS transistor, they are relevant for both low-frequency analog design and RF design.

In the next chapter, the transition is made to real submicron CMOS RF design. The combination of the two chapters should give a clear understanding of the relation between the operating point of a submicron CMOS transistor and elementary specifications of receiver circuits (particularly low noise amplifiers).

This chapter is organized as follows. First, the simple DC hand-calculation model for the drain current of a submicron CMOS device is reviewed. In the two subsequent sections, this model is used to explain the dynamics of the transconductance efficiency and the linearity as a function of the operating point and the device length. This issue has remained unaddressed in the past. It will be shown that the improvement in speed — the property which is stressed in most textbooks and essentially makes RF CMOS feasible — is nevertheless accompanied by a worse transconductance efficiency and a worse third-order linearity when compared to long channel transistors.

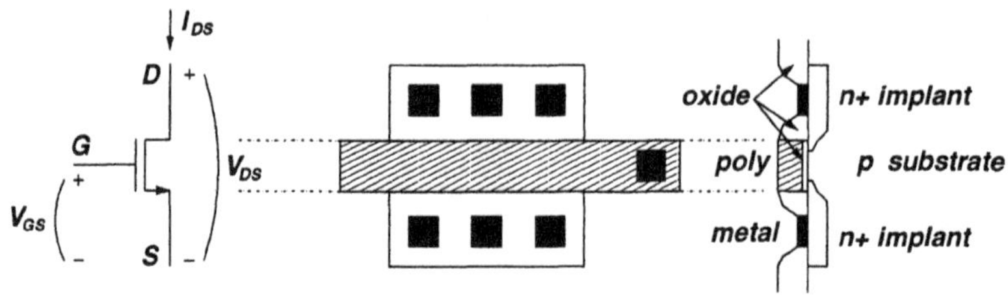

Figure 4.1: *The MOS transistor.*

4.2 Hand Calculation Model

Although simulation is essential to correctly compute the different MOS variables taking into account all interdependencies, it is good design practice to use hand-calculation models in the early design phase (using Matlab, Maple, etc.). Even the incremental design optimization in a circuit simulator can be supported by insight provided by the hand calculation models. Several requirements must be fulfilled for a hand calculation model to be workable:

insight It must provide insight into the basic dynamics of a MOS variable (e.g. I_{DS}) as a function of the most important controllable or observable variables (like e.g. V_{GS}, V_{DS} and the W/L ratio). The same applies to variables that can directly be derived from this quantity (like e.g. g_m).

manageable The model must be manageable, i.e. it must describe the dynamics of the quantity in a particular region of operation with a minimum number of parameters.

accurate A hand calculation model must support calculations with a for design purposes acceptable accuracy of e.g. 5 to 10 percent. A higher accuracy would be an overkill because the process variations are generally larger anyway.

intuitive Although it is not required that there is a direct, quantitative relationship between the physical reality and the hand calculation parameters or model, it is desirable that the model can still be intuitively mapped onto the physics of the MOS device.

Since the birth of CMOS, a multitude of transistor models have been developed for inclusion in circuit simulators. Nowadays, the models that are implemented in simulators have reached a very high degree of complexity, like e.g. the Bsim family and the MOS level 9 model. These models require several tens of parameters that need to be supplied by the foundry. Hence, none of them complies to the above-mentioned requirements for a good hand calculation model. But what could one expect. The simulation model is supposed to model all peculiarities of submicron CMOS transistors with a maximum accuracy, while the hand calculation model is supposed to help the design engineer to see the wood for the trees.

For this reason it is chosen to map the I_{DS} characteristic onto a hand calculation formula that is strongly related to the SPICE level 2 and SPICE level 3 models for a MOS in saturation [HSpice]:

$$I_{DS} = K\frac{W}{L}V_{GST}^2\frac{1}{(1-\Lambda V_{DS})}\frac{1}{(1+\Theta V_{GST})}. \tag{4.1}$$

Greek capital letters are used to distinguish between the SPICE model parameters (e.g. $\lambda, \theta, \ldots$). The symbol V_{GST} should be read as $V_{GS} - V_T$, and is only introduced to reduce formula size. In the text itself, $V_{GS} - V_T$ will continue to be used.

Apart from the quadratic factor in $V_{GS} - V_T$, describing the main dynamics of the current in strong inversion, two additional factors are present. First of all, mobility degradation due to both the transversal and the longitudinal electrical field is being modeled by the $1 + \Theta V_{GST}$ component in the denominator. This formulation is derived from a simplified version of the SPICE level 3 mobility model, given by

$$\mu_{eff} \cong \frac{\mu_0}{1 + \theta V_{GST} + (\frac{\mu_0}{v_{sat} L_{eff}}) V_{DSmin}} \cong \frac{\mu_0}{1 + (\theta + \frac{\mu_0}{n v_{sat} L_{eff}}) V_{GST}}. \tag{4.2}$$

In this equation, θ models the transversal mobility degradation due to surface scattering. Longitudinal mobility degradation is taken into account by the $\mu_0 / n v_{sat} L_{eff}$ term, v_{sat} being the saturation speed of the carriers. The $1 + \Theta V_{GST}$ component linearizes the quadratic factor, yielding a super-linear I_{DS} characteristic. At high $V_{GS} - V_T$ values, the transistor starts operating in velocity saturation and the drain current becomes proportional to $V_{GS} - V_T$. The model is therefore suitable for hand calculations in both strong inversion and velocity saturation. The transit between strong inversion and velocity saturation occurs at a $V_{GS} - V_T$ value of

$$V_{si-vs} = \frac{1}{\Theta}. \tag{4.3}$$

The relative amount of velocity saturation is thus expressed by

$$r_{vs} = \Theta V_{GST} = \frac{V_{GST}}{V_{si-vs}}. \tag{4.4}$$

These definitions will prove helpful in the subsequent sections.

The term in Λ (V^{-1}) also needs some additional explanation. Typically, in hand calculations this term is found in the numerator of the I_{DS} expression, under the form $1 + \Lambda V_{DS}$. This formulation is inherited from the SPICE level 1 MOS model. In (4.1) a somewhat different expression is used, i.e. $1 - \Lambda V_{DS}$ in the *denominator*. The same notation is also used in the SPICE level 2 MOS model. Basically it is a coarse, linearized[1] model of the channel length reduction due to an increasing drain potential. In practice, both notations are largely equivalent since

$$1 + x \sim \frac{1}{1 - x} \qquad \text{for } x \ll 1.$$

Here the second expression is chosen, for reasons that will become clear in Section 4.4.

Table 4.1 summarizes the hand calculation parameters for minimum length transistors in different submicron CMOS technologies. The parameter extraction is performed by the [MOSCAL][2] tool using a least-squares fit of the simulated I_{DS} characteristic of the respective MOS transistors in the saturation region. The Λ parameter is extracted at a $V_{GS} - V_T$ of 0.4 V and for V_{DS} ranging from 0.7 V to 1.5 V. The K, Θ and V_T parameters are extracted at a V_{DS} of 1.5 V and for $V_{GS} - V_T$ between 0.15 V and 0.9 V. As a result, the extracted parameters are valid for $V_{GS} - V_T$

[1] In reality the Λ parameter is voltage–dependent.

[2] The MOSCAL tool (MOS CALculator) is an in-house wrapper above SPICE.

values between 0.15 V and 0.9 V and for V_{DS} values between 1.5 V and the current V_{DSsat}. The worst-case deviation between the extracted hand calculation model and the simulated characteristic is approximately 2 percent, even near 0.15 V i.e. in the transition region between strong and weak inversion . At the cost of an increased deviation of about 10%, the parameters can be used until a $V_{GS} - V_T$ value of 0.1 V. Below this value, a dedicated hand calculation model for weak inversion becomes mandatory.

Table 4.1 illustrates shows that mobility degradation, incarnated by Θ, strongly affects nMOS transistors whereas it is much less pronounced in pMOS transistors. This is due to the significantly higher mobility of the electrons in the nMOS transistors compared to the holes in pMOS transistors. Additionally, the electrons in nMOS transistors experience less deceleration owing to the lower channel doping level. The Θ parameter is also a strong function of the channel length: in a 0.5 μm technology, Θ is still 1.3 V^{-1} for a minimum length transistor, while it becomes as high as 4.5 V^{-1} in the 0.25 μm technology mentioned in the table (effective length of 0.19 μm). Note that this implies that the transit voltage V_{si-vs} is barely 220 mV! This dynamics could have been foreseen by looking at (4.2). Indeed, the θ value in (4.2) is in general smaller than the term representing the longitudinal mobility degradation effect. Consequently, the Θ term is — in first order — inversely proportional to the effective channel length. Even though this reasoning is — strictly speaking — only valid within a given technology, i.e. for a fixed mobility and a fixed saturation speed v_{sat}, it can normally be extrapolated across different technologies because carrier mobility and saturation speed are only a weak function of the technology; A 0.5 μm MOS transistor in a 0.25 μm technology normally has a Θ value which is quite comparable to a 0.5 μm transistor in a 0.5 μm technology.

However, this observation can not always be made from a set of extracted parameters; One must not forget that the transistor structure itself is affected by the scaling (e.g. the drain engineering) and the processing know-how of the foundry. Moreover, the numerical simulation model that is supported by the foundry (Bsim3v3, Bsim1, MOS level 9,...) often changes from generation to generation: In a 0.7 μm technology, SPICE level 2 was current, while nowadays Bsim3v3 is considered a standard. Hence, one can not expect effects that indeed become pronounced in 0.25 μm but weren't visible in the 'larger' technology nor supported by the models available at that time, to be incorporated. Also, the accuracy of the parameter extraction by the manufacturer is strongly related to the competence of the modeling people and their insight in the importance of subtle effects for analog designers. But, what can one say if 90 percent of the people (read: the digital designers) are satisfied with the model.

type	nMOS						pMOS					
parameter	L_{eff}	K_n	V_{Tn}	Θ_n	Λ_n	err	L_{eff}	K_p	V_{Tp}	Θ_p	Λ_p	err
unit	μm	$\mu A/V^2$	V	V^{-1}	V^{-1}	%	μm	$\mu A/V^2$	V	V^{-1}	V^{-1}	%
MTC 0.5	0.47	85	0.60	1.3	0.06	2	0.46	22	0.52	0.56	0.05	2
TSB 0.4	0.4	71	0.56	0.78	0.1	0.2	0.4	23	0.65	0.38	0.1	0.2
TSB 0.25	0.19	218	0.61	4.15	0.11	1	0.25	44	0.71	1.1	0.10	2

Table 4.1: *Hand calculation parameters for different submicron CMOS technologies. The parameters have been derived for minimum channel length transistors in the respective technologies and connect the I_{DS}, $V_{GS} - V_T$ and the $(W/L)_{eff}$ of the devices.*

Thus, when making hand-calculations one should always work with the Θ value extracted for the current process, either directly extracted for that particular channel length or under the form

$$\Theta_{eff} = \Theta_{eff,min}(L_{eff,min}/L_{eff}). \tag{4.5}$$

The effect of Θ on the magnitude of the current is — at first sight — fully compensated by the higher K value. However, it is shown in the following sections that one should not jump to conclusions.

4.3 Transconductance and Transconductance Efficiency

The active character of a MOS transistor is expressed by its transconductance. This parameter is closely related to thermal noise and amplifier gain-bandwidth for a given load. The expression for the transconductance can be derived from (4.1) and is given by

$$g_m \triangleq \frac{\partial I_{DS}}{\partial V_{GS}} = 2K\frac{W}{L}V_{GST}\frac{1}{(1-\Lambda V_{DS})}\frac{1}{(1+\Theta V_{GST})}\left(\frac{1+\frac{\Theta}{2}V_{GST}}{1+\Theta V_{GST}}\right). \tag{4.6}$$

The first two factors consists of the well-known expression for g_m in the classical, long channel model. The transconductance is further degraded by two factors. The denominator of the second last factor, $1+\Theta V_{GST}$, represents the decrease in transconductance due to mobility degradation. Ultimately, at large values of $V_{GS}-V_T$, this factor removes the dependency of g_m on $V_{GS}-V_T$. Hence, in this region of operation, the transconductance is solely controlled by the W/L ratio. The last factor embodies the changing dynamics of the drain current I_{DS}, i.e. the fact that the I_{DS} versus $V_{GS}-V_T$ characteristic ultimately becomes *linear* instead of quadratic. Its only effect — owing to the mathematical fact that $\frac{\partial v^2}{\partial v}=2v$ and $\frac{\partial v}{\partial v}=1$ — consists in gradually eliminating the coefficient of two in the expression for g_m. For all purposes, this factor acts like a pole-zero pair in the $V_{GS}-V_T$ domain.

Based on (4.6) and (4.1), the transconductance to current ratio or *transconductance efficiency* is defined as

$$\eta_{g_m} \triangleq \left(\frac{g_m}{I_{DS}}\right) = \frac{2}{V_{GST}}\left(\frac{1+\frac{\Theta}{2}V_{GST}}{1+\Theta V_{GST}}\right). \tag{4.7}$$

Again, the first factor represents the classical expression for the transconductance efficiency. This component already indicates that the efficiency drops rapidly with increasing $V_{GS}-V_T$, irrespective of the channel length. Furthermore, the pole-zero effect that is present in (4.6) is inherited as well. As explained in the previous paragraph, this factor is caused by the changing dynamics of the current. Note that a lower mobility in itself does *not* result in a decreasing transconductance efficiency since this effect is present both in I_{DS} and g_m.

Fig. 4.2(a) plots the expression for the transconductance efficiency given in (4.7) against the gate over-drive voltage $V_{GS}-V_T$ for various values of Θ, hence for different channel lengths. The efficiency for long channel transistors, i.e. for $\Theta \to 0\ \mathrm{V}^{-1}$, is represented by the dashed line. From the plot it is clearly seen that short channel transistors exhibit a lower efficiency than

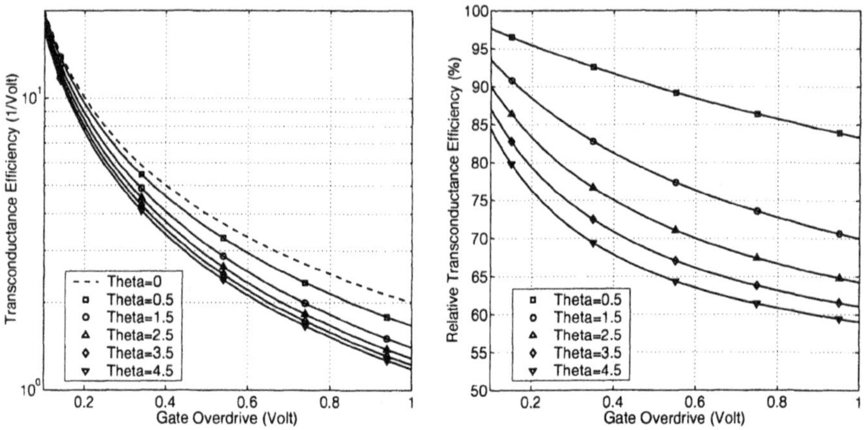

(a) Transconductance efficiency (η_{g_m}). (b) Relative transconductance efficiency.

Figure 4.2: *Transconductance efficiency* η_{g_m} *as a function of gate over-drive and* Θ.

long channel transistors. The drop in efficiency becomes even more clear from Fig. 4.2(b), where the ratios between the different efficiencies and the long channel efficiency are depicted. It can easily be seen that a loss in efficiency of 20 to 30 percent with respect to long channel efficiencies is not uncommon at practical values of $V_{GS} - V_T$. The efficiency for $\Theta = 4.5\ \mathrm{V}^{-1}$ at a $V_{GS} - V_T$ of 0.3 V is even as low as the efficiency at a $V_{GS} - V_T$ of 1 V for $\Theta = 1.5\ \mathrm{V}^{-1}$! At large values of ΘV_{GST}, the loss tends asymptotically to 50 percent.

It becomes clear that $V_{GS} - V_T$ must be kept as low as possible to efficiently generate the required transconductance and that this consideration becomes even more imperative when designing in deep submicron CMOS technologies.

It may seem evident to keep a low $V_{GS} - V_T$, but one must not forget that this may affect bandwidth and especially linearity. On the other hand, using a non-minimum channel length also improves the efficiency but compromises the frequency performance. How the $V_{GS} - V_T$ value and the channel length influence transistor linearity is dealt with in the next section.

4.4 Distortion and Intermodulation

A transistor is essentially a non-linear voltage to current converter; When a single frequency f is applied to the gate terminal, harmonic components at $2f$, $3f$, etc. are being generated apart from the linear (fundamental) output current at f, $g_m V(f)$. A two-tone input at f_1 and f_2 will not only generate harmonics of the respective frequencies, but also intermodulation products at $f_1 \pm f_2$, $2f_1 \pm f_2$, $2f_2 \pm f_1$, etc. Because linearity is an important issue in receiver design, it is essential to have an idea about the underlying mechanisms of transistor non-linearities and their relation to the operating point.

The linearity of long channel MOS devices has already been analyzed extensively in the past [Sans99]. Unfortunately, the results can not directly be transferred to deep submicron devices since other effects (like e.g. mobility degradation) affect the linearity. This work focuses pre-

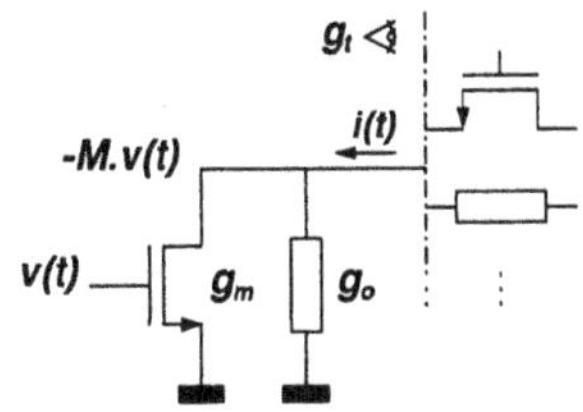

Figure 4.3: *Drain termination condition for MOS nonlinearity calculations.*

cisely on these previously unaddressed issues.

Consider the MOS transistor in Fig. 4.3. The device is biased in saturation at a given $V_{GS}-V_T$ and at a certain V_{DS}. In the frequency range of interest, the drain of the transistor is assumed to be terminated into a resistive impedance (like e.g. a resistor, a cascoding device, a resonating LC-tank, etc.). This assumption is usually valid because a non-resistive impedance is generally related to a pole, and hence unwanted loss. In addition, parasitic device capacitances have been neglected since they only contribute to second-order feedback effects. An AC signal, $v(t)$, is applied at the gate terminal. As a consequence of this excitation, a small-signal current is being generated within the transistor. The current is largely flowing through the reference plane at the drain and is transferred to a possible load (a small amount of current is lost in the output conductance). In first order there is a *linear* relationship between $i(t)$ — the net generated AC current through the reference plane — and $v(t)$, given by

$$i(t) = -g_m \frac{g_l}{g_l + g_o} v(t). \tag{4.8}$$

Hence, the AC voltage at the drain is — again in first order — also proportional to $v(t)$, and is given by

$$v_{ds}(t) = -\frac{g_m}{g_l + g_o} v(t) \triangleq -Mv(t). \tag{4.9}$$

where M is defined as the absolute value of the Miller amplification factor under the given drain termination condition. This formulation will prove to be very useful since it effectively decreases the equation depth. Besides, the M factor is typically known beforehand because it is intimately connected with e.g. the sizing rule of the cascoding device (see (6.56)). In the other case, M simply equals the low frequency amplification factor.

Using (4.9) and (4.1), the net current through the reference plane can be approximated by

$$i(t) = \frac{K\frac{W}{L}(V_{GST} + v(t))^2}{(1 + \Theta(V_{GST} + v(t)))(1 - \Lambda(V_{DS} - Mv(t)))}. \tag{4.10}$$

Note that this equation already incorporates the current through the finite output resistance by using the linear estimation for the drain voltage (4.9) in the denominator. The application of this linear approximation gives rise to some inaccuracies, yet it is imperative to remove the implicit nature from the formula; Otherwise the problem would have been unsolvable. The formulation also reflects that a small increase in the over-drive voltage slightly increases mobility degradation since the V_{DSsat} becomes temporarily higher.

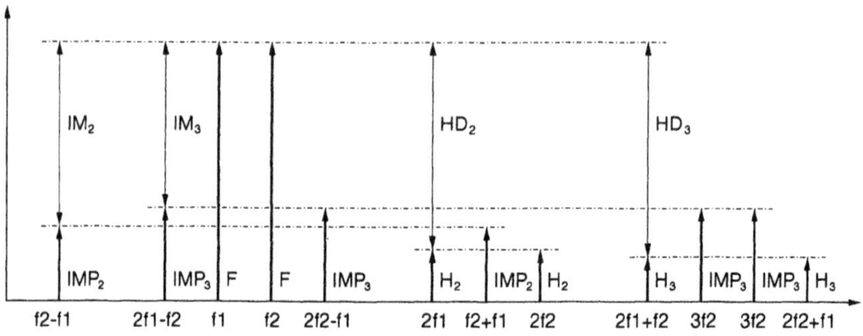

Figure 4.4: *Intermodulation and harmonic distortion during a two-tone test.*

The denominator of (4.10) can be factored out and written as a second-order polynomial in $v(t)$,

$$(1+\Theta V_{GST})(1-\Lambda V_{DS})+v(t)\cdot\big((1+\Theta V_{GST})\Lambda M+(1-\Lambda V_{DS})\Theta\big)+\Theta\Lambda M\cdot v^2(t). \quad (4.11)$$

Although it is very tempting to remove the term in $v^2(t)$ — which is allowed from a mathematical point of view since $\lim_{v\to 0} v/v^2 = 0$ — it is absolutely required to keep it there! Indeed, this term may (and will) still give rise to second and third-order non-linearities. To simplify the succeeding calculations, (4.10) is written under the form

$$i(t) = \frac{c_0(c_1+v(t))^2}{c_2+c_3v(t)+c_4v^2(t)} \quad (4.12)$$

with the following auxiliary parameters:

$$c_0 = K(W/L) \quad (4.13)$$
$$c_1 = V_{GST} \quad (4.14)$$
$$c_2 = (1+\Theta V_{GST})(1-\Lambda V_{DS}) \quad (4.15)$$
$$c_3 = (1+\Theta V_{GST})\Lambda M + (1-\Lambda V_{DS})\Theta \quad (4.16)$$
$$c_4 = \Theta\Lambda M. \quad (4.17)$$

Linearity is generally expressed in terms of ratios that are extracted at the hand of a two-tone test. They are defined as (see Fig. 4.4):

$$IM_2 \triangleq \frac{IMP_2}{F} \quad \text{and} \quad IM_3 \triangleq \frac{IMP_3}{F} \quad (4.18)$$

$$HD_2 \triangleq \frac{H_2}{F} \quad \text{and} \quad HD_3 \triangleq \frac{H_3}{F} \quad (4.19)$$

where F is the amplitude of the fundamental output tone and IMP_i and H_i represent the amplitudes of the ith order intermodulation product and harmonic, respectively. IM_i and HD_i stand for the ith order intermodulation ratio and harmonic distortion ratio, respectively.

The expressions for the different intermodulation and distortion ratios are deduced in Appendix B by expanding (4.12) in a Taylor series and substituting $v(t)$ by a two-tone input signal,

$$v(t) \mapsto U\cos(\omega_1 t) + U\cos(\omega_2 t). \quad (4.20)$$

This results in the following expressions for IM_2, IM_3, HD_2 and HD_3:

$$IM_2 = 2HD_2 = \left| \frac{(-c_2 + c_1c_3)^2 - c_1^2c_2c_4}{c_1c_2(-2c_2 + c_1c_3)} \right| \cdot U \tag{4.21}$$

$$IM_3 = 3HD_3 = \left| \frac{3}{4} \frac{(-c_2 + c_1c_3)((-c_2 + c_1c_3)c_3 - 2c_1c_2c_4)}{c_1c_2^2(-2c_2 + c_1c_3)} \right| \cdot U^2. \tag{4.22}$$

where U denotes the input amplitude (in V) of one of the component tones in (4.20). Because the expressions for the harmonic distortion ratios only differ from the respective intermodulation ratios by a factor of two or three, we will confine the discussion to the intermodulation behavior.

Substituting the c_i values from (4.13)–(4.17) into (4.21) and (4.22) still leads to a complicated expression which is hardly interpretable. Therefore, the analysis is done in two steps. First of all, the impact of a non-zero Θ value on IM_i is analyzed separately as a function of $V_{GS} - V_T$. During this analysis the drain termination is assumed to be an ideal voltage source, which corresponds to setting the Miller factor M to zero. Secondly, the effect of a non-zero Miller factor is investigated as a function of $V_{GS} - V_T$, but at a fixed value of Θ.

Setting M to zero yields the following expressions for IM_i:

$$IM_2 = \frac{1}{V_{GST}(2 + \Theta V_{GST})(1 + \Theta V_{GST})} \cdot U \tag{4.23}$$

$$IM_3 = \frac{3}{4} \frac{\Theta}{V_{GST}(2 + \Theta V_{GST})(1 + \Theta V_{GST})^2} \cdot U^2. \tag{4.24}$$

Recognizing the ith order intercept point, IP_i, to be the U value at which IM_i extrapolates to one, (4.24) can be rewritten as

$$IM_2 = \frac{U}{IP_2[\mathrm{V}]} \quad \text{and} \quad IM_3 = \frac{U^2}{IP_3^2[\mathrm{V}]}. \tag{4.25}$$

where IP_2 is given by

$$\begin{aligned} IP_2[\mathrm{V}] &= V_{GST}(2 + \Theta V_{GST})(1 + \Theta V_{GST}) \\ &= V_{GST}(2 + r_{vs})(1 + r_{vs}) \end{aligned} \tag{4.26}$$

and IP_3 is given by

$$\begin{aligned} IP_3[\mathrm{V}] &= \sqrt{\frac{4}{3} \frac{V_{GST}}{\Theta}(2 + \Theta V_{GST})(1 + \Theta V_{GST})^2} \\ &= \sqrt{\frac{4}{3} V_{GST} V_{si-vs}(2 + r_{vs})(1 + r_{vs})^2}. \end{aligned} \tag{4.27}$$

Until now, the IP_i values have been expressed in V (amplitude, not RMS value). Indeed, a MOS transistor essentially is a *voltage* driven device and the natural way to express the IP_i is in V, or in dBV. On the other hand, the power-oriented RF community rather likes to express the IP_i in dBm, 0 dBm being defined as a *power* of 1 mW. Hence, it would be useful to have a unit that stands midway between dBV and dBm: close enough to dBV to accommodate the analog community and close enough to dBm to allow a fast conversion to dBm, especially when the transistor is matched to the standard 50 Ω value.

The answer consists in expressing the IP_i value relative to 316 mV (amplitude) by using the unit dBVref$_{[316\text{ mV}]}$ instead of dBV; Then, in a 50 Ω environment, the input power (in dBm) can directly be related to the gate-source voltage excursions (in dBVref$_{[316\text{ mV}]}$) via the Q factor of the matching network:

$$v_{gs}[\text{dBVref}_{[316\text{ mV}]}] = 20\log_{10}(Q_M) + P_{in}[\text{dBm}]. \tag{4.28}$$

The reason for this is that an amplitude of 316 mV ($\sqrt{50\cdot 0.001}\sqrt{2}$ V) across 50 Ω corresponds precisely to 0 dBm.

Since the unit designation 'dBVref$_{[316\text{ mV}]}$' is somewhat bulky, we prefer to define a synonym 'dBm$_V$'. As the name indicates, one can work with this unit as if it was dBm— it even *is* dBm if the voltage is measured in 50 Ω— but the subscript indicates that it is still a dB unit based on *voltages* and not on power. Whenever the original unit dBm is used, *power* is aimed at. It is important to note that in order to express an RMS voltage in dBm$_V$, one should use 'dBVref$_{[223\text{ mV}_{rms}]}$' instead of 'dBVref$_{[316\text{ mV}]}$'.

For a terminal at impedance level R_S, the relation between dBm$_V$ and dBm is given by:

$$x\text{ dBm} \Leftrightarrow \left(x - 10log_{10}\left(\frac{50}{R_S}\right)\right)\text{dBm}_V. \tag{4.29}$$

Converting (4.26)-(4.27) from V to dBm$_V$ leads to

$$\begin{aligned} IP_2[\text{dBm}_V] &= 20\log_{10}\left(\frac{V_{GST}(2+r_{vs})(1+r_{vs})}{\sqrt{50\cdot 0.001}\sqrt{2}}\right) \\ &= 10 + 20\log_{10}(V_{GST}(2+r_{vs})(1+r_{vs})) \end{aligned} \tag{4.30}$$

and

$$\begin{aligned} IP_3[\text{dBm}_V] &= 20\log_{10}\left(\frac{\sqrt{\frac{4}{3}V_{GST}V_{si-vs}(2+r_{vs})(1+r_{vs})^2}}{\sqrt{50\cdot 0.001}\sqrt{2}}\right) \\ &\cong 11.25 + \mathbf{10}\log_{10}\left(V_{GST}V_{si-vs}(2+r_{vs})(1+r_{vs})^2\right). \end{aligned} \tag{4.31}$$

Plots of (4.30) and (4.31) are shown in Fig. 4.5(a) and Fig. 4.5(b), respectively.

Let us first consider Fig. 4.5(a), showing the IP_2 as a function of $V_{GS}-V_T$ at different values of Θ (hence different channel lengths). Obviously, the linearity at a given Θ value drastically increases with increasing $V_{GS}-V_T$ (about 20-40 dB/V). The reason behind this is the linearization of the quadratic characteristic taking place when moving towards the velocity saturation regime. Naturally, this also implies that the growth is accompanied by a decreasing transconductance efficiency (Fig. 4.2). Hence, for a given channel length, there is a direct trade-off between the IP_2 value and the power consumption.

From Fig. 4.5(a) one can also conclude that at a given $V_{GS}-V_T$, the IP_2 value increases with increasing Θ (decreasing channel length); Hence, at the same power supply voltage, short channel transistors can offer better linearity performance. On the other hand, the $V_{GS}-V_T$ value can be taken lower for the same IP_2. However, this does not mean that the linearity is available at a better efficiency. On the contrary, in general submicron transistors offer the same IP_2 at a 10 to 15 percent *lower* efficiency (Fig. 4.2) in spite of the lower $V_{GS}-V_T$!

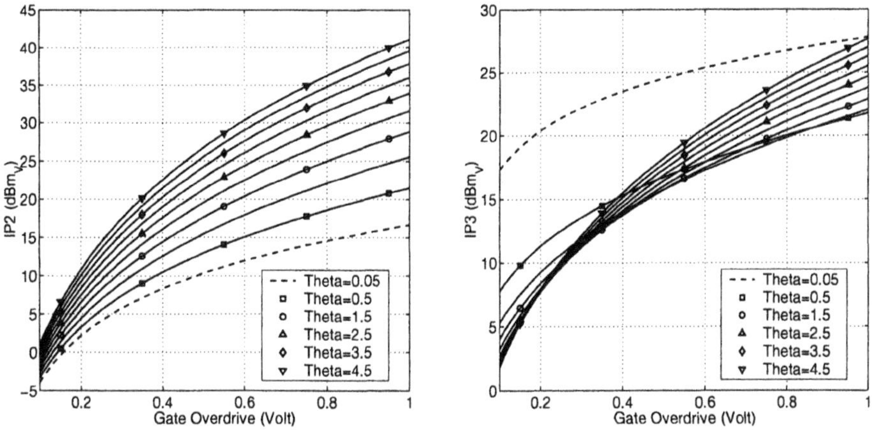

(a) Second-Order Intermodulation Point (IP_2). (b) Third-Order Intermodulation Point (IP_3).

Figure 4.5: *Linearity as a function of gate over-drive and* Θ.

The picture becomes somewhat more complicated when looking at the IP_3 plot in Fig. 4.5(b). The IP_3 value still significantly increases as a function of $V_{GS}-V_T$ but the slope of IP_3 is clearly halved on a dB scale: it is only 15-20 dB/V (4.31). At a fixed Θ value, the growing linearity as a function of $V_{GS}-V_T$ is again accompanied by a decreasing transconductance efficiency.

Compared to a long channel transistor (modeled by taking[3] $\Theta = 0.05\ \text{V}^{-1}$ and represented by the dashed line in Fig. 4.5(b)), the IP_3 value for short channel transistors is significantly *lower* especially at low values of $V_{GS}-V_T$ [Jans98a]. This behavior is caused by the super-linear nature of (4.1). Indeed, a long channel transistor ideally exhibits a quadratic I_{DS} characteristic, and hence does *not* produce third-order intermodulation products. It is precisely the gradual transition towards a linear characteristic that comes at the cost of third-order intermodulation.

At practical values of Θ, two regions in the $V_{GS}-V_T$ domain can be distinguished in the IP_3 plot; At high $V_{GS}-V_T$ values (velocity saturation), deep submicron transistors (exhibiting a large Θ, e.g. 0.25 μm) clearly exhibit a better linearity than plain submicron transistors (exhibiting lower Θ values, e.g. 0.7 μm) because they are *already* operating in the velocity saturation regime. Hence, short channel transistors offer a maximum amount of linearity at a given power supply voltage and require the minimum $V_{GS}-V_T$ for a given IP_3 (yet, still at an efficiency penalty of about 5 to 10 percent). On the other hand, at low $V_{GS}-V_T$ values (strong inversion) and again at practical values of Θ, short channel transistors perform slightly worse than long channel ones (1-2dB) (at a 10 percent lower efficiency).

Until now, the impact of Θ on both IP_2 and IP_3 has been analyzed for a fixed drain potential ($M = 0$). In the next paragraphs, it is illustrated how a nonzero Miller gain influences the IP_i values. Now, there is no escape from using (4.21) and (4.22).

Fig. 4.6 shows the IP_i values as a function of $V_{GS}-V_T$ for different values of M. Θ and Λ are set to 1.3 V^{-1} and 0.06 V^{-1}, respectively, corresponding to the MTC05 technology in Table 4.1. For the V_{DS} parameter, a value equal to the over-drive voltage $V_{GS}-V_T$ is chosen. This ensures

[3] If Θ is zero, IP_3 becomes infinite!

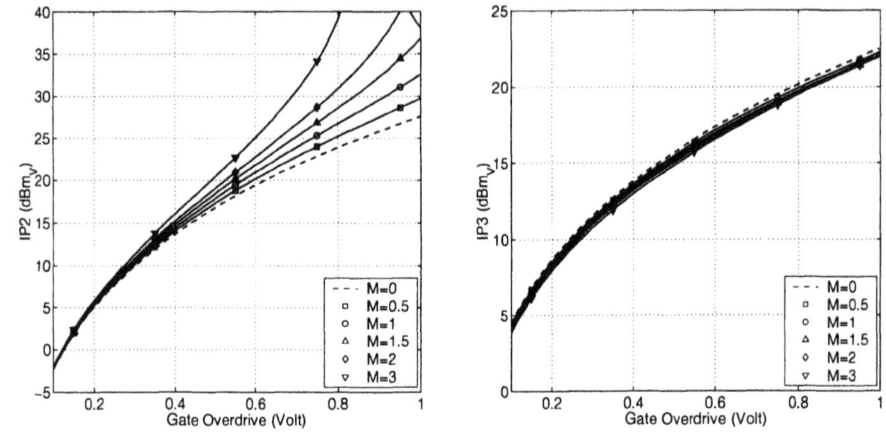

(a) Second-Order Intermodulation Point (IP_2). (b) Third-Order Intermodulation Point (IP_3).

Figure 4.6: *Linearity as a function of gate over-drive and M* ($\Theta = 1.3\ \mathrm{V}^{-1}$,$\Lambda = 0.06\ \mathrm{V}^{-1}$)

that the transistor is always biased in the saturation region since $V_{DS} = V_{GS} - V_T \cong nV_{DSsat}$. By the way, taking a different value for V_{DS} does not influence the conclusions.

A can be seen from Fig. 4.6(a), IP_2 increases with increasing Miller effect, implying that the extra second-order intermodulation terms that are generated by the presence of a nonzero M are of opposite sign. Theoretically, there even exists a $V_{GS} - V_T$ value at which the contribution of Θ and the contribution of a finite M neutralize each other. In Fig. 4.6(a) this occurs at a $V_{GS} - V_T$ of 0.9 V for $M = 3$. In practice, this must be somewhat relativized; After all, (4.10) is only a first-order approximation because in reality the Λ factor is voltage-dependent and (4.9) is only a rough simplification. Additionally, the analysis relies on an exact phase difference of 180 degrees between the gate and the drain voltage. These — and other — effects will actually damp the 'resonance'. It is thus not wise to rely on this singularity during design.

The M factor has a small impact on the IP_3 characteristic (Fig. 4.6(a)): the IP_3 is uniformly degraded by maximally 1 dB. Making (4.31) an adequate model. If one is interested in having a first-order approximation of the 'ΛM' effect, one could use

$$IP_3[\mathrm{dBmv}] \cong 11.25 + \mathbf{10}\log_{10}\left(\frac{V_{GST}}{\Theta + \Lambda M}(2 + r_{vs})(1 + r_{vs})^2\right). \tag{4.32}$$

It is important to note that the impact of ΛM on the linearity is *not* caused by the non-linear nature of the output impedance: the phenomenon is equally present[4] when using $1 + \Lambda v_{DS}$ in the numerator instead of $1 - \Lambda v_{DS}$ in the denominator. Instead, the non-linearities are generated by modulating the drain voltage of a transistor when *at the same time* modulating the 'quadratic' characteristic.

[4]both models give the same results, provided that Λ has been extracted for the respective DC model.

4.5 Conclusion

Although moving towards a smaller channel length drastically improves the intrinsic speed of a MOS device — which is essentially the enabler of CMOS RF design —, not all parameters evolve in the right direction:

- Short channel transistors exhibit a *lower* transconductance efficiency than long channel transistors. A loss in efficiency of 20 to 30 percent with respect to long channel efficiencies is not uncommon at practical values of $V_{GS} - V_T$. At large values of $V_{GS} - V_T$ the loss even tends to 50 percent.

- Compared to a long channel transistor, the IP_3 value for short channel transistors is significantly *lower*, especially at low values of $V_{GS} - V_T$ [Jans98a]. This behavior is caused by the super-linear nature of the DC current equation which models velocity saturation. Indeed, a long channel transistor ideally exhibits a quadratic I_{DS} characteristic, and hence does *not* produce third-order intermodulation products. It is precisely the gradual transition towards a linear characteristic that comes at the cost of third-order intermodulation.

Hence, if speed is not absolutely required, it may be better not to opt for minimum length transistors.

Chapter 5

RF CMOS Design for Analog Designers

5.1 Introduction

The design of high-frequency wireless front-ends involves cooperation between people from three distinct worlds: the world of analog microelectronics — for the IF and baseband circuits —, the RF world — for the RF circuits and interfaces —, and the microwave world — for the antenna. As illustrated in Table 5.1, each design community uses its own proper vocabulary and looks at the same problem from a different perspective. Yet, nowadays more and more 'traditional' analog designers move in to RF design. This especially applies to CMOS analog designers; Whereas in the past only exotic and/or expensive technologies (like e.g. GaAs) were capable of providing the required RF performance, the ever increasing speed of submicron CMOS *enables* mainstream CMOS to become a viable, low-cost alternative for implementing systems in the low-GHz domain. In addition, CMOS is already the technology of choice for the mixed-signal back-end so that there is an even stronger *drive* to integrate the RF circuits in CMOS, and if possible, on the same die.

Most of the time, a lot of fundamental questions arise when a CMOS designer is confronted with the power-oriented view of RF designers; First of all, a CMOS analog designer is used to look at a CMOS transistor as a voltage-to-current converter, leading to questions like: "Why do I need power transfer when a CMOS transistor only needs the voltage?" and "What's the use of 50 Ω in a CMOS environment". Secondly, an analog designer is especially interested in the relation between transistor sizing and circuit performance. After all, a CMOS transistor features two degrees of freedom (e.g. I_{DS} and V_{GST} (or g_m, W/L)) which must be filled in. This is somewhat different from classical RF design, where traditionally more emphasis lies on the design of passives and matching networks around characterized, off-the-shelf transistors, or, where a designer can only choose from a limited set of devices, each being targeted at a specific

	Typical statement	Vocabulary
Analog designer	"Capacitance means signal loss"	$v, i, Z, g_m, A_0, \ldots$
RF designer	"An ideal capacitor is a lossless element"	$S_{11}, S_{21}, \ldots$
Microwave designer	"There is no such thing as a capacitor"	$\vec{E}, \vec{H}, \vec{S}, \ldots$

Table 5.1: *Three different kinds of designers.*

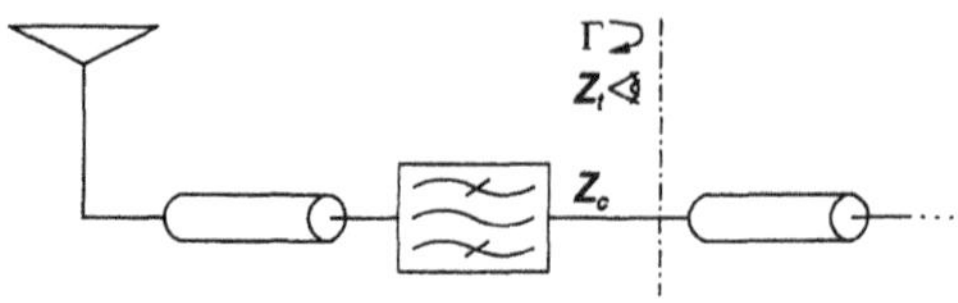

Figure 5.1: *Termination at a component's reference plane.*

performance like e.g. low noise or high gain.

For this reason, this chapter discusses RF design from an analog designer's perspective. It is especially aimed at analog designers who want to familiarize themselves with the fundamental and more advanced aspects of RF CMOS design. It provides answers to commonly asked questions like e.g. "Does a matched MOS device always provide more output signal than an unmatched device?" and "Does a matched MOS device really exploit the signal power it is provided with?".

The chapter is organized as follows; First, the exact definitions of power matching, impedance matching and noise matching are reviewed — concepts which are often mixed up. The main part of this chapter concentrates on the performance of input-matched MOS devices; Gain, noise and linearity (IP_3) are analyzed in full detail as a function of transistor sizing and as a function of the effective source impedance. Along with this, the role of the non-quasi static gate resistance is exposed and it is shown how this resistance puts a fundamental limit on the achievable noise figure and the gain, even at low-GHz radio frequencies. The insights and analytical formulas developed in this chapter will form the basis of the systematic LNA design methodology described in Chapter 6.

5.2 Impedance-, Power- and Noise Matching

5.2.1 Termination

One speaks about termination whenever an interface or reference plane of a component is required to be 'terminated' into a specific impedance value (the characteristic impedance) in order to guarantee the nominal performance.

The quality of a termination is expressed by its reflection coefficient Γ, defined as

$$\Gamma = \frac{Z_t - Z_c}{Z_t + Z_c}, \tag{5.1}$$

where Z_c is the characteristic impedance of the reference plane, indicated by the dash-dotted line, and Z_t the value of the terminating impedance (Fig. 5.1). In RF subsystems, a Z_c value of 50 Ω is commonly used. This value is c compromise between maximum power handling capability — corresponding to a characteristic impedance of 30 Ω— and minimum attenuation — corresponding to a characteristic impedance of about 77 Ohm.

The system performance can be strongly affected by the quality of a termination. Indeed, if a termination is not sufficient, the reflected power stream may give rise to multiple reflections. Ultimately, this may result in destructive interference at the input of the destination component.

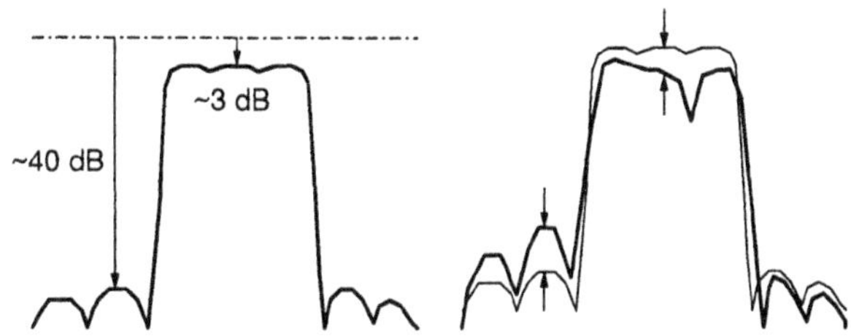

(a) Adequately terminated. (b) Badly terminated.

Figure 5.2: *The antenna filter under different termination conditions.*

Fig. 5.2 shows the relevance of correctly terminating an RF filter. Fig. 5.2(a) indicates the nominal characteristic of a typical antenna filter, corresponding with a Γ below -10 dB. The filter exhibits a relatively flat pass–band and the side–lobe levels in the stop–band are several tens of dB's below. When the reflection coefficient Γ becomes too high, the filter starts deviating from its nominal transfer function, as shown in Fig. 5.2(b). First of all, the side–lobes increase, compromising the attenuation of the out of band blocking levels. Additionally, the pass band loss may increase and even notches may arise. These effects are highly unwanted in mobile receivers, the first leading to system overloading and the second decreasing the sensitivity.

Note that termination is *not* concerned with maximum power transfer (explained in Subsection 5.2.2), but only concerned with keeping the power flow at the characteristic impedance level. Under these conditions, the terminated building block shows no interaction with the succeeding blocks and can be considered uni-directional.

In the foregoing discussion we have silently assumed that the antenna is a 50 Ω component which is connected to the 50 Ω blocking filter using a 50 Ohm transmission line. In reality, however, the antenna impedance can considerably change depending on the exact environment, like e.g. nearby metal. It is important to understand that this does not change the required termination impedance of the antenna filter: the filter still requires a 50 Ω termination to ensure a normal filter characteristic. From the point of view of the filter, the only difference is that when the antenna had also been 50 Ω, we would have had a doubly-terminated system whereas now it is only terminated one side. Nevertheless, when the antenna impedance differs too much from the impedance of the connected transmission line, this results in less energy being transferred to the system. As a result, the sensitivity of the receiver as a whole may still be compromised as a result of the changing antenna impedance.

5.2.2 Impedance Matching versus Power Matching

Impedance matching and power matching are often mixed up. In this subsection the — sometimes subtle — difference between the two concepts is reviewed.

Impedance matching denotes the process of ensuring that a destination component presents the correct impedance level that is required for proper *termination* of the output terminal of a component or a transmission line; The reflection coefficient Γ — or equivalently the *VSWR* — is minimized by ensuring that

$$Z_l = Z_c. \tag{5.2}$$

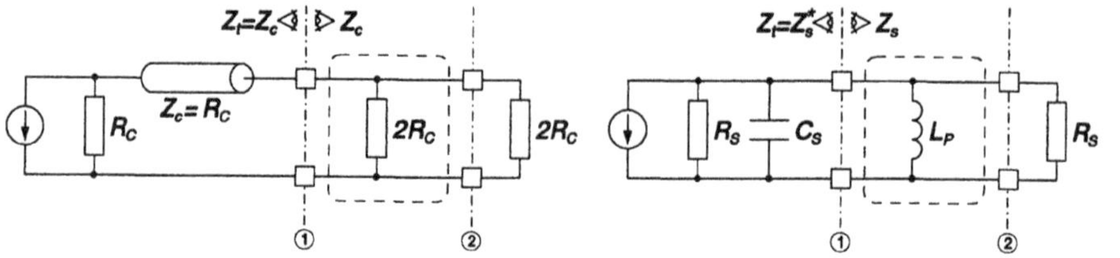

(a) Impedance matching. (b) Power matching.

Figure 5.3: *The difference between impedance matching and power matching.*

Impedance matching is illustrated in Fig. 5.3(a). This figure shows a component represented by its Norton model — a current source shunted by R_c — whose signal flows to the load though a transmission line with $Z_C = R_C$. For the sake of simplicity we will assume that the load impedance — i.e. the input impedance of a building block which needs to be connected to the transmission line — is two times larger than Z_C. Impedance matching at reference plane ① can now simply be done by shunting the load impedance $2R_c$ by a resistor with value $2R_c$. The impedance seen by the transmission line now equals its characteristic impedance.

Power matching, on the contrary, aims at transferring a maximum amount of power from the source to a given load. It tries to transform the load impedance into the complex conjugate of the source impedance or equivalently, transform the source impedance into the complex conjugate of the load impedance. Hence, in contrast to impedance matching, the target value of Z_l becomes

$$Z_l = Z_s^*, \tag{5.3}$$

where Z_s is the source impedance. Consider for instance the band-limited source in Fig. 5.3(b), represented by its Norton equivalent consisting of a current source shunted with R_s and C_s. Let us assume that the target load has the same value as R_s. A power match can now be ensured by adding inductor L_p. In this particular case, the matching condition is of course only fulfilled at a single frequency.

The difference between the two concepts is obvious from the figure. Impedance matching is clearly not concerned about the amount of power that is transferred into the load impedance; In Fig. 5.3(a), the impedance at the left side of load reference plane ② ($R_C//2R_c$) is clearly *not* the complex conjugate of the load impedance, $2R_C$. Half the power is lost in the added resistor. Thus, impedance matching is only concerned with providing the correct termination. It is also worth noticing that the power matching constraint applies to both reference planes in Fig. 5.3(b), whereas in Fig. 5.3(a), the impedance matching constraint is only valid at the component reference plane; Power matching is bi-directional — the load is adapted to the source and the source is adapted to the load — while impedance matching is mostly uni-directional.

Even though (5.3) guarantees a maximum power transfer to the input of a system, nothing is known about how *efficiently* the power is used by that particular system. More specifically, the question 'does maximum power transfer to a system always lead to maximum output power?' remains unaddressed... In the case of purely passive systems, there is no discussion: maximum power transfer to a passive system — by construction — always means maximum signal power in the load. However, when transistors are involved, this is not necessarily true any longer. This peculiarity will be dealt with in the next sections.

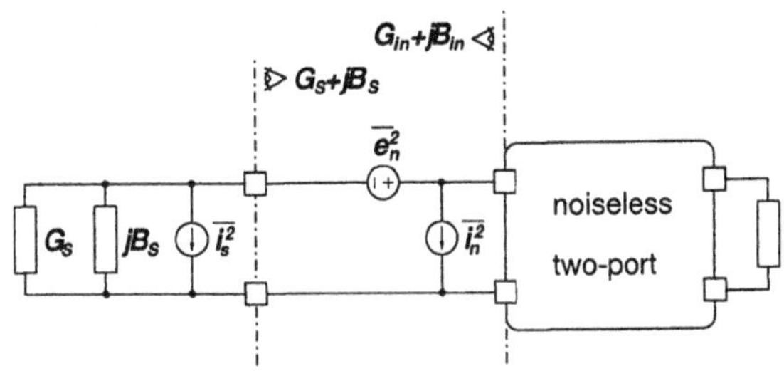

Figure 5.4: *Equivalent noise model of a two-port and its signal source.*

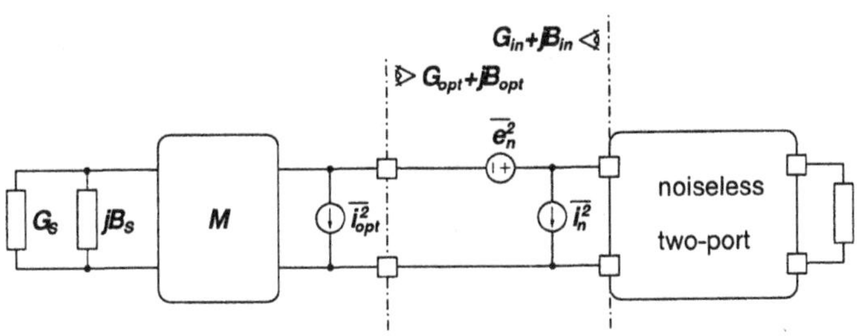

Figure 5.5: *Noise matching.*

5.2.3 Noise Matching

Let's consider the two-port in Fig. 5.4. All internal noise sources have been externalized and are represented by the equivalent input noise sources $\overline{e_n^2}$ and $\overline{i_n^2}$. The noise current associated with the real part of the source admittance is represented by $\overline{i_s^2}$.

According to two-port noise theory ([Haus60, Chan91, Lee98], restated in Appendix C), the noise figure of a two-port can be stated in terms of the source admittance, Y_S, and a number of parameters describing the magnitude and the correlation between the two-port noise sources $\overline{e_n^2}$ and $\overline{i_n^2}$. More important, the theory predicts that there exists an optimum source admittance that minimizes the noise figure of the system, denoted by Y_{opt} (i.e. $G_{opt} + jB_{opt}$). By choosing Y_{opt} as source admittance, the best compromise is made between (1) balancing the effect of the different uncorrelated noise sources and (2) maximally exploiting the correlation between the correlated noise components. As to the second aspect, the fraction of the current i_n that is fully correlated with e_n *and* orthogonal to e_n is even completely canceled!

The term 'noise matching' then refers to the process of transforming a given source admittance Y_S — by means of a noiseless matching network — into the optimum source admittance Y_{opt} (Fig. 5.5). Indeed, directly connecting the source to the two-port almost never yields an optimum *SNR* since the source impedance generally has a different value (e.g. 50 Ω).

It is important to understand that a noise match is completely different from an impedance match or a power match; Most of the time Y_{opt} considerably differs from Y_{in} or Y_{in}^*. Probably the best example is that for a purely passive two-port with input admittance Y_{in}, $Y_{opt} \to \infty$; The optimum source with respect to noise figure is a voltage source. It is a pity however to see that under these conditions no power is transferred to the passive network. After all, under noise matching conditions the voltage driving the passive two-port goes to zero, since it is given by

$$v_S \cong 2\sqrt{\frac{P_{av}}{G_S \to \infty}} \to 0, \tag{5.4}$$

where P_{av} is the available source power.

5.3 MOS Power Matching by Inductive Source Degeneration

5.3.1 Matching Principle

A well-known method [Shae97, Lee98] to match the inherently capacitive MOS device to a known source impedance R_S consists in inductively degenerating the transistor's source terminal (Fig. 5.6). In this matching scheme, the interaction of the ground inductance L_s with the g_m and the C_{gs} of the transistor generates a real part in the input impedance, given by

$$\Re(Z_{in}) = \frac{g_m L_s}{C_{gs}} \tag{5.5}$$

$$= \omega_T L_s, \tag{5.6}$$

where ω_T denotes the cut-off frequency of the transistor — the frequency at which the current gain is unity. The frequency-dependent imaginary part,

$$\Im(Z_{in}) = \frac{1}{j\omega C_{gs}} + j\omega(L_g + L_s), \tag{5.7}$$

can be neutralized around the operating frequency ω_0 by appropriately choosing the element values. By making the value of the real part equal to the desired source impedance R_S and adjusting the gate inductance L_g to yield series resonance at ω_0, i.e. adjusting L_g so that

$$\omega_0 = \frac{1}{\sqrt{(L_g + L_s)C_{gs}}}, \tag{5.8}$$

a good match is achieved around ω_0:

$$Z_{in} \cong \Re(Z_{in}) = \omega_T L_s \longmapsto R_S. \tag{5.9}$$

For reasons of clarity, non-idealities like the bulk transconductance, g_{mb}, the finite resistance of the inductors and the parasitic capacitances C_{gd} and C_{sb} are not incorporated in the foregoing formulae. Because this scheme matches the transistor to the source in a single step, we will refer to this scheme as the 'direct' source matching scheme.

5.3.2 Effective Transconductance and Power-to-Current Conversion

The effective transconductance, $g_{m,eff}$, defined in Fig. 5.6, is given by

$$g_{m,eff} \stackrel{\Delta}{=} \frac{i_{out}}{v_s} = \frac{\omega_T}{j\omega_0}\frac{1}{2R_S}. \tag{5.10}$$

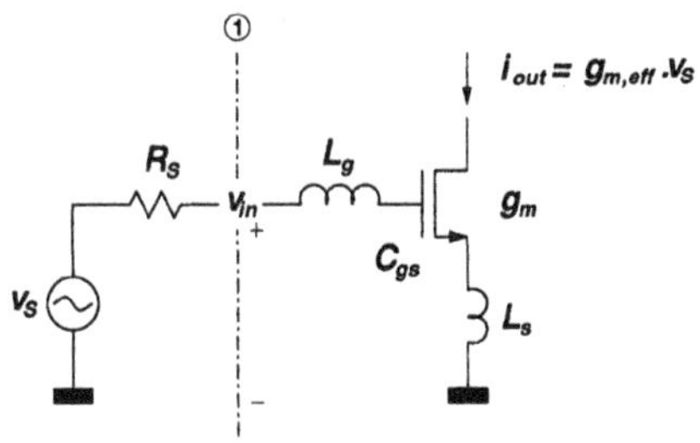

Figure 5.6: *MOS matching by inductive source degeneration.*

It is worth pointing out that $g_{m,eff}$ depends neither on the biasing current nor on the device transconductance. $g_{m,eff}$ only depends on the ratio between the cut-off frequency ω_T (proportional to $V_{GS}-V_T$), the operating frequency ω_0, and from the source resistance R_S. The better the intrinsic frequency performance of the MOS transistor with respect to the operating frequency, the higher the gain.

Restating (5.10) in terms of the original device transconductance g_m leads to

$$|g_{m,eff}| = Q_M g_m \tag{5.11}$$

$$Q_M \triangleq \frac{v_{gs}}{v_s} = \frac{1}{2\omega_0 R_S C_{gs}} = Q_{in}. \tag{5.12}$$

Hence, before controlling the gate-source terminal of the device, the source voltage v_s is multiplied by the quality factor of the matching network, Q_M. One can also look at (5.12) in another way; The amount by which the bandwidth of the unmatched system, given by

$$\omega_{3dB} = \frac{1}{R_S C_{gs}}, \tag{5.13}$$

exceeds the operating frequency ω_0, is converted into extra gain.

The reason why it suffices to consider i_{out} is that the load impedance is usually much lower than the output impedance, making it unnecessary to specify the output impedance value; For all purposes, the amplifier core behaves as a current source with respect to the load impedance. One of the main reasons behind this is that the quality factor of the load network (and therefore its R_p) is either limited by the quality factor of the passives — and hence by technology and the available silicon area — or constrained by the gain flatness specification. In addition, a cascode transistor is generally placed on top of the device to ensure a large output impedance.

Until now, the gain has been represented by the $g_{m,eff}$, i.e. the voltage to current conversion. However, from an RF point of view it is interesting to look at the the power-to-current conversion or *PCC*, defined as

$$PCC \triangleq \frac{i_{out}^2}{P_{av}}, \tag{5.14}$$

where i_{out} is the output current of the device and P_{av} is the available source power. The $g_{m,eff}$ and the *PCC* are closely related since the source voltage v_s directly relates to the available input power P_{av} by

$$v_s = 2\sqrt{P_{av} R_S}. \tag{5.15}$$

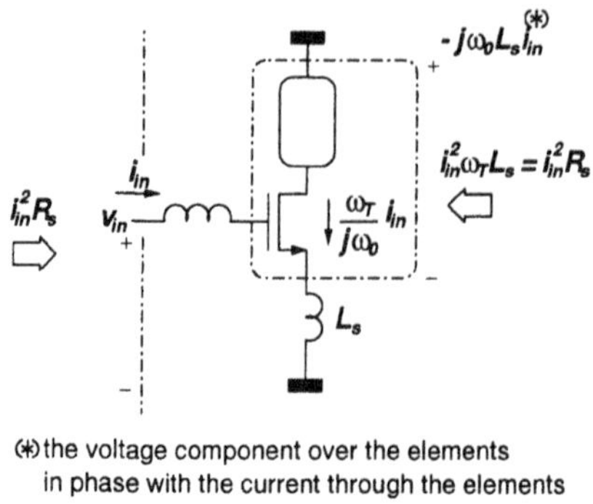

Figure 5.7: *The power flow in the classical matching scheme.*

The PCC in terms of $g_{m.eff}$ is thus given by

$$PCC(\Omega^{-1}) = 4g_{m.eff}^2 R_S \tag{5.16}$$

$$= \left(\frac{\omega_T}{\omega_0}\right)^2 \frac{1}{R_S}. \tag{5.17}$$

Hence, when comparing two schemes that match towards an identical source resistance R_S, it is sufficient to compare their respective $g_{m.eff}$ values; The PCC can then be derived from (5.16). On the other hand, the impact of the source resistance on the gain of a particular matching scheme can only be evaluated by looking at the PCC.

Of course, the $g_{m.eff}$ and the PCC only tell part of the story, namely: "How efficiently is the device used to generate gain?". In the end, however, i_{out} will be used either to generate voltage across or power in a certain load network. Later on, this will lead to the definition of the so-called power-to-voltage conversion (PVC) and the transducer power gain (G_t).

5.3.3 Analysis of the Power Flow

The origin of the real part in the input impedance has been explained in Subsection 5.3.1. However, it has not yet been clarified where the source power is flowing to, nor how efficiently this power is used to generate gain.

Since the matching elements themselves are lossless, the source power must inevitably flow into the transistor and all elements that follow. This is illustrated in Fig. 5.7. The signal power dissipation in the enclosed area is given by

$$P_D = v_{over} i_{ds}^* = j\omega_0 L_s i_{in} \frac{\omega_T}{j\omega_0} i_{in} \tag{5.18}$$

$$= \omega_T L_s i_{in}^2, \tag{5.19}$$

where v_{over} denotes the voltage component that is in phase with the current through the enclosed area. If the drain of the transistor is at a fixed potential, which is the case when an ideal cascoding device is connected to the transistor's drain, P_D is entirely dissipated in the transistor.

The fact that there is indeed maximum power transfer to the transistor does not mean that it is effectively used nor that it is the steering variable... According to (5.17), the output current and the input power are related by

$$i_{out}^2 = \left(\frac{\omega_T}{\omega_0}\right)^2 \frac{P_{in}}{R_S}. \tag{5.20}$$

Obviously, the output current not only depends on the input power, P_{in}, but also on the impedance level of the input reference plane; When the reference plane impedance R_S is lower, the same source power leads to a higher output current. In fact, in this matching scheme the *current* through the reference plane rather than the input power is the controlling variable:

$$i_{out} = \frac{\omega_T}{j\omega_0} i_{in}. \tag{5.21}$$

For all purposes, the matched device behaves as a bipolar transistor with

$$r_\pi = R_S \tag{5.22}$$

$$\beta = \frac{\omega_T}{\omega_0}. \tag{5.23}$$

The fact that the transferred power in itself is not important is further illustrated by the following calculations. Suppose that the input impedance of the matching section is different from the source impedance, i.e. $\Gamma \neq 0$. Additionally, the input impedance is assumed to be resistive — corresponding to a 'wrong' $\omega_T L_s$ value — so that $\Gamma \epsilon \mathbb{R}$. The net power that travels over the reference plane, P_{in}, and the input impedance, R_{in}, is then given by

$$P_{in} = P_{av}(1 - |\Gamma|^2) \tag{5.24}$$

$$R_{in} = R_S \left(\frac{1+\Gamma}{1-\Gamma}\right). \tag{5.25}$$

Combining the formula

$$i_{out}^2 = \left(\frac{\omega_T}{\omega_0}\right)^2 \frac{P_{in}}{R_{in}} \tag{5.26}$$

with the above expressions and applying the definition of the *PCC* given by (5.14), yields the *PCC* as a function of Γ:

$$PCC = PCC_{\Gamma=0}(1 - \Gamma)^2. \tag{5.27}$$

Fig. 5.8 shows the *PCC* enhancement — the ratio between the *PCC* when $R_{in} \neq R_S$ ($\Gamma = (R_{in} - R_S)/(R_{in} + R_S) \neq 0$) and the *PCC* when the MOS is matched to R_S— as a function of Γ. Astonishingly, the highest *PCC* occurs at full reflection, i.e. when there is *no* net power transfer to the circuit at all[1]. The fact that the maximum gain occurs at $\Gamma = -1$ shows once more that only the input current is relevant; The source sees a short and simply dumps a maximum amount of current into the circuit. Note that a Γ of -1 corresponds to a simple series-resonance at the input! In practice, the termination requirement of the antenna filter limits the allowable $|\Gamma|$ to about -10 dB. This is indicated by the dash-dotted lines in Fig. 5.8. Nevertheless, the

[1] In the next section we will show that due to the non-quasi static gate resistance there always is some power transferred to the MOS which makes that Γ can never become exactly -1. However, if $r_{g,NQS} \ll R_S$, Γ can come very close to it.

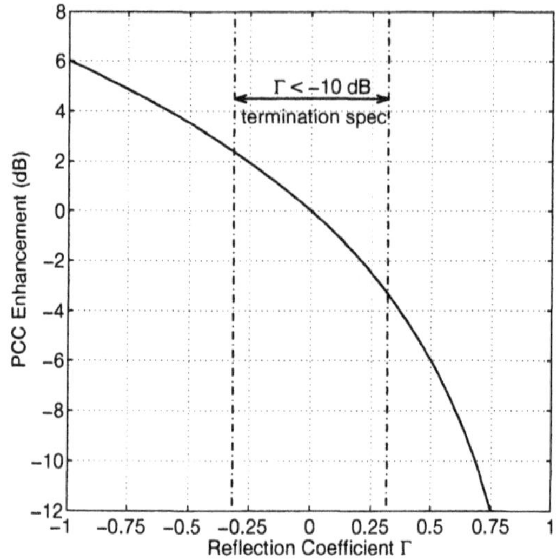

Figure 5.8: *The PCC enhancement as a function of the reflection coefficient* Γ. *The PCC enhancement is defined as the ratio between the PCC when* $R_{in} \neq R_S$ ($\Gamma \equiv (R_{in} - R_S)/(R_{in} + R_S) \neq 0$) *and the PCC when* $R_{in} = R_S$ ($\Gamma = 0$).

preceding calculations demonstrate that $|\Gamma|$ can be traded for gain: about 2 dB extra gain can be obtained while still complying to the constraint on $|\Gamma|$. Hence, the answer to the questions in Section 5.1, "Does a matched MOS device always provide more output signal than an unmatched device?" and "Does a matched MOS device really exploit the signal power it is provided with?", is negative; at least that's what follows from the above discussion. Before we can refine this point of view, it is necessary to introduce the non-quasi static effect.

5.4 The Non-Quasi Static Effect

5.4.1 Origin of the Non-Quasi Static Effect

At low frequencies, the inversion layer in a MOS transistor can be considered to be in equilibrium with the terminal voltages. This so-called *quasi-static* assumption allows one to make abstraction of the channel charge and to simply think of the v_{gs} controlling the i_{ds}. In addition, the input impedance behaves purely capacitive, i.e.

$$q_g = v_{gs} C_{gs}. \tag{5.28}$$

In reality however, the gate-source capacitance is not active between the gate and source terminals — like in most HSpice models — but acts in fact as a distributed capacitance to the channel. After all, the inversion layer charge stretches out over the full gate length. Depending on the distance along the channel, every infinitesimal gate-channel capacitor (transistor section) sees an equivalent conductance to the source. The value of this conductance is a function of the

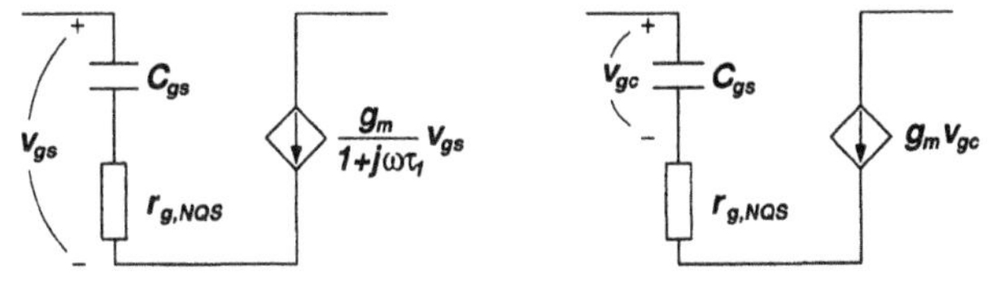

Figure 5.9: *A first order non-quasi static small-signal model for transistors operating in strong inversion: (a) valid both in the linear and the saturation region (b) approximate model, valid in the saturation region only.*

local charge distribution, i.e. the charge itself and the gradients in the channel.

It is precisely this physical reality that gives rise to the *non-quasi static effect*; When the gate electrode is excited at very high frequencies — in the order of several ω_T' — the finite channel conductance limits the speed of the inversion layer build-up so that it takes some time for the channel charge to get in equilibrium with the terminal voltages. For all purposes, the channel acts as a lossy transmission line. Since the i_{ds} is determined by the 'thickness' of the inversion layer, and since there is no inversion layer until the charging wave reaches the drain, also the drain current is delayed with respect to v_{gs}.

5.4.2 First Order Non-Quasi Static Model

[Tsiv87] models the non-quasi static effect — more specifically, the phase lag in the channel charge build-up — by adding a non-quasi static gate resistance $r_{g,NQS}$ in series with the C_{gs}, whose value corresponds to the effective resistance that is seen by the C_{gs}. Because of the distributed nature of the gate-channel capacitance, $r_{g,NQS}$ boils down to a certain fraction of the physical channel resistance. Apart from the inertia of the channel charge, [Tsiv87] models the delay of i_{ds} versus $g_m v_{gs}$ with a time constant τ_1. The corresponding small-signal equivalent circuit is shown in Fig. 5.9(a). In order not to overload the picture, the standard parasitics (like e.g. C_{gd}, C_{db}, g_{mb}, etc.) and the more exotic parasitics (like e.g. the drain inductance) are not shown. It is worth pointing out that — apart from the gate charge — the non-quasi static effect also affects all other terminal charges, giving rise to similar RC time constants at each terminal. Again, for reasons of compactness, the elements modeling this effect are not depicted in Fig. 5.9.

In strong inversion, the expression for the non-quasi static gate resistance is given by

$$r_{g,NQS} = \frac{\tau_{g,NQS}}{C_{gs}} = \frac{\tau_1 - \tau_2}{C_{gs}}, \tag{5.29}$$

where τ_1 and τ_2 are given by

$$\tau_1 = \frac{2}{5}\frac{1}{\omega_T} f_1(\frac{V_{DS}}{V_{DSsat}}) \tag{5.30}$$

$$\tau_2 = \frac{1}{5}\frac{1}{\omega_T} f_2(\frac{V_{DS}}{V_{DSsat}}), \tag{5.31}$$

and ω_T is the cut-off frequency of the MOS transistor. f_1 and f_2 are continuous functions of V_{DS}/V_{DSsat} that make the expressions valid in both the linear and the saturation region. In the saturation region, f_1 and f_2 simply amount to 1, so that

$$r_{g,NQS} = \frac{1}{5g_m} \tag{5.32}$$

$$\tau_{g,NQS} = \frac{1}{5\omega_T}. \tag{5.33}$$

Apparently, the time constant with which the drain current settles (τ_1) is two times larger than the time constant with which the gate charge settles ($\tau_{g,NQS} \stackrel{\Delta}{=} \tau_1 - \tau_2$). The reason for this is that the drain current only starts to increase when the charging wave of electrons reaches the end of the channel, or in other words, when most of the channel charge has already settled.

In practice, the operating frequency is much lower than the frequency where the NQS effect is active, so that $\omega\tau_1 \ll 1$. Consequently, the $\omega\tau_1$ term in Fig. 5.9(a) might as well be omitted. If one wants to retain some delay on the v_{gs} to i_{ds} transfer function, Fig. 5.9(b) is a good candidate model. This model is obtained by replacing the time constant τ_1 in Fig. 5.9(a) by the (smaller) non-quasi static time constant of the gate $\tau_1 - \tau_2$ and by recognizing that this time constant can be realized by taking the voltage v_{gc} over the C_{gs} as steering variable for the active part of the transistor. Apart from being more handy, this model gives more insight: instead of v_{gs}, the v_{gc} controls the channel charge, which on his turn controls the i_{ds}. The only difference between Fig. 5.9(a) and Fig. 5.9(b) is that in Fig. 5.9(b) the electrical length of the transistor is assumed to be zero for electrons traveling through the channel.

(5.32) has been derived under the long channel approximation; Velocity saturation and several short channel effects have not been taken into account to calculate the distribution of the channel charge. A more general formulation is obtained by replacing the factor 5 in the denominator of (5.32) by the so-called Elmore constant ε which defaults to 5 [Tsiv87]. This variant is included in the BSIM3v3 MOS model description [BSIM3v3]. To avoid any misinterpretation — ε is also the symbol for the electrical permittivity — in this text instead of ε the symbol κ is used for the Elmore constant:

$$r_{g,NQS} = \frac{1}{\kappa g_m}. \tag{5.34}$$

If a non-quasi static model is not available in the simulator — e.g. the non-quasi static extension of BSIM3v3 is not implemented in the HSpice BSIM3v3 model [HSpice, p. 16–239] — the correct resistance can either be put in series with the gate terminal or transformed into an equivalent parallel conductance,

$$g_{g,NQS} = \frac{1}{r_{g,NQS}(Q^2+1)}, \tag{5.35}$$

where Q is the quality factor of the gate-channel capacitor at the operating frequency ω_0, given by

$$Q = \frac{\kappa\omega_T}{\omega_0}. \tag{5.36}$$

Of course, this equivalent model is only valid at frequencies near ω_0. When this model is implemented in a netlist, is is recommended to put a large capacitor in series with the conductance to prevent DC current from flowing to the capacitor.

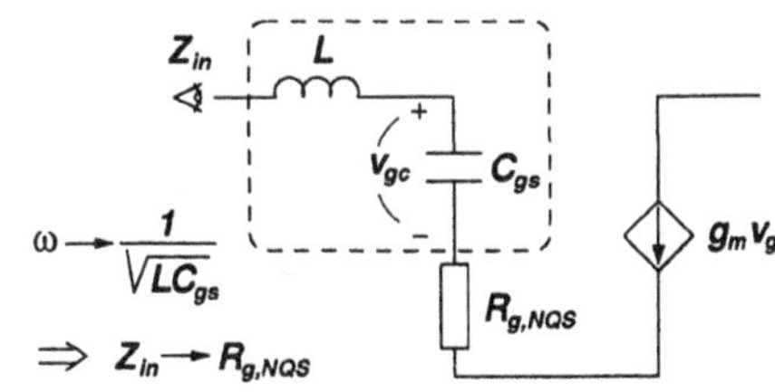

Figure 5.10: *Exposing $r_{g,NQS}$ by resonating out the gate-channel capacitance.*

5.4.3 Importance of the Non-Quasi Static Effect in the Low GHz Range

In general, it is believed that the non-quasi static effect only becomes important (read: 'necessary to model') at very high frequencies, e.g. at some fraction of the cut-off frequency ω_T. Indeed, the time constant of the phenomenon, $\tau_{g,NQS}$, is in the order of several ω_T's. Transposing this reasoning to the input impedance, this means that at normal frequencies ($< \omega_T/2$, e.g. < 10 GHz) the imaginary part (the capacitive part) of the input impedance is several orders of magnitude larger than the real part (the resistive part):

$$\frac{1}{\omega C_{gs}} \gg r_{g,NQS}. \tag{5.37}$$

Hence, it is very tempting to remove $r_{g,NQS}$ from the model.

However, when an inductor (e.g. a bondwire) is put in series with the gate, at a certain frequency the capacitive part of the input impedance of the MOS transistor is canceled by a series resonance phenomenon. At this frequency, the total input impedance behaves purely resistively; For all purposes, the oxide capacitance is made transparent to the signal so that the channel is directly exposed, as shown in Fig. 5.10. This phenomenon has been measured in [Jans98a]. It is worth pointing out that this frequency can be made *arbitrarily* low, just by changing the value of the series inductor! Thus, although the operating frequency is much lower than ω_T or $1/\tau_{g,NQS}$, a non-quasi static model is needed to correctly describe the MOS transistor.

The configuration in the thought-experiment of Fig. 5.10 is not purely theoretical. In fact, the matching technique described in Section 5.3 uses exactly the same configuration. This implies that the real part of Z_{in} in the classical matching scheme is no longer given by (5.6) but by

$$\Re(Z_{in}) = r_{g,NQS} + \omega_T L_s. \tag{5.38}$$

Not taking into account $r_{g,NQS}$ can significantly degrade the input match: for a g_m of 10 mS (0.2 V $V_{GS} - V_T$, 1 mA I_{DS}), a perfect 50 Ω match according to (5.9) results in reality in 70 Ω. Further, it will be shown in Section 5.5 that $r_{g,NQS}$ imposes a fundamental limit on power matching.

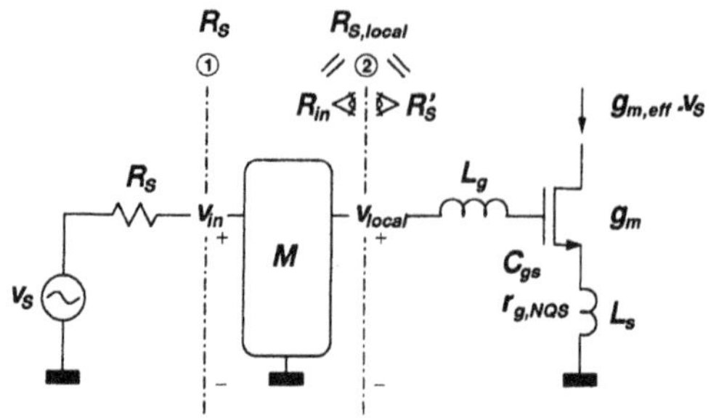

Figure 5.11: *Generalization of the classical MOS matching scheme by removing the impedance constraint on the input current.*

5.5 Optimum MOS Power Matching

5.5.1 Indirect Matching Principle

The matching approach in Section 5.3 suffers from the basic shortcoming of fixing the impedance of the current through C_{gs} at the value of the source resistance R_S [Jans99a]. Indeed, precisely due to matching, the current through reference plane ① is *by construction* at impedance level R_S. Its impact is clearly reflected in (5.17): the *PCC* is inversely proportional to R_S. Hence, the gain in this 'direct' source matching scheme is fundamentally limited by the value of the source impedance.

As shown in Fig. 5.11, the impedance constraint on the input current can be removed by splitting up the reference plane into two separate ones [Jans99a]: an input reference plane ① and a local reference plane ②. Now, the impedance level at ② (i.e. $R_{S,local}$ [2]) can be lowered while still keeping the source match at ①. The matching network M is designed to transform $R_{S,local}$ into R_S around ω_0 to preserve source matching at ①. The matching network can be implemented straightforwardly by means of a two-element matching scheme.

Using the power conservation theorem, the following expression for $g_{m.eff}$ is derived:

$$g_{m.eff} = g_{m.eff}[\text{eqn.(5.10)}]\sqrt{\frac{R_S}{R_{S,local}}} \quad (5.39)$$

$$= \frac{\omega_T}{j\omega_0}\frac{1}{2\sqrt{R_S R_{S,local}}}. \quad (5.40)$$

[2] $R_{S,local}$ is defined as the impedance level at the local reference plane ② under matched conditions. Whenever R_{in}, the input impedance of the MOS input section, equals R'_S, the effective source impedance, this common value of R_{in} and R'_S can be denoted by $R_{S,local}$. In case the matching network fails to transform R_S into $R'_S = R_{in}$ or in case the quality of the match is not known, we will explicitly use the variables R_{in} and R'_S.

The alternative expression for $g_{m,eff}$, (5.11), remains valid but (5.12) now becomes

$$Q_M = Q_{M[\text{eqn.}(5.12)]}\sqrt{\frac{R_S}{R_{S,local}}} \tag{5.41}$$

$$= \frac{1}{2\omega_0\sqrt{R_S R_{S,local}}C_{gs}}. \tag{5.42}$$

The expression for the *PCC* is now given by

$$PCC = PCC_{[\text{eqn.}(5.17)]}\frac{R_S}{R_{S,local}} \tag{5.43}$$

$$= \left(\frac{\omega_T}{\omega_0}\right)^2 \frac{1}{R_{S,local}}. \tag{5.44}$$

(5.39) and (5.43) clearly indicate that lowering $R_{S,local}$ with respect to R_S results in considerably larger $g_{m,eff}$ and *PCC* values than in the original matching technique, (5.10) and (5.17). A gain improvement of 6 to 10 dB can easily be obtained by equalizing $R_{S,local}$ to $0.25R_S$ and $0.1R_S$, respectively. Of course, in practice $R_{S,local}$ is limited by the inductor loss, the resistance of the gate fingers and the non-quasi static gate resistance. This will be covered in Subsection 5.5.3. It also follows from the formulae that the gain solely depends on the ω_T of the device (i.e. the $V_{GS} - V_T$); It neither depends on the current nor on the g_m. From an analog designer's perspective this is quite amazing, yet, it's a fact.

It is worth pointing out that the original source together with the matching network M acts as a composite source with source resistance $R_{S,local}$. Since the matching network M is lossless, the available power of the composite source is equal to that of the original source. Hence, as far as gain is concerned, it is always advantageous to have a low-ohmic source. The rest of the story — i.e. the relation with noise figure and IP_3 — is covered in Subsection 5.7.2 and Section 5.8, respectively.

5.5.2 How Leaving Out the Matching Network M Affects the *PCC*

Fig. 5.12 explains what it's all about; In principle, the modified matching technique requires an extra matching network M to implement the impedance transformation between R_{in} and the fixed source resistance R_S. However, one could ask oneself how the *PCC* is affected by leaving out this extra matching network. In the next paragraphs it will be shown that, under certain circumstances, M can indeed be left out without causing any loss in *PCC*.

To prove this, (5.43) — which is the gain when M is present — is first restated in terms of the parameter Υ (Upsilon), defined as

$$\Upsilon \triangleq \frac{R_{in} - R_S}{R_{in} + R_S}, \tag{5.45}$$

yielding the following expression for the *PCC* under matched conditions:

$$PCC_M = PCC_{[\text{eqn.}(5.17)]}\left(\frac{1-\Upsilon}{1+\Upsilon}\right). \tag{5.46}$$

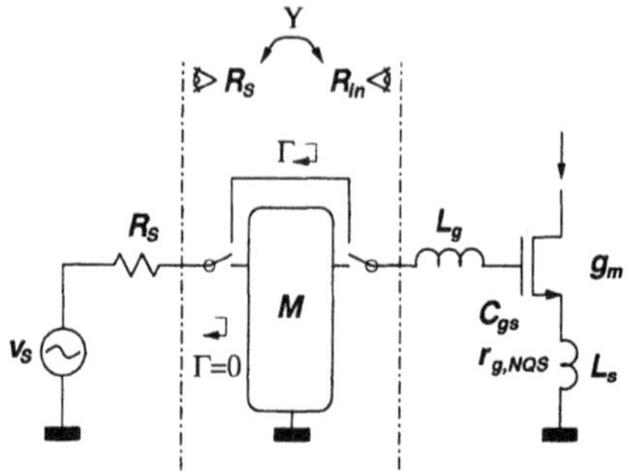

Figure 5.12: *Examining the impact of omitting M on the PCC.*

It is worth pointing out that in this context the parameter Υ has nothing to do with the reflection coefficient Γ, which is zero because of the 'exact' source match. On the other hand, when the matching section would be omitted, the *PCC* in terms of Υ is given by (5.27):

$$PCC_{\neg M} = PCC_{[\text{eqn.}(5.17)]}(1 - \Upsilon)^2. \tag{5.47}$$

In this case, Υ *does* stand for the reflection coefficient Γ. Note that for a given transistor, $(r_{g,NQS} - R_S)/(r_{g,NQS} + R_S)$ is the lowest attainable Υ value because R_{in} can never become lower than $r_{g,NQS}$.

Fig. 5.13 plots expressions (5.46) and (5.47) — actually, the respective *PCC* enhancement — versus Υ. Clearly, when Υ is smaller than 0.25 (i.e. -12 dB), or equivalently $0.6R_S < R_{in} < 1.67R_S$, the difference between both expressions is negligable. After all, based on (5.46) and (5.47) it can be shown that

$$\frac{PCC_{\neg M}}{PCC_M} = 1 - |\Gamma|^2. \tag{5.48}$$

This indicates that under these conditions the extra matching network M might as well be omitted. Of course, leaving out M means that the reflection coefficient Γ is no longer zero, but becomes equal to Υ. Yet, if a reflection coefficient of -10 dB is still tolerable, one can decide to omit the matching network in the region marked by the dash-dotted lines. Outside of this region, a matching section is always required.

It is important to note that when the matching network M is not simply omitted but merely fails to transform R_S into an effective source impedance $R'_S = R_{in}$, the above formulae are still valid when 'R_S' is replaced by 'R'_S'. Hence, the above formulae can also be used whenever the input network features a finite Γ. The effect of a finite Γ is further discussed in Subsection 6.12.1.

5.5.3 Fundamental Power Matching Limit

In Subsection 5.5.1 it is shown that the *PCC* can dramatically be improved by lowering $R_{S,local}$ with respect to R_S. Nevertheless, the *PCC* can not be made arbitrarily high; According to (5.38),

$$\Re(Z_{in}) = r_{g,NQS} + \omega_T L_s, \tag{5.49}$$

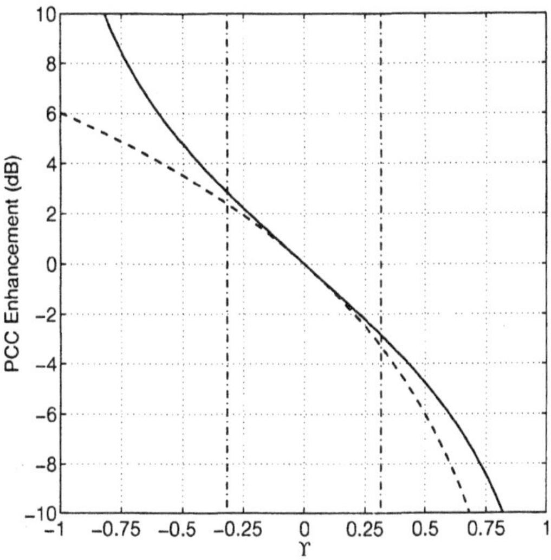

Figure 5.13: *The impact of leaving out the extra matching network M. The figure shows the PCC enhancement **(1)** when the extra matching section is present (solid line, ϒ (Upsilon) is related to R_{in} by equation (5.45)) and **(2)** when the extra matching section is omitted (dashed line, ϒ is the actual reflection coefficient Γ). The enhancement is calculated with respect to the PCC in case the circuit had been directly matched to R_S. It's worth noting that for a given transistor, $(r_{g,NQS} - R_S)/(r_{g,NQS} + R_S)$ is the lowest attainable ϒ value because R_{in} can never become lower than $r_{g,NQS}$.*

so that

$$R_{S,local} \geq r_{g,NQS}. \tag{5.50}$$

For a given MOS device, i.e. a fixed ω_T (only a function of $V_{GS} - V_T$) and a fixed g_m (a certain I_{DS} for a given $V_{GS} - V_T$), the maximum PCC is thus obtained for $L_s \rightarrow 0$, and is asymptotically given by

$$PCC_{max} = \left(\frac{\omega_T}{\omega_0}\right)^2 \frac{1}{r_{g,NQS}}. \tag{5.51}$$

Hence, the gain of a MOS transistor is maximized when all the available power is directed to $r_{g,NQS}$, i.e. the absolute minimum $R_{S,local}$. While all other ohmic parasites, like e.g. the resistance of the gate fingers and the inductors, can be made arbitrarily low either by increasing the number of fingers or increasing the silicon area, the non-quasi static gate resistance is always present.

People that are still not convinced that in order to really maximize the output signal of a transistor, the non-quasi static gate resistance needs to be *matched* to the source, should be by the following reasoning. Maximum output current requires a maximum amount of gate drive, i.e. a maximum v_{gs}. This requires dumping the maximum possible current in the input impedance, which in his turn boils down to maximizing the power delivered to the non-quasi static gate resistance; After all, since $r_{g,NQS}$ stands in series with C_{gs}, the power dissipation in $r_{g,NQS}$ equals

$$P_D = i_{in}^2 r_{g,NQS}. \tag{5.52}$$

Hence, it directly follows that the output signal is maximized when $r_{g,NQS}$ is matched to the source impedance. One of the simplest ways to implement this match is to resonate the C_{gs} against an inductor — exposing $r_{g,NQS}$— and transforming the source impedance R_S (down) to $R_{S,local} = r_{g,NQS}$. It is exactly this scheme that has been discussed in this subsection.

Using (5.34) and using approximations of the well-known relations between ω_T, g_m, I_{DS} and $V_{GS} - V_T$ for a MOS transistor in saturation,

$$\omega_T \cong K_T V_{GST} \tag{5.53}$$

$$g_m \cong \frac{2I_{DS}}{V_{GST}}, \tag{5.54}$$

the maximum PCC can be related to the minimum DC power dissipation in the device [Jans99a]:

$$PCC_{max} \simeq 2\kappa K_T^2 \frac{I_{DS} V_{GST}}{\omega_0^2} \tag{5.55}$$

$$= C_{techn.}^{si} \frac{P_{DC,min}}{\omega_0^2}. \tag{5.56}$$

The transition from (5.55) to (5.56) is made by recognizing that

$$V_{GST} = nV_{DSsat} = nV_{DSmin}. \tag{5.57}$$

Hence, it follows that the maximum gain is only limited by technology constants, the power dissipation in the transistor and the operating frequency.

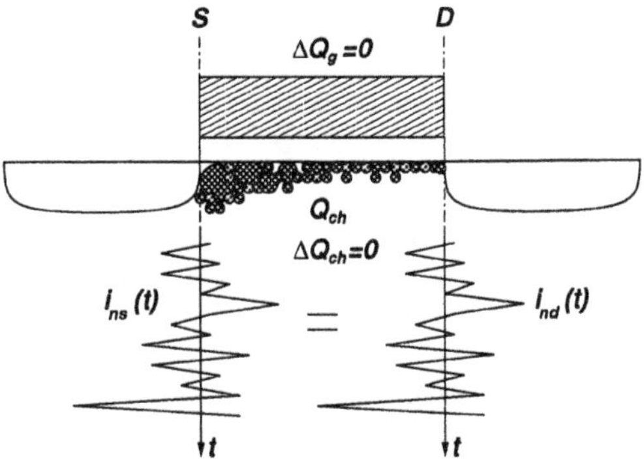

Figure 5.14: *The basic assumptions in the classical MOS noise model.*

5.6 Noise Sources in MOS Devices

5.6.1 Classical Channel Thermal Noise

Thermal noise in a MOS transistor is mainly caused by the agitation of the electrons in the inversion layer: due to the Brownian motion, electrons randomly cross the source and drain boundaries, giving rise to the noise currents $i_{nd}(t)$ and $i_{ns}(t)$. The classical MOS noise model is based on the assumption that the total channel charge remains constant, or in other words, that there is current continuity throughout the channel (see Fig. 5.14):

$$i_{nd}(t) = i_{ns}(t). \tag{5.58}$$

or in other words, that the total channel charge remains constant. Since the channel charge remains untouched, the noise current can be modeled by a current source between the transistor's source and drain terminals. In [Tsiv87], it is shown that

$$\overline{i_{nd}^2} = 4kT\frac{\mu}{L^2}Q_N\Delta f, \tag{5.59}$$

where μ is the carrier mobility, L is the channel length and Q_N is the total inversion layer charge.

For long channel transistors, (5.59) boils down to:

$$\overline{i_{nd}^2} = 4kT\gamma g_{ds0}\Delta f, \tag{5.60}$$

with g_{ds0} the zero-V_{DS} drain-source conductance and γ a bias-dependent parameter (function of the ratio V_{DS}/V_{DSsat}) [vdZi86, Tsiv87, Fox93, Mank99]. Through γ, (5.60) is valid both in the saturation region and in the linear region; For long channel transistors in saturation γ is calculated to be 2/3. In the linear region its value gradually increases from 2/3 to 1 for V_{DS} values between V_{DSsat} and zero, respectively. This standard noise model is used in the HSpice simulator [HSpice] provided that the *NLEV* parameter is set to 3.

In today's devices, however, short channel effects make the drain noise current somewhat larger. There are several reasons for this:

- In short channel transistors, the L parameter in (5.59) must be replaced by the effective *electrical* channel length, L_{eff}, to include channel length modulation. L_{eff} is defined as

$L - \Delta L$, where L is the channel length between the drain and the source (drawn channel length minus drain and source lateral diffusions) and ΔL is the reduction in electrical channel length due to the extension of the drain depletion region into the channel when $V_{DS} > V_{DSsat}$. The ΔL value strongly depends on technology (channel length, doping levels, etc.), V_{DS} and V_{DSsat}.

As a result of channel length modulation, the noise of a short channel device in saturation increases with increasing V_{DS}. This behavior occurs in both pMOS and nMOS devices. Due to this effect, the noise figure can be 20 to 30 percent larger than predicted by (5.60), and this even at moderate V_{DS} values (like e.g. a V_{DS} of $2V_{DSsat}$). A noise model incorporating channel length modulation as well as velocity saturation can be found in [Wang94]. This more advanced model is incorporated in the BSIM3v3 model definition [BSIM3v3], and can be activated in HSpice by setting *NLEV* = 0 and *NOIMOD* = 2 [HSpice].

- Under certain conditions, the high electric field at the drain of a short channel transistor may produce hot electrons with a (noise) temperature significantly above the lattice temperature. Due to this effect, γ might even become as high as two to three [Jind86, Abid86]. In addition, when the electrons can acquire enough energy between scattering events, these high energy electrons might cause impact ionization, leading to noisy substrate currents that can be coupled to the drain current through the g_{mb} [Jind85, Tsiv87]. Again, this effect strongly depends on technology (the channel length, the LDD structure, the doping levels, etc.) and on the applied stress factors (like e.g. the $[I_{DS}, V_{DS} - V_{DSsat}]$ vector).

The only way to prevent the noise in short channel transistors from increasing too much, is to keep the V_{DS} relatively close to V_{DSsat} (e.g. $V_{DSsat} + V_{ptp}$) and to foresee a lot of substrate straps grounding the bulk.

In the saturation region, it is more convenient to formulate (5.60) in terms of the transconductance:

$$\overline{i_{nd}^2} = 4kT\frac{\gamma}{\alpha}g_m\Delta f, \tag{5.61}$$

where α is defined as

$$\alpha \triangleq \frac{g_m}{g_{ds0}} \tag{5.62}$$

$$\cong \frac{1}{n} \triangleq \frac{g_m}{g_m + g_{mb}}. \tag{5.63}$$

Sometimes, the drain noise current is referred to the gate-source terminal, where it leads to two simultaneously active noise sources $\overline{e_n^2}$ and $\overline{i_n^2}$, given by

$$\overline{e_n^2} = \frac{\overline{i_{nd}^2}}{g_m^2} \tag{5.64}$$

$$\overline{i_n^2} = \overline{i_{nd}^2}\left(\frac{\omega}{\omega_T}\right)^2. \tag{5.65}$$

However, since e_n and i_n are *fully* correlated — $i_n = j\omega C_{gs}e_n$ — a lot of care must be taken when performing calculations with these variables.

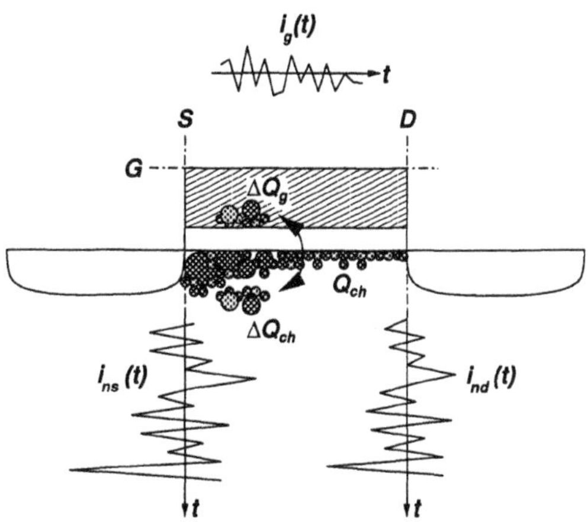

Figure 5.15: *The origin of the gate noise current.*

5.6.2 Non-Quasi Static Gate Noise Current

As stated in the previous subsection, the classical noise model is only valid under the assumption that the total channel charge stays constant in time. At low frequencies this model holds perfectly well. However, as charge *can* be stored in the channel, the momentary source and drain noise currents need not necessarily be identical. Since charge conservation tells us that each additional electron in the channel repulses a corresponding electron from the gate terminal, an extra gate noise current is induced (see Fig. 5.15), given by [vdZi86]:

$$\overline{i_{ng}^2} = 4kT\delta\frac{(\omega C_{gs})^2}{\kappa g_{ds0}}\Delta f. \tag{5.66}$$

Note that $\overline{i_{ng}^2}$ is completely different from the input-referred drain noise current, given by (5.65)!

In the long channel approximation, δ is calculated to be 4/3 [vdZi86]. As is the case with the γ parameter, in the short channel regime the exact value of δ depends both on the technology and the applied stress [Mank99]. Nevertheless, since the induced gate noise and the classical channel noise stem from the same phenomenon, it is very tempting to postulate the relation $\delta = 2\gamma$ [3], as is done in [Lee98]. The gate noise is also partly correlated with the drain noise. The correlation is embodied by a non-zero correlation coefficient c, defined as

$$c = \frac{\overline{i_{ng}\, i_{nd}^*}}{\sqrt{\overline{i_{ng}^2}\;\overline{i_{nd}^2}}}. \tag{5.67}$$

In the long channel regime, c equals $j0.395$ [vdZi86] while in the short channel regime its value is somewhat different [Mank98, Mank99]. The reason why the correlation coefficient is purely imaginary, can be attributed to the capacitive nature of the coupling between the channel and the gate.

[3] δ is probably somewhat larger because the closer to the drain, the tighter the coupling to the gate becomes... and it is precisely at the drain where the local electron temperature is the highest.

Although it may not immediately be obvious from (5.62), the gate noise current is intimately related to the non-quasi static gate resistance. To grasp this, (5.66) is first transformed into an alternative form using (5.34) and (5.62), yielding

$$\overline{i_{ng}^2} = 4kT\delta\alpha\left(\frac{\omega}{\kappa\omega_T}\right)^2\frac{1}{r_{g,NQS}}. \tag{5.68}$$

This noise current can also be thought of originating from a voltage noise source in series with the gate, whose magnitude is given by

$$\overline{e_{ng}^2} = \left(\frac{1}{\omega^2 C_{gs}^2} + r_{g,NQS}^2\right)\overline{i_{ng}^2} \tag{5.69}$$

$$\cong \frac{\overline{i_{ng}^2}}{\omega^2 C_{gs}^2} \tag{5.70}$$

$$= 4kT\delta\alpha r_{g,NQS}. \tag{5.71}$$

It immediately follows that, apart from the factor $\delta\alpha$, the gate noise simply boils down to the noise of the non-quasi static gate resistance.

5.6.3 Exotic and Parasitic Noise Sources

In the remainder of this text, only the classical noise and the non-quasi static gate noise will be taken into account. Nevertheless, there exist other noise sources that may play an important role if one does not take care during layout. For the sake of completeness, they are mentioned here.

In the previous subsection the non-quasi static gate noise current has been introduced as if it was *the* missing link between $i_{nd}(t)$ and $i_{ns}(t)$. However, one must not forget that a MOS device is in reality a four terminal structure, that also includes a bulk terminal. This means that extra channel charge not only gives rise to a change in the gate charge, but also affects the bulk charge, inducing a bulk noise current $\overline{i_{nb}^2}$. In a good approximation,

$$\overline{i_{nb}^2} \cong (n-1)\,\overline{i_{ng}^2}, \tag{5.72}$$

where n is defined by (5.63) [Tsiv87]. To keep this current from generating bulk voltages that modulate the back gate of the MOS device, it's important to reduce the resistance seen by the bulk terminal during layout. This can be done by dividing the transistor in several parallel sections and by foreseeing a lot of substrate straps around the transistor islands. In this way the mean distance between the bulk material underneath the gate and the ground terminal — and therefore the total bulk resistance — is reduced.

These countermeasures have some other beneficial effects. First of all, reducing the bulk resistance also cuts down its own voltage noise. Consequently, the back gate is less modulated so that less extra drain noise is induced:

$$\overline{i_{nd,extra}^2} = 4kTR_b g_{mb}^2 \Delta f. \tag{5.73}$$

Secondly, the resistance of the gate fingers decreases quadratically with the number of parallel sections

$$\overline{e^2_{ng,extra}} = 4kT\frac{R_g}{3}\Delta f, \tag{5.74}$$

where R_g is the DC gate resistance, given by

$$R_g = \frac{\rho_{poly}W}{n^2L}, \tag{5.75}$$

with n the number of fingers. The factor three in (5.74) takes into account the distributed nature of the gate resistance [Jind84, Chan91, Raza94].

There also exists bulk-induced noise which increases with the number of fingers[4]. Intuitively, the existence of this noise can be explained as follows; Near the edges of each finger, the oxide thickness gradually changes from t_{ox} to the thickness of the field oxide (i.e. the so-called "bird's beak"). In that region, the threshold voltage increases so that there is no longer a channel at this position. However, due to the small oxide thickness there is still capacitive coupling (via C_{gb}) to the noisy bulk material that lies underneath. [Kish99] models this noise source by an additional gate-to-substrate noise current, given by

$$\overline{i^2_{n,gb}} = \frac{4kT}{R_b}\left|\frac{j\omega R_b C_{gb}}{1+j\omega R_b C_{gb}}\right|^2 \Delta f. \tag{5.76}$$

As can be seen from the above equation, this noise source has a frequency-behavior similar to the non-quasi static gate noise current in (5.68). The measurement results in [Kish99] confirm that the noise is independent of the transistor bias conditions — which is consistent with the intuitive explanation — and strongly depends on device geometry and the C_{gb}. Whereas this noise source can be reduced by flanking the transistor by a lot of substrate straps, it does not improve with the number of fingers. The noise current increases since the time constant R_bC_{gb} is independent of the number of fingers and R_b is inversely proportional to the number of fingers. In this work, it is assumed that the classical back-gate noise in (5.73) is dominant, so that increasing the number of fingers does pay off.

One of the parasitic noise sources that can not be decreased by proper layout is the source resistance; Because during processing no difference is made between the source and the drain terminal of a MOS device, the LDD (lowly doped drain diffusion) is equally present at the source[5]. The resistance value of the 'lightly doped source' can be written as

$$R_{LDS} = \frac{\rho_{LDS}}{W}, \tag{5.77}$$

where ρ_{LDS} is the source resistance for a 1 μm wide transistor (in $\Omega\mu$m) and W is de width of the transistor (in μm). A typical value for ρ_{LDS} is 400 to 500 $\Omega\mu$m. In practical LNA designs, the transistor width is relatively large so that the source resistance is in the order of a few Ohms.

5.6.4 Performing Advanced Noise Simulations in a Classical Simulator

When there is no non-quasi static model available in the simulator — as is the case in the HSpice simulator, where the non-quasi static extension of the BSIM3v3 MOS model is simply not im-

[4]This is not the same as the transistor-width!

[5]in future, advanced devices this shortcoming may be solved.

plemented [HSpice, p. 16–239] (neither the resistance nor the noise) — or when the internal noise calculations employ the standard long-channel γ value of 2/3, a work-around needs to be found to correctly simulate the noise.

The first step in correctly simulating the noise is to account for the possible excess noise that is generated within a submicron CMOS transistor — expressed by a γ value larger than 2/3. The correct noise factor can then be calculated as follows:

$$F = (F_{sim} - 1)\frac{\gamma}{\gamma_{sim}} + 1, \tag{5.78}$$

where F_{sim} denotes the simulated noise factor and γ_{sim} is the γ value implemented in the simulator.

However, there is still a problem since (5.78) multiplies the noise of all the resistive elements in the circuit by γ/γ_{sim} as well. This problem can be alleviated by decreasing the temperature of all the resistive elements beforehand by a factor γ/γ_{sim}:

$$T_{resistor} = T_{sim}\frac{\gamma_{sim}}{\gamma}, \tag{5.79}$$

where T_{sim} is the internal temperature of the simulator, which is generally set to 300 K.

Since the intrinsic MOS device is often not accessible, the non-quasi static effect and its associated noise needs to be modeled explicitly by putting a (noisy) conductance — whose value is given by (5.35) — in parallel with the gate-source terminals. Of course, this model is only valid around the operating frequency ω_0. In order to implement the non-quasi static noise current — whose expression is given by (5.68) — the temperature of the conductance needs to be adjusted to:

$$T_{conductance} = T_{sim}\frac{\alpha\delta\gamma_{sim}}{\gamma}. \tag{5.80}$$

Note that this equation does not yet take into account the correlation between the gate noise and the drain noise current. However, as will be shown later on, the effect of the correlation coefficient c can easily be included by multiplying (5.80) by $1 - |c|^2$.

5.7 The Noise Figure of an Input-Matched MOS Device

This section presents a detailed analysis of the noise performance of a source-matched MOS device[6]. The noise equations derived in this section form the basis of the LNA design methodology described in Chapter 6. There, they will be adapted to incorporate all kinds of non-idealities, like e.g. the C_{gd}, the parasitic input capacitance due to the bondpad and its protection, etc.

After a short review of the noise performance of a noise-matched MOS device, the noise figure of an input-matched MOS device is derived and discussed. The impact of the source-matching scheme (or equivalently, the source impedance value) on the magnitude of the non-quasi static gate noise and the classical drain noise is thoroughly examined. Among others, it will be shown that by adjusting the source impedance, the noise figure of an input-matched device can be made extremely close to the minimum noise figure F_{min}, i.e. the noise figure when there is a noise match. The trade-off between *PCC* and noise figure is examined as well. The section concludes with some considerations on noise optimization.

[6] A stripped-down version of this section can be found in [Jans99b].

	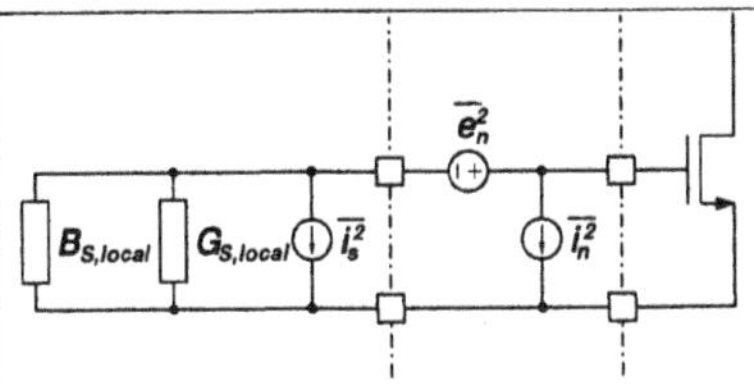 Noise Matching	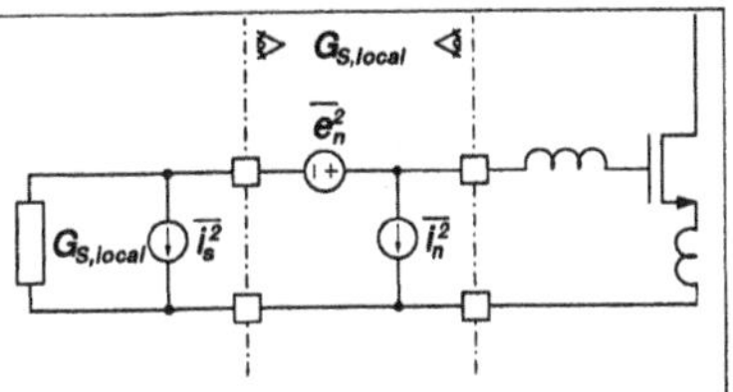 Source/Power Matching
$G_s \triangleq \frac{\overline{i_s^2}}{4kT\Delta f}$	$\frac{1}{R_{S,local}}$	
$R_n \triangleq \frac{\overline{e_n^2}}{4kT\Delta f}$	$\frac{\gamma}{\alpha}\frac{1}{g_m}$	$\frac{1}{4}\frac{\gamma}{\alpha}\left(\frac{\omega}{\omega_T}\right)^2\left(R_{S,local}+\frac{1}{\kappa g_m}\right)^2$
$G_u \triangleq \frac{\overline{i_u^2}}{4kT\Delta f}$	$\frac{\alpha\delta}{\kappa}g_m\left(\frac{\omega}{\omega_T}\right)^2(1-\lvert c\rvert^2)$	$\frac{\alpha\delta(1-\lvert c\rvert^2)}{\kappa g_m R_{S,local}^2}$
$Y_c \triangleq \frac{i_c}{e_n}$	$g_m\frac{j\omega}{\omega_T}\left(1+\lvert c\rvert\alpha\sqrt{\frac{\delta}{\gamma\kappa}}\right)$	$\frac{1}{R_{S,local}}\left(1-c\alpha\sqrt{\frac{\delta}{\gamma\kappa}}\right)$ $(c \in \mathbb{C})$

Table 5.2: *Two-port noise parameters under different matching conditions.*

5.7.1 Noise Figure under Noise Matching Conditions

According to Subsection 5.2.3, the noise figure of any two-port can be minimized by transforming the source conductance into $G_{opt} + jB_{opt}$, an optimum conductance which only depends on the magnitude and the correlation of the equivalent input noise sources of the two-port. G_{opt} and B_{opt} can be calculated using the conventional two-port noise theory ([Haus60, Lee98], restated in Appendix C), which makes use of some auxiliary variables: Y_s, R_n, G_u and Y_c.

In case of a single MOS device, the Y_s, R_n, G_u and Y_c parameters can be found in the left part of Table 5.2. The calculations that lead to these expressions can be found in [Lee98]. This yields the following expression for the minimum noise figure:

$$F_{min} = 1 + 2\left(\frac{\omega_0}{\omega_T}\right)\sqrt{\frac{\gamma\delta}{\kappa}(1-\lvert c\rvert^2)}. \tag{5.81}$$

Hence, the minimum noise figure depends only on the ratio between the operating frequency and the ω_T of the device. Note that this means that for a given current, a device with a small W/L offers a lower noise figure than a device with a large W/L !

5.7.2 Noise Figure under Source Matching Constraints

In case of a source-matched MOS device, the derivation of the auxiliary parameters Y_s, R_n, G_u and Y_c is quite intricate. Since we have already performed the detailed calculation of the auxiliary parameters in [Lero99], we will confine ourselves to citing the results; The resulting expressions for Y_s, R_n, G_u and Y_c are stated in the right part of Table 5.2. Substituting these parameters in

(C.5) yields the expression for the noise factor under matched conditions:

$$F = 1 + \frac{\left(\frac{\alpha\delta(1-|c|^2)}{\kappa g_m R_{S,local}^2} + \left(\frac{4}{R_{S,local}^2} + \frac{|c|^2\alpha^2}{R_{S,local}^2}\frac{\delta}{\gamma\kappa}\right)\frac{1}{4}\frac{\gamma}{\alpha}\left(\frac{\omega_0}{\omega_T}\right)^2 g_m\left(R_{S,local} + \frac{1}{\kappa g_m}\right)^2\right)}{\frac{1}{R_{S,local}}}. \tag{5.82}$$

This expression can be reorganized to

$$F = 1 + (F_d - 1) + (F_{g,u} - 1) + (F_{g,c} - 1), \tag{5.83}$$

where F_d, $F_{g,u}$ and $F_{g,c}$ are defined by

$$F_d - 1 = \frac{\gamma}{\alpha}\left(\frac{\omega_0}{\omega_T}\right)^2 \frac{g_m}{R_{S,local}}\left(R_{S,local} + \frac{1}{\kappa g_m}\right)^2 \tag{5.84}$$

$$F_{g,u} - 1 = \frac{\alpha\delta(1-|c|^2)}{\kappa g_m R_{S,local}} \tag{5.85}$$

$$F_{g,c} - 1 = \frac{|c|^2\alpha^2\delta}{4\gamma\kappa}\frac{\gamma}{\alpha}\left(\frac{\omega_0}{\omega_T}\right)^2 \frac{g_m}{R_{S,local}}\left(R_{S,local} + \frac{1}{\kappa g_m}\right)^2. \tag{5.86}$$

The $F_d - 1$, $F_{g,u} - 1$ and $F_{g,c} - 1$ terms represent the respective contributions of the drain noise, the uncorrelated and the correlated gate noise. Because

$$\frac{F_{g,c} - 1}{F_d - 1} = \frac{|c|^2\alpha^2\delta}{4\gamma\kappa} \ll 1, \tag{5.87}$$

$F_{g,c} - 1$ can be neglected compared to $F_d - 1$, so that

$$F \cong 1 + (F_d - 1) + (F_{g,u} - 1). \tag{5.88}$$

The error is hardly 1.5%. After some manipulations, (5.88) becomes:

$$F \cong 1 + \frac{\gamma}{\alpha}\left(\frac{\omega_0}{\omega_T}\right)^2\left(\sqrt{g_m R_{S,local}} + \frac{1}{\kappa\sqrt{g_m R_{S,local}}}\right)^2 + \frac{\alpha\delta(1-|c|^2)}{\kappa g_m R_{S,local}}. \tag{5.89}$$

Obviously, g_m and $R_{S,local}$ can be weighed against one another since the product '$g_m R_{S,local}$' is treated as a single unit in this expression. The dynamics in (5.89) can be explained as follows:

$F_d - 1$: Lowering the g_m value — while maintaining a certain ω_T— leads to a lower drain noise contribution because the device's drain noise current *decreases* while the *PCC* stays constant. This in contrast with the input-referred noise of a simple common source structure which *increases* as g_m decreases. The reason is that in the case of a matched device, the noise transfer function is virtually independent from the g_m of the device[7] and only depends on the *PCC* (i.e. ω_0, ω_T and $R_{S,local}$).

A lower $R_{S,local}$ value decreases the input-referred drain noise — because the *PCC* of the matched device is inversely proportional to $R_{S,local}$— while the noise power of the signal

[7] The non-quasi static gate resistance, which depends on g_m, slightly affects the noise transfer function.

source remains the same (kTB). Again, the noise factor behaves differently than the noise factor of a common source structure, which *increases* as $R_{S,local}$ becomes lower.

$F_{g,u} - 1$: The larger the g_m, the lower the value of the parasitic non-quasi static resistance becomes, so that less non-quasi static noise is induced at the source reference plane. On the other hand, when $R_{S,local}$ increases, the induced noise at the source reference plane becomes relatively lower compared to the noise of $R_{S,local}$.

$F_d - 1$ vs. $F_{g,u} - 1$: $F_d - 1$ rapidly decreases as the frequency performance of the device improves. When $\omega_0 \ll \omega_T$ — which is almost always the case in submicron CMOS technologies — and at normal values of the source impedance, the noise is dominated by the non-quasi static gate noise, $F_{g,u} - 1$, so that often $F \cong F_{g,u}$. In the next section, the balance between these two noise contributions will be illustrated with a numerical example.

5.7.3 Impact of the Source Matching Scheme on Noise Figure and *PCC*

In this subsection the noise figure of the scheme described in Subsection 5.5.1 is compared with the noise figure of the direct matching scheme described in Subsection 5.3.1. It will be shown that the extra degree of freedom offered by the indirect matching technique — the impedance of the local reference plane — enables us to cover the whole spectrum from maximum gain (i.e. a non-quasi static match) to minimum noise figure (very close to a noise-match!), and this while keeping the source match ($\Gamma = 0$)! Since the only difference between the two matching schemes is the value of $R_{S,local}$, it suffices to analyze the dynamics of the noise for a fixed device as a function of $R_{S,local}$.

Because (5.89) contains two terms with opposite dynamics, there exists an optimum noise factor; Differentiating this expression with respect to $R_{S,local}$, yields the $R_{S,local}$ value that optimizes noise for the given device:

$$R_{S,opt} = \frac{1}{\kappa g_m}\sqrt{1 + \frac{\alpha\delta(1 - |c|^2)}{\Phi}} \tag{5.90}$$

where

$$\Phi \triangleq \frac{\gamma}{\alpha\kappa}\left(\frac{\omega_0}{\omega_T}\right)^2. \tag{5.91}$$

The optimum noise factor under source matching constraints is then given by

$$F_{opt} = 1 + 2\left(\Phi + \sqrt{\Phi\alpha\delta(1 - |c|^2) + \Phi^2}\right). \tag{5.92}$$

If the operating frequency is significantly smaller than ω_T, F_{opt} is very close to the minimum noise factor F_{min}, given by (5.81). Under these conditions,

$$\Phi \ll \alpha\delta(1 - |c|^2), \tag{5.93}$$

α	κ	γ	δ	c	$\omega_0/2\pi$	$\frac{\omega_0}{\omega_T}$	g_m	C_{gs}	W/L
0.85	5	$2\cdot\frac{2}{3}$	$2\cdot\frac{4}{3}$	$j0.4$	1.8 GHz	$\frac{1}{12}$	50 mS (~ 5 mA at 0.15 V)	370 fF	300/0.25

Table 5.3: *The MOS device used in the numerical examples. The device parameters are representative for the amplifying device in a 0.25* μm, *1.8* GHz *LNA.*

so that F_{opt} can indeed be approximated by

$$F_{opt} \cong 1 + 2\sqrt{\Phi\alpha\delta(1-|c|^2)} \tag{5.94}$$

$$= 1 + 2\left(\frac{\omega_0}{\omega_T}\right)\sqrt{\frac{\gamma\delta}{\kappa}(1-|c|^2)} \tag{5.95}$$

$$= F_{min}. \tag{5.96}$$

This clearly shows that matching a given device to the source impedance is not at the expense of noise figure as long as the intermediate impedance at reference plane ② in Fig. 5.11 , $R_{S,local}$, is chosen correctly. Again, apart from some technology constants, F_{opt} only depends on the operating frequency and the ω_T (i.e. the V_{GST}) of the device. There is no relation with the I_{DS} or the g_m. The same conclusion as for the noise-matched device in Subsection 5.7.1 applies here: for a given current, a small device features a lower noise figure than a large device.

The solid line in Fig. 5.16 plots the noise figure *NF* ($10\log_{10}(F)$) against $R_{S,local}$, for the MOS device described in Table 5.3. The optimum noise figure NF_{opt} of 0.54 dB is obtained at an $R_{S,local}$ value of 118 Ω (Fig. 5.16(d)). Since (5.93) applies, NF_{opt} is indeed very close to NF_{min}, which equals 0.53 dB. If $R_{S,local}$ is chosen to be 50 Ω, which corresponds to applying the classical, direct matching technique in a 50 Ω environment, the noise figure increases to about 0.74 dB (Fig. 5.16(c)). At low $R_{S,local}$ values, the noise figure is clearly dominated by the non-quasi static gate noise current. Ultimately, at the minimum $R_{S,local}$ value, given by

$$R_{S,local} = r_{g,NQS} = \frac{1}{\kappa g_m}, \tag{5.97}$$

the expression for the noise factor becomes

$$F = 1 + \underbrace{\frac{4\gamma}{\alpha\kappa}\left(\frac{\omega_0}{\omega_T}\right)^2}_{\ll 1} + \alpha\delta(1-|c|^2). \tag{5.98}$$

For the MOS device in Table 5.3, $r_{g,NQS}$ equals 4 Ω. This yields a noise figure of 4.6 dB (Fig. 5.16(a)). Note that if δ had been equal to its long channel value, the non-quasi static noise and the noise from the source resistance would have balanced, yielding a noise figure of about 3 dB.

One could ask oneself what would have been the conclusion if the non-quasi static gate noise current had been neglected, i.e. when the classical MOS noise model had been used. This would

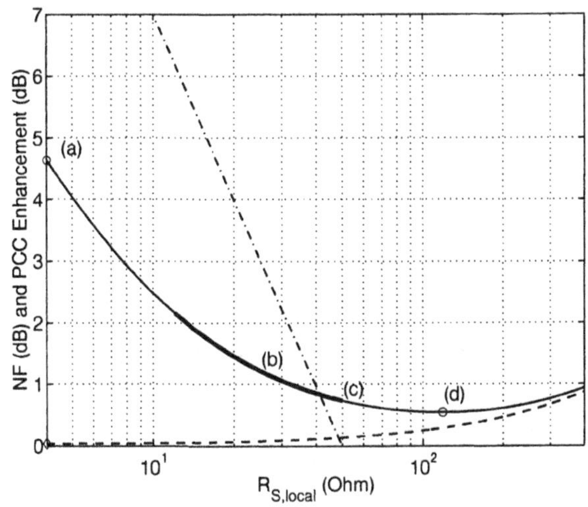

Figure 5.16: *The noise figure and the PCC enhancement as a function of $R_{S,local}$ for the device in Table 5.3. The noise figure including the non-quasi static noise current (solid line) is contrasted against the noise figure predicted by the classical noise model (dashed line). The dash-dotted line represents the PCC enhancement with respect to a direct match to a 50 Ω source.*

have lead to the following set of equations:

$$F = 1 + \frac{\gamma}{\alpha}\left(\frac{\omega_0}{\omega_T}\right)^2 \left(\sqrt{g_m R_{S,local}} + \frac{1}{\kappa\sqrt{g_m R_{S,local}}}\right)^2 \tag{5.99}$$

$$R_{S,opt} = \frac{1}{\kappa g_m} = r_{g,NQS} \tag{5.100}$$

$$F_{opt} = 1 + 4\Phi. \tag{5.101}$$

Hence, one would have concluded that the noise figure is minimized at the same $R_{S,local}$ value that maximizes the power gain (PCC). Additionally, NF_{opt} would have been predicted to be extremely low, in the order of 0.04 dB. The noise figure predicted by the classical noise model is represented by the dashed line in Fig. 5.16.

To clarify the trade-off between noise figure and PCC, the dash-dotted line in Fig. 5.16 plots the PCC enhancement as a function of $R_{S,local}$ for the usual 50 Ω source. As expected, the PCC increases rapidly at the cost of a progressively degrading noise figure. Still, an excellent compromise between noise performance and PCC can be obtained by selecting trajectory (b) [Jans99b]; By lowering $R_{S,local}$ from 50 Ω down to $3r_{g,NQS}$ (i.e. 12 Ω for the device in Table 5.3), the gain increases by 6 dB. The noise figure on the other hand raises by 1.4 dB, yielding a total noise figure of only 2.1 dB. Such a noise figure still complies to the most stringent cellular receiver specifications. Besides, in CMOS low-IF receivers the $1/f$ noise of the down-conversion mixers often dominates the system noise figure anyway. The trade-offs are summarized in Table 5.4.

Matching	NQS Match	Indirect Source Match	Direct Source Match	Near Noise Match
Source Match	Yes	Yes	Yes	Yes
$R_{S,local}$	$r_{g,NQS}$ (a)	$> 4r_{g,NQS}$ but $< R_S$ (b)	R_S (c)	$R_{S,opt}$ (d)
PCC	Maximum	High	Moderate	Low
NF	> 3 dB	Low	Low $\sim NF_{min} + .1 \cdots .5$dB	Optimum $\sim NF_{min}$

Table 5.4: *Trading noise figure against PCC for a given device.*

5.7.4 Some Early Considerations on Noise Optimization

In the previous subsection, we always assumed a fixed device (i.e. $V_{GS} - V_T$ and I_{DS} fixed) while $R_{S,local}$ has been used as a degree of freedom to optimize the noise figure. However, since '$g_m R_{S,local}$' behaves as an indivisible quantity in (5.89), we could as well have differentiated (5.89) with respect to $g_m R_{S,local}$ instead of just $R_{S,local}$. Of course, this still requires that ω_T— or equivalently $V_{GS} - V_T$— is kept fixed. This would have lead to the same result, be it in a slightly different form:

$$(g_m R_{S,local})_{opt@V_{GST}} = \frac{1}{\kappa}\sqrt{1 + \frac{\alpha\delta(1 - |c|^2)}{\Phi}}. \tag{5.102}$$

If (5.93) holds — which is typically the case —,

$$(g_m R_{S,local})_{opt@V_{GST}} \cong \frac{1}{\kappa}\sqrt{\frac{\alpha\delta(1 - |c|^2)}{\Phi}} \tag{5.103}$$

$$= \frac{\omega_T}{\omega_0}\sqrt{\frac{\alpha^2\delta(1 - |c|^2)}{\gamma\kappa}}. \tag{5.104}$$

Since

$$\omega_T \triangleq \frac{g_m}{C_{gs}} = \frac{3g_m}{2WLC_{ox}}, \tag{5.105}$$

it follows that

$$W_{opt@V_{GST}} \cong \frac{1}{\omega_0 R_{S,local}}\left(\frac{3}{2LC_{ox}}\sqrt{\frac{\alpha^2\delta(1 - |c|^2)}{\gamma\kappa}}\right). \tag{5.106}$$

For a CMOS technology with the noise parameters in Table 5.3, and for an operating frequency and a source resistance of 1.8 GHz and 50 Ω, respectively , it can be shown that $W_{opt@V_{GST}}$ is approximately 700 μm. Note that this value does not depend on $V_{GS} - V_T$!

At any given $V_{GS} - V_T$ (and hence any ω_T), the optimum noise factor F_{opt} in (5.95) — which depends only on the ratio of ω_0 and ω_T— can always be achieved as long as W and $R_{S,local}$ are related by (5.106). This leads to the following remarkable observation: If $R_{S,local}$ is a degree of freedom, $W_{opt@V_{GST}}$ — and hence I_{DS}— can be made *arbitrarily* low by increasing $R_{S,local}$. Consequently, the optimum noise factor in (5.95) can in principle be obtained at an extremely

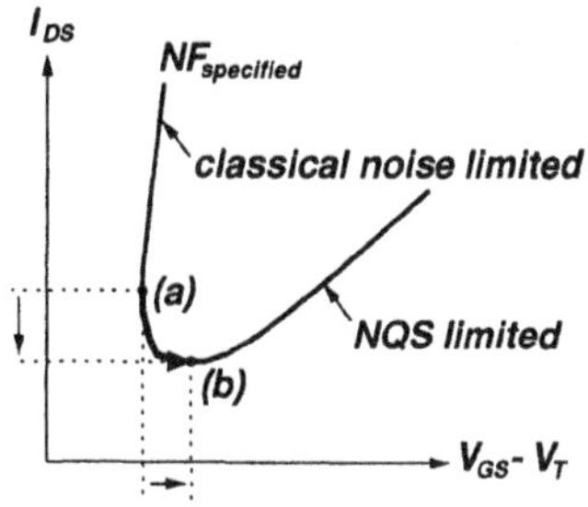

Figure 5.17: *A typical noise figure contour in the I_{DS}/V_{GST} design space.*

low I_{DS}! Of course, the pitfall is that too large $R_{S,local}$ values come at the cost of a low PCC, requiring a load with a very large quality factor when this device would be used as the input device of an amplifier.

When $R_{S,local}$ is no degree of freedom, the $V_{GS} - V_T$ can be determined so that F_{opt} equals $F_{specified}$. By subsequently setting the device width to $W_{opt@V_{GST}}$, we have obtained the transistor that offers the specified noise figure at the lowest possible $V_{GS} - V_T$ value (see Fig. 5.17(a)); After all, according to (5.95) the minimum noise figure value is always larger at lower $V_{GS} - V_T$ values.

Still, it should be clear that this solution does not guarantee a minimum power consumption; Optimization of the power consumption would require that the specified noise figure is offered at the minimum possible I_{DS} ... which is of course very different from the guarantee that the noise figure is offered at a minimum $V_{GS} - V_T$. What we do know however, is that the width of the transistor that optimizes the power consumption for a given noise figure is necessarily smaller than the $W_{opt@V_{GST}}$ in (5.106): since $V_{GS} - V_T$ *must* be larger there is simply no other choice than to go after a significantly lower W in order to pursue a lower I_{DS} (see Fig. 5.17(b)). As a consequence, when the power consumption is really minimized for a given noise figure, the non-quasi static noise again becomes the limiting factor.

In contrast to the $V_{GS} - V_T$-constrained noise minimization (or the noise-constrained $V_{GS} - V_T$ minimization) an analytical solution to the power-constrained noise optimization problem — which is the same as a noise-constrained power optimization due to the nature of the contour lines — is very difficult to obtain without oversimplifying the problem. An example of this is the power-constrained noise optimization in [Shae97, Lee98], which leads to the simple expressions

$$W_{opt@I_{DS}} \cong \frac{3}{2\omega_0 L C_{ox} R_{S,local} Q} \tag{5.107}$$

$$Q = |c|\sqrt{\frac{5\gamma}{\delta}}\left(1 + \sqrt{1 + \frac{3}{|c|^2}\left(1 + \frac{\delta}{5\gamma}\right)}\right) \approx 4 \tag{5.108}$$

$$F_{min} \approx 1 + 2.4\frac{\gamma}{\alpha}\frac{\omega_0}{\omega_T}. \tag{5.109}$$

For a CMOS technology with the noise parameters in Table 5.3, and for an operating frequency and a source resistance of 1.8 GHz and 50 Ω, respectively, $W_{opt@I_{DS}}$ is approximately 350 μm, about two times less than $W_{opt@V_{GST}}$. The main limitation of this set of expressions is that they are derived under the assumption that $\Theta(V_{GS} - V_T) \ll 1$, an assertion which is no longer true in deep submicron CMOS technologies; For a 0.25 μm device at a $V_{GS} - V_T$ of 0.25 V, $\Theta(V_{GS} - V_T)$ already approximates 1 (see Table 4.1)! The assertion in [Shae97, Lee98] that "this inequality fails to hold only in high power circuits" is therefore not true.

There is still another reason why we must take the expressions for W_{opt} with a grain of salt:

- In practice, the minimum current consumption of e.g. an LNA is not only determined by the noise figure specification; Constraints on the total LNA quality factor, the linearity, the maximum gate inductance, etc. often put lower bounds on the power consumption which are more stringent than the noise figure specification.

- Parasites like the C_{gd}, the input capacitance C_p and the Miller effect M *dramatically* change the magnitude of and the balance between the non-quasi static noise and the classical drain noise current. These effects make the noise figure expression even more complicated than it was before.

- The noise sources in Subsection 5.6.3 have note been taken into account in this optimization. For instance, according to [Kish99], the bulk noise current in (5.76) pushes the optimum width towards larger values as an extra term is added which is inversely proportional to the square of the C_{gs}:

$$F - F_{original} \cong \left(\frac{C_{gb}}{C_{gs}}\right)^2 \frac{R_b}{R_S}\left(1 + \frac{1}{Q_{in}^2}\right). \tag{5.110}$$

  This leads to the following asymptotical dependencies:

$$\propto \frac{n}{(WL)^2} \qquad (Q^2 \gg 1) \tag{5.111}$$

$$\propto n \qquad (Q^2 \ll 1). \tag{5.112}$$

  where n is the number of fingers.

Because of this, a real power optimization can only be done by 1) taking into account all parasitic effects and transistor non-idealities 2) looking at the whole picture, i.e. including the constraints on linearity, quality factor, etc. Contour lines provide the ideal means to examine these kind of effects. The benefit of using contour lines is further illustrated by the LNA design procedure in Section 6.9. An advanced treatment of LNA noise dynamics can be found in Section 6.11.

5.8 The IP_3 of an Input-Matched MOS Device

In Section 4.4, the IP_3 of a common source MOS device has been calculated, a configuration often found on-chip. This subsection examines the IP_3 of a matched MOS device. The result is especially useful during the design of input-matched low noise amplifier.

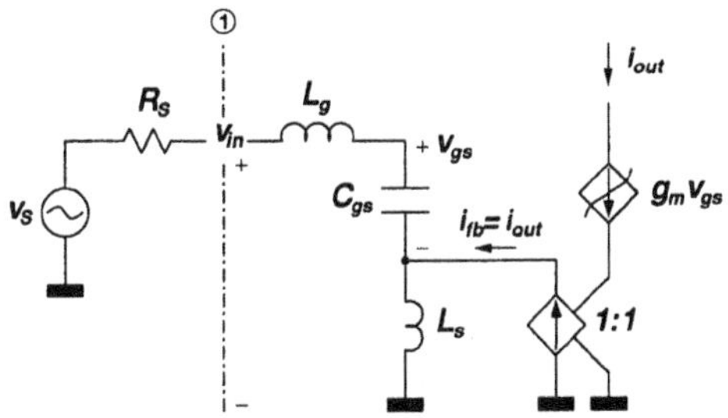

Figure 5.18: *Small signal equivalent circuit of Fig. 5.6.*

5.8.1 Impact of Feedback on Linearity

Suppose we have a system with an open loop, ith order intermodulation ratio of $IM_{i,OL}$. When this system is negatively fed back with a frequency-independent loop gain T, the closed loop IM_i is given by

$$IM_{i,CL} = \frac{IM_{i,OL}}{(1+T)^{i-1}}. \tag{5.113}$$

Since IM_i can also be rewritten as

$$IM_i = \frac{v^i}{IP_i^i}, \tag{5.114}$$

and taking into account the definition of IP_i,

$$IP_i \triangleq v(IM_i = 1), \tag{5.115}$$

the resulting closed loop IP_i can be calculated as

$$IP_{i,CL} = IP_{i,OL}(1+T). \tag{5.116}$$

The exact nature of the feedback and its associated loop gain T must be deduced from a signal flow graph.

Note that the foregoing expressions are only valid if T is both independent of frequency and a real number. That this constraint is important — even when considering intermodulation products that are close to the operating frequency — can be illustrated as follows... As we all know, third-order intermodulation between two tones at frequencies f_1 and f_2 generates a product at $2f_2 - f_1$. But, also second-order intermodulation between a *fed back* second-order harmonic at $2f_2$ and the original tone at f_1 leads to a product at $2f_2 - f_1$. Hence, even for third-order intermodulation products — whose frequency is inherently very close to the frequency of the applied tones — the amount of loop gain at frequencies much further away may play an important role.

5.8.2 Case Study: The IP$_3$ of an Input-Matched MOS Device

Fig. 5.18 depicts the small signal equivalent circuit of an input-matched MOS device. Clearly, the generated output current is reinjected into the input network, an example of shunt-series feedback.

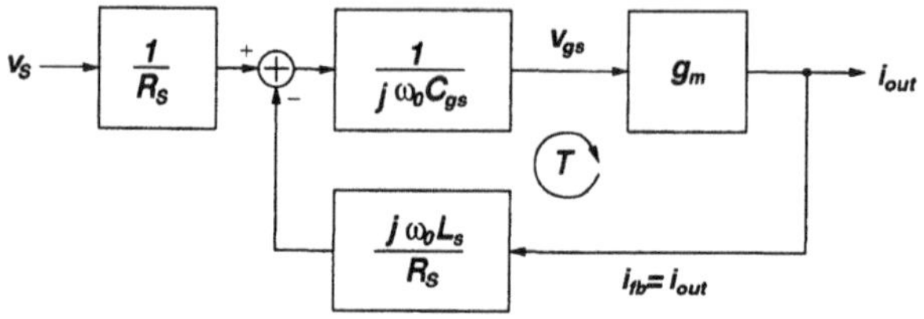

Figure 5.19: *Signal flow graph corresponding to Fig. 5.18.*

Before we move on to calculating IP_3 based on this scheme, the following reasoning tries to keep us from making a basic mistake. IP_3 is defined as the single-tone input power that needs to travel through reference plane ① to make IM_3 equal to one during a two-tone test. However, the two applied tones are not the only signals that travel through the reference plane; In fact, because of the feedback part of the generated intermodulation products also cross this plane. Therefore, the terminal voltage v_{in} at plane ① can not be considered 'clean', and consequently it can not be considered as the controlling voltage. v_s on the other hand, is clean. Hence, to find the IP_3, one is obliged to first calculate IP_3 with respect to the clean source v_s (in dBm_V) and then calculate the corresponding input power (in dBm) that effectively crosses the reference plane. Then, using (4.29), the real IP_3 (in dBm) can readily be obtained from the IP_3 with respect to v_s (in dBm_V):

$$IP_3@①[\text{dBm}] = IP_3@v_s[\text{dBm}_\text{V}] - 6 + 10\log_{10}\left(\frac{50}{R_S}\right). \tag{5.117}$$

Fig. 5.19 shows the signal flow graph of the circuit near the operating frequency ω_0. First, the input signal v_s is converted into a current by means of the internal source resistance R_S. Note that L_g, L_s and C_{gs} do not participate in the conversion since they resonate at ω_0. This current is only one of the two current components that flow through the C_{gs}; Also a fraction of the injected feedback current flows through the C_{gs}, be it in the opposite direction:

$$\left(\frac{i_{fb,C_{gs}}}{i_{fb}}\right)_{\omega_0} = \frac{\dfrac{1}{R_S - j\omega_0 L_s}}{\dfrac{1}{R_S - j\omega_0 L_s} + \dfrac{1}{j\omega_0 L_s}} \tag{5.118}$$

$$= \frac{j\omega_0 L_s}{R_S}. \tag{5.119}$$

Finally, the net current through C_{gs} generates a voltage which is amplified by the device. Near ω_0, the loop gain is thus given by

$$T(\omega_0) = \frac{1}{j\omega_0 C_{gs}} \cdot g_m \cdot \frac{j\omega_0 L_s}{R_S} \tag{5.120}$$

$$= \mathbf{1}\,! \tag{5.121}$$

Now that we have the loop gain, IP_3 can readily be derived; According to (5.116),

$$IP_3@v_s[\text{dBm}_\text{V}] \cong IP_{3,OL}@v_s[\text{dBm}_\text{V}] + 20\log_{10}(1 + T(\omega_0)), \tag{5.122}$$

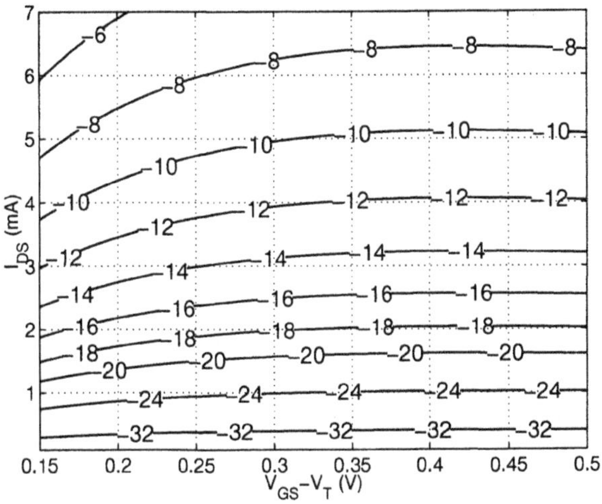

Figure 5.20: *Contour plot of IP_3 for the 0.25 μm technology in Table 4.1. R_S and the operating frequency are assumed to be 50 Ω and 1.8* GHz, *respectively.*

which boils down to

$$IP_3@v_s[\text{dBmv}] \cong IP_{3[\text{eqn.(4.31)}]}[\text{dBmv}] + 20\log_{10}\underbrace{(\omega_0 C_{gs} R_S)}_{1/Q_{OL}} + 6, \tag{5.123}$$

where Q_{OL} is the quality factor of the input series resonance when the loop is open. Combining (5.123) with (5.117) yields the following expression for IP_3:

$$IP_3@①[\text{dBm}] \cong IP_{3[\text{eqn.(4.31)}]}[\text{dBmv}] + 20\log_{10}(\omega_0 C_{gs} R_S) + 10log_{10}\left(\frac{50}{R_S}\right) \tag{5.124}$$

$$\cong \underbrace{IP_{3[\text{eqn.(4.31)}]}[\text{dBmv}]}_{f(V_{GST})} + 20\log_{10}\left(\omega_0 C_{gs}\sqrt{50 R_S}\right). \tag{5.125}$$

As can be seen from the formula, the IP_3 consists of two terms: an intrinsic IP_3, which only depends on $V_{GS} - V_T$ according to (4.31), and a term caused by the extra voltage boost due to the matching network. For instance, for the 0.15 V $V_{GS} - V_T$ MOS transistor in Table 5.3, the intrinsic IP_3 can be calculated to be +5.8 dBm (using Table 4.1 and (4.31)). If this 370 fF transistor is matched to 50 Ω at 1.8 GHz, the intrinsic IP_3 is degraded by 13.6 dB, resulting in a net IP_3 of about -7.8 dBm.

Fig. 5.20 shows a contour plot of the IP_3 as a function of power (I_{DS}) and $V_{GS} - V_T$ (proportional to ω_T) for the 0.25 μm technology in Table 4.1. In this example, R_S and the operating frequency are assumed to be 50 Ω and 1.8 GHz, respectively. Obviously, the overall IP_3 behaves completely different than the intrinsic IP_3; the overall IP_3 is only a very weak function of $V_{GS} - V_T$ whereas the intrinsic IP_3 only depended on $V_{GS} - V_T$! Additionally, the overall IP_3 does not follow the dynamics of the second term either. Otherwise the contour lines had to be parabolae since $I_{DS} \propto C_{gs}V_{GST}^2$. The relatively horizontal behavior of IP_3 as a function of $V_{GS} - V_T$ can be understood as follows. At any given current, a larger $V_{GS} - V_T$ value leads to a larger intrinsic linearity. However, the transistor needs to be made considerably smaller to keep the same current level, increasing the quality factor of the input section, and hence decreasing the second term. In the end, both dynamics more or less cancel out. Hence, by approximation, the linearity can only be improved at the cost of a higher power consumption[8].

There is still one thing which is quite remarkable: A given linearity spec can in principle be obtained at a near-zero current when the transistor is biased at extremely low $V_{GS} - V_T$ values, i.e. in weak inversion[9]. In this region, it can be shown that the intrinsic IP_3 equals

$$IP_{3,intrinsic} = \sqrt{8}\,\frac{kT}{\alpha}, \tag{5.126}$$

which generally boils down to a value of about -12 dBm$_V$. Consequently,

$$IP_3@①[\mathrm{dBm}] \cong -12 + 20\log_{10}\left(\omega_0 C_{gs}\sqrt{50R_S}\right), \tag{5.127}$$

so that IP_3 becomes a function of the transistor width (C_{gs}) only. Hence, when $V_{GS} - V_T \ll 0.15$ V, the contour lines in Fig. 5.20 become exponentials. Of course, there are also some disadvantages to this design point: a low $V_{GS} - V_T$ is — among others — associated with a low input quality factor and thus a low PCC, which increases the required quality factor of the load in an LNA.

Note that (5.122), and hence also (5.125), is only an approximate expression for IP_3. The reason is that, strictly speaking, (5.116) is only valid if T is frequency-independent while in this particular case the loop gain is a function of frequency. Still, (5.125) does give a good idea about the IP_3 under matched conditions; Since a two-tone signal centered around ω_0 gives rise to third-order intermodulation products around ω_0, at least the third-order intermodulation products themselves experience the loop gain $T(\omega_0)$. In addition, it can be shown that although $T(2\omega_0)$ *is* not unity, it asymptotically *tends* to unity for large ω_0 values.

5.9 Conclusions

In this chapter, analytical expressions have been derived for the gain, the noise figure and the linearity of an input-matched MOS device; The dynamics of these specifications have been thoroughly analyzed and explained as a function of transistor parameters and the effective source impedance seen by the transistor. It is shown that some of the performance limitations at radio frequencies can be tracked down to fundamental properties at the MOS device level, like e.g. the non-quasi static gate resistance, and its noise. The insights and formulas developed in this chapter will form the core of the systematic LNA design methodology described in the Chapter 6.

The three most important conclusions of this chapter are as follows:

- The conventional matching method of inductive degeneration is indeed an efficient way to match the predominantly capacitive MOS device to a known source impedance; The

[8] At any given $V_{GS} - V_T$, the overall IP_3 is proportional to C_{gs} and hence to I_{DS}.

[9] The only reason why the curves are not drawn explicitly for $V_{GS} - V_T < 0.15$ V is that the values in Table 4.1 are only valid for $V_{GS} - V_T \geq 0.15$ V.

inductor synthesizes the required filter termination impedance, while the lossless nature of the match ensures a maximum power transfer to the device. However, this work has shown that the transferred signal power itself is *not* the basic steering variable. As a matter of fact, in this matching scheme, the transistor's output current is controlled by the *current* delivered by the source — which happens to be the current that flows into the transistor and generates the v_{gs}— and not by the transferred power itself:

$$|i_{out}| = \frac{\omega_T}{\omega_0}|i_{in}|. \tag{5.128}$$

Since the delivered current highly depends on the impedance of the signal source, the gain differs from case to case — even though exactly the same power is transferred to the given MOS device. Hence, a good quality input match is not so much essential to generate gain, as it is to satisfy the filter termination requirement.

- This work has shown that the classical matching method — a direct source match using inductive source degeneration — suffers from the basic shortcoming that the gain is fundamentally limited by the value of the source impedance because the impedance level of the power flowing into the transistor is constrained to precisely this value. In this work, a modified 'indirect' matching scheme has been proposed which solves this problem by transforming the physical source into a source with a lower impedance, and matching the MOS device towards this value. In this way, the input current for a given signal power — and hence the gain — can be increased. It is shown that the extra degree of freedom offered by this matching technique — the impedance of the local reference plane — enables us to cover the whole spectrum from maximum gain (i.e. a non-quasi static match) to minimum noise figure (very close to a noise-match!), and this while keeping the source match ($\Gamma = 0$)!

- This work has demonstrated that even at operating frequencies far below the transistor's cut-off frequency (ω_T), a non-quasi static model is needed to correctly describe the MOS transistor. In fact, it has been shown that the non-quasi static gate resistance puts a fundamental limit on the achievable performance:

  - Inductive input-matching schemes directly expose the non-quasi static gate resistance. Apart from affecting the real part of the input impedance, it is also the absolute minimum input impedance. The non-quasi static resistance therefore puts a fundamental limit on the achievable gain: the maximum gain occurs when all the signal power is directed to the non-quasi static resistance, or in other words, when all the available power is used to charge and discharge the channel capacitance.

  - When the source impedance is low, the noise figure of an input-matched MOS device is largely determined by the non-quasi static noise. For instance, at source impedance values lower than 50 Ω, the noise figure of a 5 mA I_{DS}-0.15 V $V_{GS} - V_T$ device is simply dominated by non-quasi static noise! At lower current levels and/or larger $V_{GS} - V_T$ values, this effect becomes even more pronounced. As a result, any noise optimization necessarily requires a trade-off between classical noise and non-quasi static noise.

Chapter 6

Systematic CMOS LNA Design

6.1 Introduction

In the previous chapter, analytical expressions have been derived for the gain, the noise figure and the linearity of input-matched CMOS devices. In this chapter, the gathered knowledge is further refined and applied to the analysis and the design of CMOS low noise amplifiers. The ultimate goal: to be able to systematically design an LNA for a given set of specifications (e.g. NF, IP_3 and gain) while complying to the imposed constraints (e.g. maximum values for input reflection, total quality factor and external elements). In the end, this chapter will enable us to answer fundamental questions like: "What is the actual importance of non-quasi static noise in a real-life CMOS LNA?", "Is input matching really essential to obtain a good LNA performance?", and "50 Ω sources in a RF CMOS context, should we like them or not?".

The chapter is organized as follows. First, Section 6.2 and Section 6.3 introduce the topology which is used as a demonstration vehicle throughout this chapter: a cascode LNA providing either power-to-voltage or power-to-power conversion. The basic performance dynamics of the cascode LNA are covered in Section 6.4. The impact of several important non-idealities on the LNA performance forms the subject of the next sections. This analysis results in a bank of analytical equations in Section 6.7, describing all important LNA specifications — noise figure, IP_3, gain and quality factor — in terms of the parameters of the involved devices.

In Section 6.9, the step is made from analysis to real LNA design; A systematic LNA design strategy is presented which employs the above-mentioned analytical model in conjunction with contour lines to find the best point in the design space. The proposed method clearly identifies un-realizable, inactive and binding constraints and reveals the precise impact of an arbitrary constraint on the minimum LNA power consumption. The methodology is illustrated by means of a case study — the design of a low noise amplifier for the DCS–1800 application. In Section 6.10, the same methodology is applied to look for the source resistance which minimizes the LNA power consumption. Section 6.11 covers some fallacies and pitfalls regarding the LNA noise figure dynamics. Section 6.12 analyzes the impact of a finite input reflection coefficient on LNA performance. Finally, the case study is resumed one last time in Section 6.13 to help digesting Section 6.12.

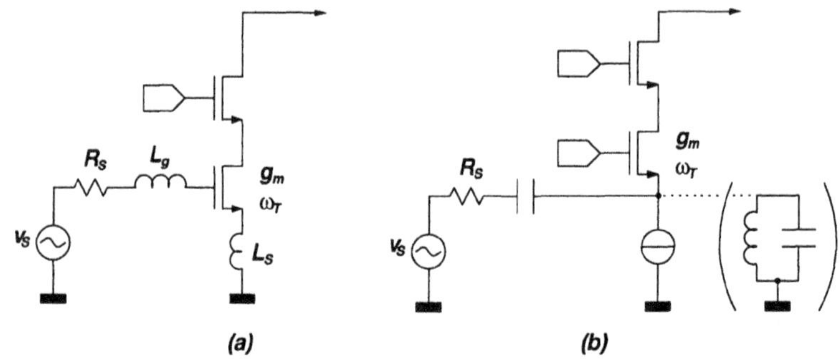

Figure 6.1: *The common-source and the common-gate configuration.*

6.2 Narrow-band Low Noise Amplifier Topologies

Basically, any amplifier with a reasonably large gain and a reasonably low noise figure could be designated as a low-noise amplifier. As a result, a multitude of topologies exist that fulfill these requirements. This is especially true for wide-band LNAs; In this case, virtually all functional combinations of transistors and resistors could yield a suitable wide-band amplifier. Topology examples are the transimpedance-type common-source-based LNA in [Jans97], the multi-stage stacked-transistor common-source LNA in [Jans98b], common-gate LNAs and the less conventional two-transistor configurations in [Bruc00]. Unfortunately, in wide-band amplifiers a low noise figure and a high gain often come at the cost of a relatively large power consumption. For this reason, highly sensitive wireless receivers commonly use narrow-band low-noise amplifiers. Narrow-band low-noise amplifiers can be optimized more effectively as they only need to achieve the performance in the band of interest. For instance, inductors can be employed to cancel out the capacitive parasitics, to generate a noise match, or to implement impedance transformation networks, etc.

In practice only two amplifier cores are used: the common-source configuration (Fig. 6.1(a)) and the common-gate configuration (Fig. 6.1(b)). Of these two topologies, the common-source configuration is definitely the best candidate for building low-power high-performance low-noise amplifiers:

- As indicated by (5.21), the common-source input stage of Fig. 6.1(a) implements a current amplification by a factor ω_T/ω_0. This not only lowers the input-referred channel noise of the MOS transistor but also decreases the load impedance required to implement a certain gain. This is not the case in the common-gate input stage of Fig. 6.1(b); There the output current is simply identical to the input current. As the f_T is clearly of no importance, this topology does not exploit the frequency capabilities of the technology.

- As shown in Subsection 5.7.3, the common source topology is able to offer an extremely low noise figure (a quasi noise-match) whereas the noise figure of the common-gate topology is by construction limited to about 2.2 dB (i.e. $10\log_{10}(1+2/3)$).

- In case of a common-source stage, the value of the source impedance can be adjusted in order to obtain an optimum trade-off between a low noise figure an a large gain. The local

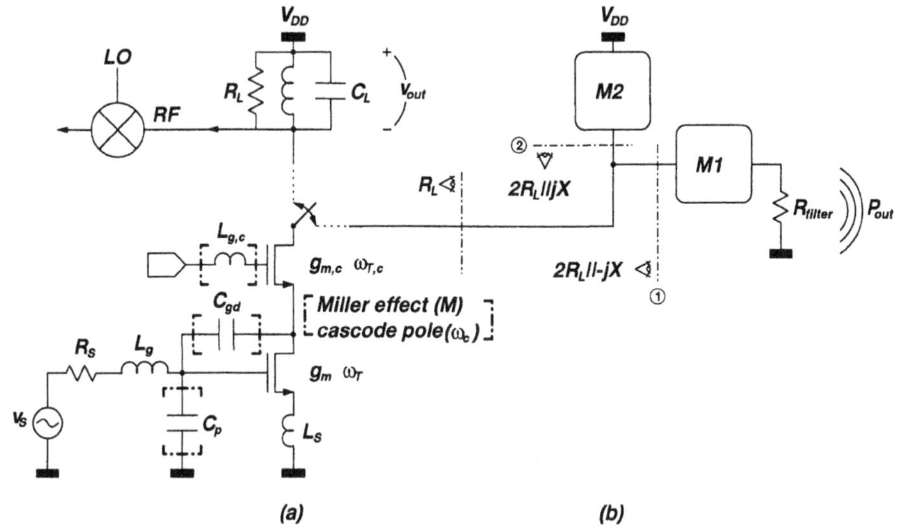

Figure 6.2: *Two single-stage CMOS LNA templates sharing the same LNA core: (**a**) power-to-voltage LNA (**b**)* power-to-power LNA.

source impedance acts as a real degree of freedom. In case of a common-gate configuration, gain and noise figure are much more entangled. In addition, a lower source impedance has a direct impact on the power consumption as the transconductance of the input device must be set to the conductance of the source.

6.3 Cascode Low Noise Amplifiers

The task of an LNA invariably consists in transferring the available antenna power towards the succeeding building block while minimally degrading the *SNR*. Depending on the system architecture, the LNA output signal taken into consideration is either a voltage (e.g. over the capacitive input impedance of the next block) or a power[1] (e.g. in the resistive input impedance of the next block). As an example, Fig. 6.2 shows two single-stage CMOS LNA templates sharing the same LNA core. The only difference lies in their output structure: the left output structure is intended for driving on-chip CMOS mixers (or building blocks with a large, predominantly capacitive input impedance) while the right one is suitable for driving an off-chip filter (or building blocks with a resistive input impedance). We will come back to this shortly.

The LNA core in Fig. 6.2 consists of a cascoded, inductively degenerated common source input stage that converts the available antenna power into a current. This topology allows a good quality input match with a a low noise figure, a large power-to-current conversion (*PCC*) and a reasonable linearity; The conversion principle of this stage, along with its noise and linearity

[1] In case of a resistive load, the signal power in the load directly relates to the voltage *over* and the current *in* that load. If one knows which one of these three quantities will be sensed by the next building block, one could as well consider this quantity as the wanted output signal (instead of just the output power).

performance, has already been discussed in detail in Section 5.3 to Section 5.8. The cascode transistor on top of the amplifying device is inserted for a number of reasons:

- The cascode device lowers the output conductance by a factor $g_m r_o$, so that the gain is completely controlled by the load network.

- The cascode device reduces the Miller effect on the C_{gd} by ensuring a low impedance at the drain of the amplifying device. This keeps the Miller effect from degrading the *PCC* as well as from increasing the input referred noise (see Section 6.6). Note that this is only true provided that $R_L < r_o$ (which mostly holds).

- The cascode improves the reverse isolation of the LNA, highly reducing the LO leakage when the LNA is embedded in a receiver. A second advantage is that — due to the increased I/O isolation — the design of the input matching network and the load network is decoupled as well. For the same reason, unwanted reflections at the output no longer result in amplifier instability.

- The cascode device sets the drain potential of the amplifying device to a value close to $V_{GS} - V_T \ll V_{DD}$. By reducing the reverse voltage over the drain junction diode, the noise temperature of the electrons at the drain is kept under control.

- Optimization of the cascode pole often leads to a device which is smaller than the amplifying device. Since this results in a lower capacitive load at the output, for a given R_L value, an inductor with a lower quality factor can be used, while for a given inductor, the gain can be increased.

The *RLC* resonant load (Fig. 6.2, top left) lends itself perfectly for fully integrated CMOS receiver systems because CMOS mixers often have a purely capacitive RF input and are generally voltage-driven. The voltage over the capacitor is maximized by resonating the capacitor against an inductor. The effective parallel resistance R_L— and hence the gain — is determined by the achievable — or allowable — quality factor of the inductor. This topology is used in the LNA which is embedded in the 2V DCS-1800 transceiver front-end described in Chapter 9.

In case the LNA needs to drive an off-chip image filter, like e.g. in a heterodyne receiver system, the output *power* is the relevant quantity, which requires an output matching network (Fig. 6.2, top right). M2 represents the part of the matching network that is directly connected to V_{DD} and carries the DC current of the LNA. Therefore, M2 generally contains an inductor. M1 subsequently transforms the impedance, formed by the LNA core and M2, down to the impedance of the filter. The fact that the impedances at reference planes ① and ② in Fig. 6.2 are designated by '$2R_L//jX$' and '$2R_L//(-jX)$', respectively, deserves some further explanation. Suppose that M1 and M2 had both been lossless. In that case, $\Re$(*Z*@②) would be infinite while $\Re$(*Z*@②) would be set to the output impedance of the LNA core, the condition for maximum power gain. However, because of the finite quality factor of the (on-chip) inductor in M2, $\Re$(*Z*@②) will be limited to the R_p of the inductor, so that *Z*@② = $R_p//jX$. The condition for maximum power gain then requires that M1 transforms the filter resistance into *Z*@① = $R_p//(-jX)$. Unlike M2, M1 can be made quasi-lossless since it can be implemented with a set of high-Q metal-metal (or metal-silicided poly) capacitors. This topology is used in the 0.8 dB *NF*, 20 dB power gain LNA in Chapter 8.

The next subsections will cover the performance of these two LNAs in detail, along with the non-idealities introduced by C_p, C_{gd}, $L_{g,c}$, M and ω_c. Finally, based on the complete set of design equations, a systematic design strategy is developed.

6.4 Gain and Noise Figure

6.4.1 From PCC to Gain

The process of generating gain can be broken up into two parts: The first part, i.e. the conversion of available source power in output current, is done in the LNA core and can be quantified by the *PCC* defined in (5.14). The second part, i.e. the conversion of the generated current in either voltage or power, is done by the output network.

When driving CMOS building blocks, most of the time the output *voltage* is the parameter of interest. Hence, the gain of the LNA should quantify how efficiently the antenna power is converted into an output voltage. This leads to the definition of the *PVC* or the power-to-voltage conversion:

$$PVC \triangleq \frac{v_{out}^2}{P_{av}}. \tag{6.1}$$

For the *RLC* resonant LNA in Fig. 6.2(a), the expression for the *PVC* becomes

$$PVC(\Omega) = PCC\, R_L^2 \tag{6.2}$$

$$= \left(\frac{\omega_T}{\omega_0}\right)^2 \frac{R_L^2}{R_S}. \tag{6.3}$$

It is important to note that, in contrast to a voltage gain or a power gain, the *PVC* is *not* dimensionless: the *PVC* carries the unit 'Ω'. For example, an LNA that converts a power of 0.01 mW into 223 $\mathrm{mV_{rms}}$ should feature a *PVC* of 5kΩ since

$$\frac{(0.223)^2\ V_{rms}^2}{0.00001\ W} = 5000\ \Omega. \tag{6.4}$$

Strictly speaking, the *PVC* can not be expressed in dB. However, this problem can be circumvented by proper normalization; An LNA that generates an output voltage of 0 $\mathrm{dBm_V}$ (i.e. 223 $\mathrm{mV_{rms}}$) at an incident power of 0 dBm (i.e. 1 mW) could for instance be defined as an LNA featuring a *PVC* of '0 dB'. Not by coincidence this requires a division by 50 Ω:

$$PVC(\mathrm{dB}) \triangleq 10\log_{10}\left(\frac{PVC(\Omega)}{50\ \Omega}\right). \tag{6.5}$$

In fact, it can be shown that in case R_S equals 50 Ω, the evaluation of (6.5) boils down to

$$20\log_{10}\left(\frac{v_{out}}{v_s/2}\right), \tag{6.6}$$

where $v_s/2$ is the input voltage of the LNA if it were perfectly matched to 50 Ω, or equivalently, the voltage over a 50 Ω termination when it is connected to the power source.

When *power* is the parameter of interest, the gain should quantify how efficiently the available source power is boosted before it is dumped in the load. Exactly this is done in the definition of the the transducer power gain:

$$G_t \triangleq \frac{P_{out}}{P_{av}}, \tag{6.7}$$

where P_{out} is the signal power in the load (e.g. R_{filter} in Fig. 6.2(b)). For the output matched circuit in Fig. 6.2(b), it can be shown that

$$G_t = \frac{PCCR_L}{2}, \tag{6.8}$$

where R_L is the resistance seen by the LNA core (see Fig. 6.2(b)). The factor two in (6.8) stems from the fact that half of the power delivered to R_L is dissipated in matching network M2. Assuming that M1 introduces no loss, the other half of the power is successfully transferred to the load.

6.4.2 Noise Figure

The noise figure of both LNAs in Fig. 6.2 is given by

$$F = 1 + (F_d - 1) + (F_{g,u} - 1) + (F_c - 1) + (F_L - 1), \tag{6.9}$$

where $F_d - 1$ and $F_{g,u} - 1$ represent the noise figure contributions of the amplifying device, $F_c - 1$ is the contribution of the cascode device and $F_L - 1$ is the contribution of the load.

The first two noise components have been extensively studied in Subsection 5.7.3; Since $\omega_0^2 \ll \omega_T^2$, i.e. applying the approximation in (5.93), $F_d - 1$ and $F_{g,u} - 1$ boil down to:

$$F_d - 1 \cong \frac{\gamma}{\alpha}\left(\frac{\omega_0}{\omega_T}\right)^2 \left(g_m R_S + \frac{2}{\kappa}\right) \tag{6.10}$$

$$F_{g,u} - 1 = \frac{\alpha\delta(1 - |c|^2)}{\kappa g_m R_S}. \tag{6.11}$$

When $\omega_0^2 \ll \omega_c^2$, the noise contribution of the cascode is given by

$$F_c - 1 = 4\gamma \left(\frac{\omega_0}{\omega_c}\right)^2 \frac{g_m}{M\,PCC} \tag{6.12}$$

$$= 4\gamma \left(\frac{\omega_0}{\omega_T}\right)^2 \left(\frac{\omega_0}{\omega_c}\right)^2 \frac{g_m R_S}{M}, \tag{6.13}$$

where the Miller factor M is defined as

$$M \triangleq \frac{\alpha g_m}{g_{m,c}}. \tag{6.14}$$

Note that by using the variable M, the noise is directly formulated in terms of the g_m and the PCC of the amplifying device. It is a good starting point to design the cascode transistor such that the pole frequency ω_c is maximized (see Section 6.8). Making the transistor smaller — to increase M a little bit at the cost of ω_c — often does not pay off because the noise contribution is already small. In fact, in Subsection 6.8.2 it will be shown that there even exists some pressure to decrease M.

For the calculation of the noise contribution of the load, we must distinguish between the *RLC* resonant LNA in Fig. 6.2(a) and the power gain LNA in Fig. 6.2(b). In case of the LNA in Fig. 6.2(a), the noise figure contribution of the load resistance R_L can be written in many different ways:

$$F_{R_L} - 1 = \frac{4}{R_L PCC} = \frac{4R_L}{PVC} = \frac{4}{\sqrt{PVC}\sqrt{PCC}} \tag{6.15}$$

$$= 4\left(\frac{\omega_0}{\omega_T}\right)^2 \frac{R_S}{R_L} \tag{6.16}$$

At first sight, the equivalent formulations in (6.15) lead to conflicting conclusions. The following explanation might give a better insight in these formulae. For a given device, i.e. a given *PCC*, a given power consumption and a given device noise, *increasing* the value of R_L leads to a lower noise figure contribution; After all, the larger R_L, the lower the noise current generated by R_L. At the same time, the LNA gain (*PVC*) becomes larger and larger, lowering the noise contribution of the succeeding mixer[2]. Hence, for a given power consumption, it is always best to maximize R_L and hence the *PVC*. However, the *PVC* can not be made arbitrarily large. since the limited input range and the linearity of the mixer put a hard constraint on the allowable LNA gain (*PVC*). It is therefore more realistic to reformulate the noise contribution of R_L in terms of the *PVC* defined in (6.2). This leads to the second expression in (6.15), which reveals that for a given *PVC*, *lowering* the R_L value leads to a proportionally lower noise contribution. Although this conclusion is correct, one must bear in mind that $F_{R_L} - 1$ does not stand on its own; While a low R_L value effectively minimizes the noise contribution of R_L for a given *PVC*, it puts a lot of pressure on the *PCC*, and hence on the $V_{GS} - V_T$ of the amplifying device. Because of the large $V_{GS} - V_T$, the non-quasi static noise term $F_{g,u} - 1$ — which is inversely proportional to g_m— becomes more and more pronounced, so that a lot of current may be required to keep this term from increasing. Hence, as long as $F_{R_L} - 1$ is smaller than $F_{g,u} - 1$, a lot of power can be saved by allowing a larger R_L value while keeping the *PVC* constant (by lowering the *PCC*). Again, this can not go on forever because $F_d - 1$ and $F_c - 1$ will eventually contribute to the noise figure due to the large g_m and the low *PCC*. A large R_L value also involves a high-Q load inductor, lowering the LNA bandwidth and making the circuit more sensitive to process variations in a position where this variation can not be compensated for. This is especially the case when the LNA needs to drive large capacitive loads:

$$Q_{tank} = R_L \omega_0 C_L. \tag{6.17}$$

In this case or in case that C_L can not be resonated out by a reasonable quality inductor — e.g. if the substrate material is highly conductive — there is no other solution than to work with a limited R_L value and use the *PCC* to create gain (*PVC*). In case the noise figure of the LNA is dominated by the non-quasi static noise term $F_{g,u} - 1$, this term can only be kept under control by consuming extra current (because g_m must be kept at a certain level while $V_{GS} - V_T$ increases). Note that this action also increases the quality factor of the input section because of the smaller

[2]It can be shown that the noise contribution of the mixer is given by $\frac{200\Omega F_{E,mixer}}{PVC}$.

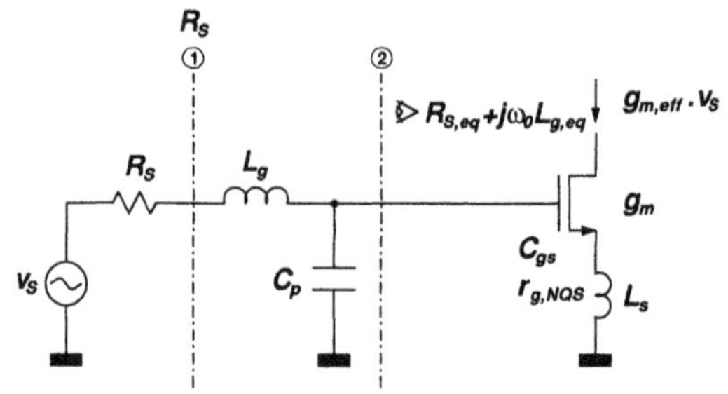

Figure 6.3: *Power matching in the presence of a parasitic input capacitance C_p.*

width of the input device.

In case of the output matched LNA in Fig. 6.2(b), it can be shown that the noise contribution of the load boils down to:

$$F_L - 1 = \frac{1}{G_t}. \tag{6.18}$$

Since G_t is generally set by the system specification, $F_L - 1$ is fixed as well.

6.5 Impact of Input Capacitance on Matching, PCC and NF

The performance of an LNA is strongly affected by the parasitic parallel input capacitance originating from the bonding pad ($\sim$ 300 fF), the protection diodes ($\sim$ 100 fF) and the C_{gd} of the device ($\sim 0.4C_{gs}$). This subsection examines the quantitative impact of the parasitic input capacitance on matching, gain and noise figure. It will be shown that this capacitance not only makes input matching more difficult but also severely degrades the *PCC*. In addition, it will be demonstrated that the parasitic input capacitance has a remarkable influence on the dynamics of the noise figure.

6.5.1 Impact on Matching

Consider for instance the device in Fig. 6.3. The device is assumed to be source-matched to R_S in the presence of a parasitic input capacitance C_p. To control the complexity of the analysis, C_p is considered ideal (i.e. without any series resistance). Though this is not the case in reality, one should always pursue this because any additional resistance injects noise.

The impedance seen at the left of reference plane ② can be written as $R_{S,eq} + j\omega_0 L_{g,eq}$, where $R_{S,eq}$ and $L_{g,eq}$ are given by

$$R_{S,eq} = \frac{R_S}{\omega_0^2 C_p^2 R_S^2 + (1 - \omega_0^2 C_p L_g)^2} \tag{6.19}$$

$$L_{g,eq} = \frac{L_g - C_p(\omega_0^2 L_g^2 + R_S^2)}{\omega_0^2 C_p^2 R_S^2 + (1 - \omega_0^2 C_p L_g)^2}. \tag{6.20}$$

For all purposes, the device can be thought of being matched to the virtual source impedance $R_{S,eq}$ by means of the virtual inductance $L_{g,eq}$. The L_g variable can be eliminated from (6.19) and (6.20) by stating that the device is power-matched in the presence of C_p:

$$R_{S,eq} = \frac{1}{\kappa g_m} + \omega_T \underbrace{\left(\frac{\omega_T}{\omega_0^2 g_m} - L_{g,eq}\right)}_{L_s}. \tag{6.21}$$

Note that the resonance constraint is incorporated in the last term. This yields the following expression for $R_{S,eq}$:

$$R_{S,eq} \cong \frac{2R_S\left(1+\frac{C_p\omega_T}{g_m}\right)^2}{\Psi+\sqrt{\Psi^2-(\Psi-1)^2\left(\frac{\omega_T}{\omega_0}\right)^2}}, \tag{6.22}$$

where Ψ is defined as

$$\Psi = 1 + \frac{2C_pR_S\omega_0^2}{\omega_T}\left(1+\frac{C_p\omega_T}{g_m}\right). \tag{6.23}$$

The only assumption made during the derivation of (6.22) is that

$$\left(\frac{\omega_0}{\omega_T}\right)^2 \ll 1. \tag{6.24}$$

Substituting the $R_{S,local}$ variable in (5.44) and (5.89) by the expression for $R_{S,eq}$ in (6.22) yields the new noise figure and PCC as a function of R_S and C_p.

Of course, only positive numbers are valid solutions for $R_{S,eq}$. Additionally, according to (6.21), it is required that $R_{S,eq} \geq r_{g,NQS}$ to ensure that $L_s \geq 0$. Yet, there still is a more important constraint: $R_{S,eq}$ and L_g must be *real* numbers. This is equivalent to stating that the terms under the square root in (6.22) need to be positive, which results in the requirement that

$$R_S \leq \frac{\omega_T}{2C_p\omega_0^2\left(\frac{\omega_T}{\omega_0}-1\right)\left(1+\frac{C_p\omega_T}{g_m}\right)} \tag{6.25}$$

or by approximation:

$$R_S \leq \frac{1}{2\omega_0 C_p\left(1+\frac{C_p}{C_{gs}}\right)}. \tag{6.26}$$

Hence, when either C_p or the operating frequency increases, it becomes progressively more difficult to match a device to a given source impedance R_S without using an extra intermediate matching section. For instance, when the 370 fF device in Table 5.3 is to be source-matched at 1.8 GHz in the presence of a C_p of $400 + 0.4 \cdot 370$ fF, the maximum achievable input impedance is barely 32.5 Ω; The device can not be matched to 50 Ω using the direct matching technique described in Section 5.3. Even by enlarging the device — to decrease the ratio between C_p and

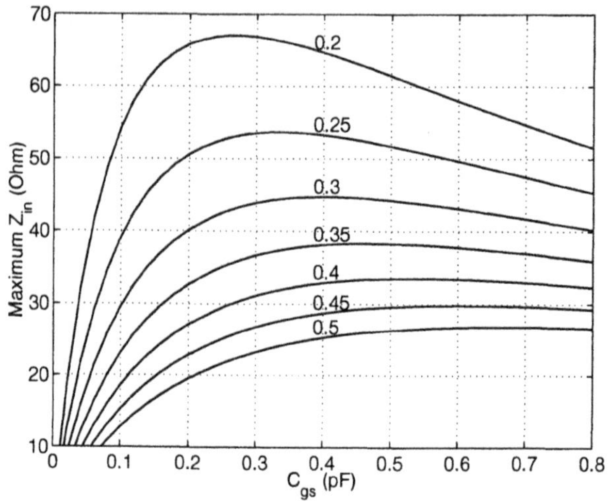

Figure 6.4: *Maximum input impedance at 1.8* GHz *as a function of C_{gs} and $C_{p,fixed}$ (in* pF*).*

C_{gs}—, the maximum achievable input impedance can hardly be increased. This is due to the fact that C_p also contains the device's C_{gd}:

$$C_p \cong C_{p,fixed} + \alpha_{gd} C_{gs} \quad \text{with} \quad \alpha_{gd} \sim 0.25 \ldots 0.4. \tag{6.27}$$

In this equation, $C_{p,fixed}$ represents the fixed part of C_p, containing the parasitic capacitance of the package, the bondpad and the input protection. This is further clarified in Fig. 6.4, which plots the maximum input impedance — i.e. the right hand side of (6.26) — at 1.8 GHz as a function of C_{gs} for a set of $C_{p,fixed}$ values (in pF). The α_{gd} parameter in (6.27) is taken to be 0.4. Using

For the above mentioned $C_{p,fixed}$ value of 400 fF, the maximum input impedance is maximized at a C_{gs} of 535 fF ... but it only equals a disappointing 33.5 Ω; In fact, according to Fig. 6.4, at 1.8 GHz an exact 50 Ω match is not feasible for $C_{p,fixed} \geq 0.27$ pF without inserting an extra matching section. Fortunately, in the interval 0.27 pF $\geq C_{p,fixed} \geq$ 0.45 pF one can still comply to the $\Gamma < -12$ dB specification without using the extra matching section M; It hardly influences the gain compared to an exact match (see Subsection 5.5.2) and it spares extra off-chip matching components. Nevertheless, the required C_{gs} rapidly increases when $C_{p,fixed}$ becomes closer to 0.45 pF, which puts a lot of pressure on the power consumption. $C_{p,fixed}$ must therefore be limited at all cost, e.g. by using small, octogonal bondpads and by inserting the smallest possible protection circuitry (Subsection 6.5.3). Fig. 6.4 also explains why one is inclined to keep the input impedance relatively low: a lower input impedance requires a lower C_{gs} and hence less power. The other alternative consists in employing an extra matching section M, to transform R_S into a lower value before connecting it to reference plane ①. This can for instance be done by adding some extra series inductance to L_g and putting an off-chip capacitor, or by decreasing L_g— if necessary, by adding an off-chip series capacitor — and putting an off-chip inductor. The last solution is probably the best one, since it allows the use of a low-value series inductor — as is the case in a packaged solution.

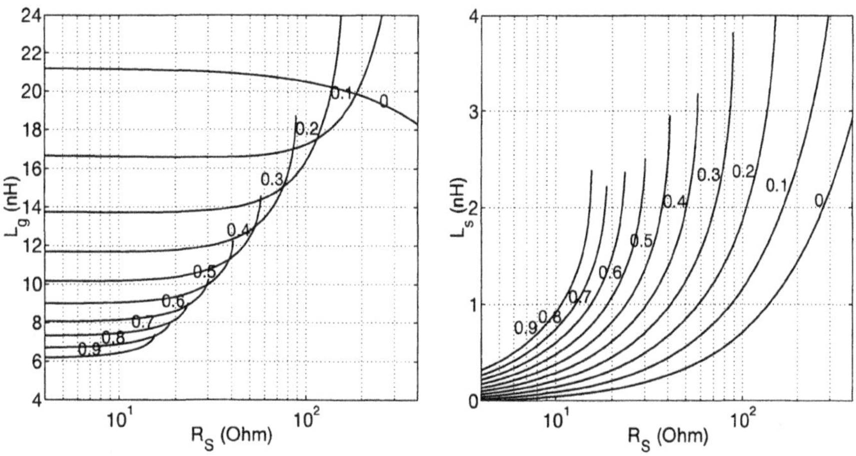

(a) Gate inductance L_g.

(b) Source inductance L_s.

Figure 6.5: *The required component values as a function of R_S and C_p (in* pF*). The plot employs the device parameters of Table 5.3.*

Fig. 6.5 shows the element values of the physical components L_g and L_s, calculated during the elimination process using the matching constraint in (6.21). As could have been predicted, L_g must decrease to keep the resonance frequency at 1.8 GHz while L_s must increase to counteract the action of the down-transformation network $C_p - L_g$. Each curve ends at the R_S value given in (6.26). As can be seen from the figure, an non-zero C_p value is not bad from a manufacturability point of view: L_g comes down from excessively high inductance values while L_s moves from sub-nH values into the nH region. Fig. 6.6(a) shows $R_{S,eq}$ as a function of R_S for different values of C_p and for the device in Table 5.3. Obviously, $R_{S,eq}$ rapidly increases with increasing C_p. This shouldn't surprise the reader, since the $L_g - C_p$ network to the left of reference plane ② acts as an impedance up-transformation network. Again, each curve ends at the R_S value given in (6.26). For completeness, Fig. 6.6(b) shows the corresponding $L_{g,eq}$ value. $L_{g,eq}$ changes much less since its value is mostly determined by the (fixed) device size.

6.5.2 Impact on PCC and Noise Figure

For all purposes, the device can be thought of being steered by a virtual source with impedance $R_{S,eq}$. Hence, the *PCC* equation in (5.44) and the equation describing the noise contribution of the amplifying device, i.e. (5.89), can be adapted to take into account the effect of C_p simply by substituting $R_{S,local}$ by $R_{S,eq}$.

Fig. 6.7(a) shows the *PCC* enhancement in the presence of a non-zero C_p as a function of R_S with respect to a normal 50 Ω match with zero C_p. Obviously, the *PCC* drops enormously because $R_{S,eq} \gg R_S$; For example, when the input is matched to 50 Ω, a C_p of 400 fF degrades the *PCC* by 8 dB! In principle, the lower *PCC* does not need to result in a lower gain (*PVC*); By increasing the value of the load impedance, R_L, the gain can still be brought to the required value. However, as mentioned in Subsection 6.4.2, a large R_L value involves a high-Q tank,

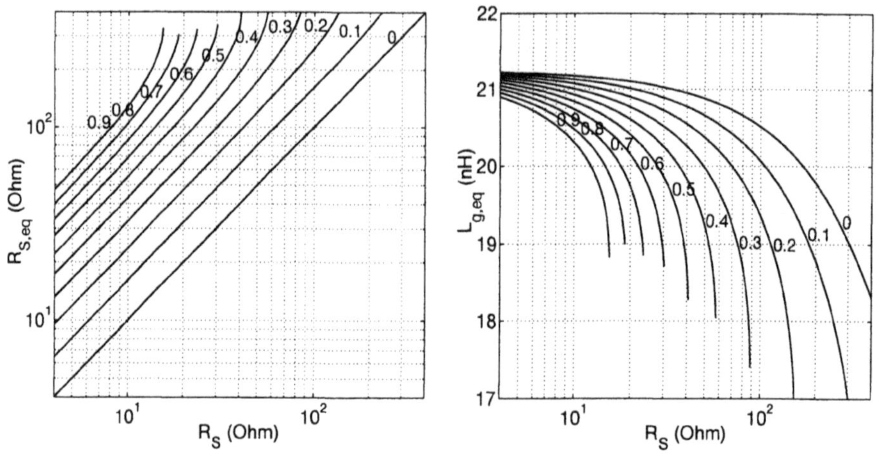

(a) $R_{S,eq}$. (b) $L_{g,eq}$.

Figure 6.6: *The virtual source impedance $R_{S,eq} + j\omega_0 L_{g,eq}$ as a function of R_S for different values of C_p (in* pF*). The plot employs the device parameters of Table 5.3.*

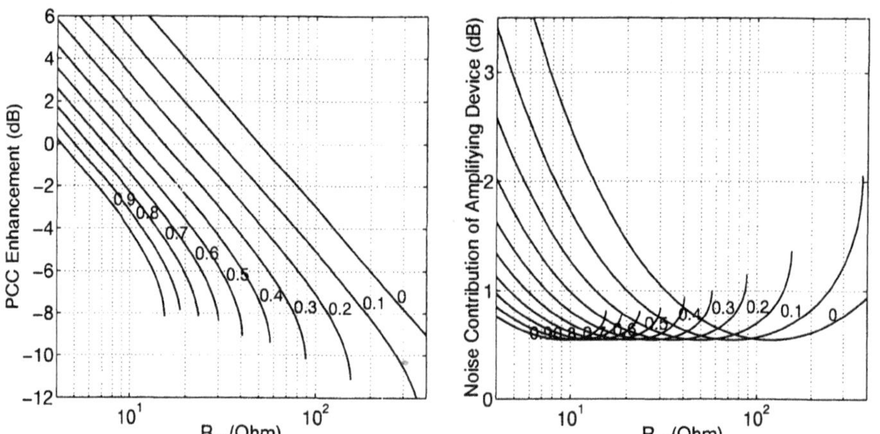

(a) *PCC* enhancement. (b) Noise contribution of amplifying device.

Figure 6.7: *The PCC enhancement and the noise figure contribution of the amplifying device as a function of $R_{S,local}$ in the presence of a non-zero C_p (in* pF*). The plot employs the device parameters of Table 5.3.*

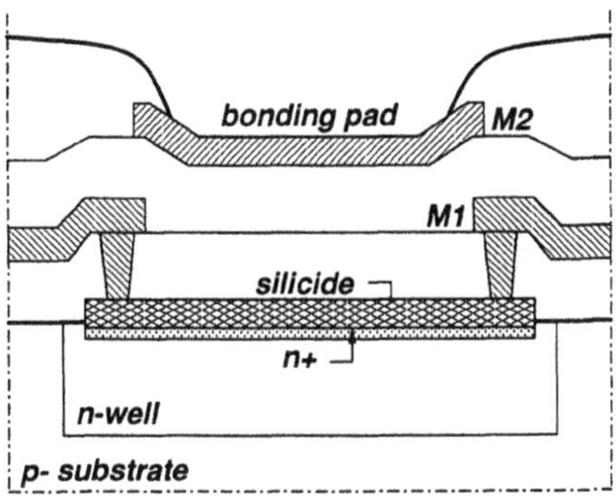

Figure 6.8: *A high-Q, low-C_p bonding pad structure in a two-level metal process.*

making the circuit more sensitive to process variations (in a position where this variation can not be compensated for) and requiring a high-Q load inductor, especially when the LNA needs to drive large capacitive loads. If one bumps into this limit or if C_L can not be resonated out by an inductor — e.g. if the substrate material is highly conductive — there is no other solution than to increase the ω_T (and hence the $V_{GS} - V_T$) of the amplifying device to boost the *PCC*.

Fig. 6.7(b) shows the noise figure contribution of the amplifying device in the presence of a non-zero C_p as a function of R_S. Since at a given R_S, a larger C_p leads to an increasing $R_{S,eq}$, NF_{min} is already reached at a lower R_S value. Non-quasi static noise is pushed to extremely low R_S values while the noise at normal R_S values is again determined by the classical drain noise. It is important to note that this means that as long as R_{opt} (i.e. (5.90)) is larger than R_S (e.g. 50 Ω), C_p can be used to transform R_S into R_{opt}, yielding a quasi perfect noise match! Unfortunately, C_p is no real degree of freedom since it already contains the bondpad capacitance and the capacitance of the protection circuitry. Most of the time, the default C_p value already increases $R_{S,eq}$ beyond R_{opt}, so that any additional C_p comes at the cost of noise figure. Additionally, $F_{R_L} - 1$ and $F_c - 1$ are inversely proportional to the *PCC*, which is strongly degraded by C_p.

6.5.3 A Low-C_p Bondpad Structure with a High Q-factor

From the above discussion it has become clear that the parasitic input capacitance C_p must be kept low for two reasons:

- A large C_p lowers the attainable input impedance, which highly complicates input matching, especially at high frequencies. The larger C_p, the larger the required power to ensure a good quality match.

- A too large C_p reduces the signal excursions at the gate, degrading both the *PCC* and the noise figure.

Apart from the transistor's C_{gd}, the most important input parasites are the parasitic capacitance of the bonding pad and the input protection circuitry. Whereas the capacitance of the protection network is a measure of its capability of withstanding ESD events, the bonding pad capacitance is a pure parasite. Therefore, one should always nibble off as much as capacitance as possible from the bonding pad structure and assigning the freed capacitance to the protection circuitry so that it can accommodate larger ESD pulses.

In order to reduce the bonding pad capacitance, the bonding pad should be implemented only in the top metal layer as to decrease its capacitance per unit area. Secondly, the size of the bonding pad should be decreased as much as possible, e.g. by moving to an octogonal structure. Though this may seem an overkill at first sight, this action decreases the capacitance by more than 17%!

However, not only the value of C_p is important; Since the second plate of the bonding pad capacitance is formed by the silicon material, the charging currents experience a significant series resistance. This series resistance can become as high as a few hundred Ohms — especially in CMOS processes that use lowly-doped, high-resistivity wafers — which significantly degrades the quality factor of the bonding pad structure. The problem can be alleviated by putting a low-ohmic shielding layer between the bonding pad and the substrate. This can for instance be done by placing a silicided n+ diffusion layer underneath the bonding pad[3] and connecting it to ground along the perimeter. In this way, the series resistance of the bondpad capacitance can be reduced to 1 or 2 Ohms. Apart from increasing the quality factor, the grounded 'bottom plate' also improves the isolation between the RF terminal and the substrate, so that the terminal picks up less disturbances coming from other parts of the circuit (like e.g. LO signals and switching noise). This property makes the bonding pad ideal for the LNA input and the TX output terminal. The ground signal can be provided by the adjacent bonding pads (e.g. in a GND-SIG-GND configuration).

Since a metal features a fifty times lower series resistance, one may ask oneself why a silicide is preferred instead of the first metal layer (like in [Rofo98a, Colv99]). The reason is that by using the silicide as ground shield, the oxide thickness is larger than when the first metal layer is used, which results in a lower capacitance per unit area. While this is no longer an issue in in today's advanced multi-level metal processes[4], it can make a huge difference in two level metal processes — which is the case in the early 0.25 μm CMOS technology used throughout the majority of the text; If the first metal layer had been used, the capacitance would have been two times larger!

It is true that a metal ground shield yields a better quality factor, yet, it is not worthwhile; It suffices that the quality factor of the bondpad is much larger than the quality factor of the 'ideal' input section, which is determined by the source resistance. Moreover, when the special bonding pad structure is used, the quality factor of the protection circuitry will most likely be less, especially when a two-diode protection structure is used.

6.6 Impact of C_{gd} and M on Matching, PCC and NF

The C_{gd} of the amplifying device has a tremendous impact on the LNA performance. This subsection examines the main degradation mechanisms originating from this capacitance.

Fig. 6.9 shows a small signal model of the LNA input section, taking into account the impedance presented by the cascode transistor. The C_{gd} can be replaced by its Y parameter

[3]Apart from upsetting the technology people — because it violates the rule that there must be a minimum distance between n+ active area and the bonding pad — this should not pose any problem.

[4]The LNA in Chapter 8, which is implemented in an advanced four-level metal 0.25 μm CMOS process, features bonding pads in the fourth metal layer, and uses the first metal layer as shielding.

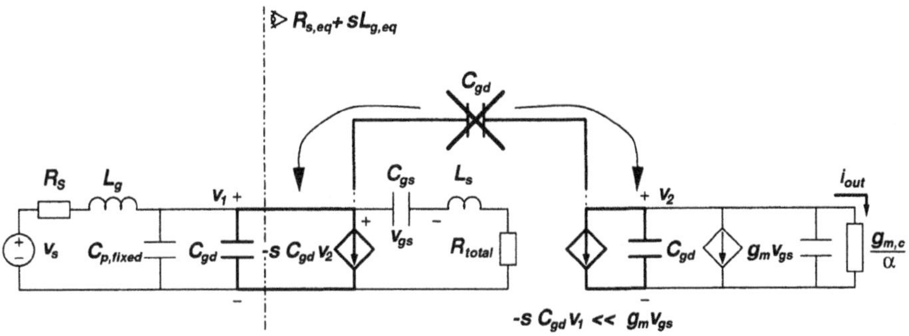

Figure 6.9: *Small signal model of the LNA input section, taking into account the impedance presented by the cascode transistor ($g_{m,c}/\alpha$) and the C_{gd} of the amplifying device (represented by its Y-parameter equivalent).*

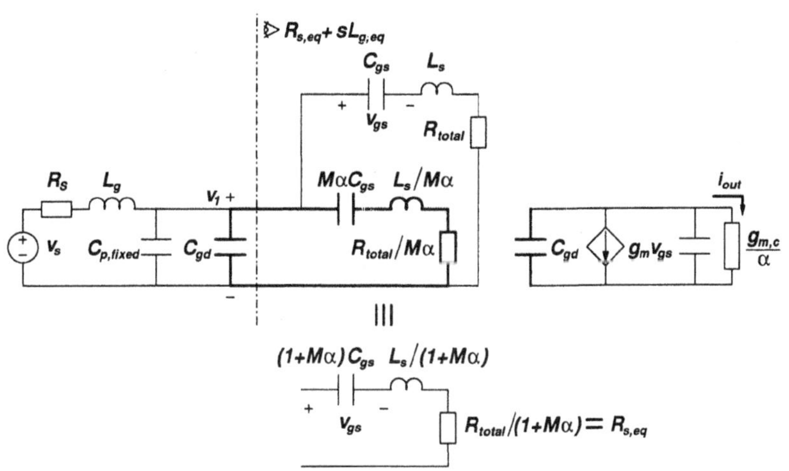

Figure 6.10: *Small signal model of the LNA input section. The capacitive feedback current on node one is modeled by an equivalent parallel impedance which is a perfectly scaled version of the 'normal' input impedance.*

equivalent, representing four different effects. First, C_{gd} contributes to the total capacitance at the gate and the drain of the amplifying device. Secondly, the C_{gd} injects a feed-forward current $sC_{gd}v_1$ into the drain, however, this contribution is negligible since $\omega_0 \ll \omega_T < g_m/C_{gd}$. The last, yet most important effect, is the feedback current $-sC_{gd}v_2$ which is drawn from the gate of the amplifying device (Miller effect).

Since

$$v_2 \cong -Mv_1 \cdot \frac{1}{1 + sC_{gs}R_{total} - \omega_0^2 L_s C_{gs}}, \tag{6.28}$$

where M is defined by (6.14) and R_{total} is given by

$$R_{total} = \omega_T L_s + r_{g,NQS}, \tag{6.29}$$

the feedback current $-sC_{gd}v_2$ can also be written in terms of v_1. The result is that the effect of the feedback current can be modeled as an extra impedance Z_p which is put in parallel to the gate of the amplifying device. Z_p is simply given by

$$Z_p = \frac{1 + sC_{gs}R_{total} - \omega_0^2 L_s C_{gs}}{sC_{gd}M} \tag{6.30}$$

$$= \frac{1}{M\alpha_{gd}}\left(\frac{1}{sC_{gs}} + R_{total} + sL_s\right), \tag{6.31}$$

where α_{gd} denotes the technology-dependent ratio between C_{gd} and C_{gs}. Note that Z_p is a perfectly scaled version of the 'normal' input impedance; all component parts are systematically $M\alpha_{gd}$ times smaller (Fig. 6.10). It is worth noting that this also implies that the controlling voltage of the transistor, v_{gs}, equally appears over the capacitive element in Z_p, $M\alpha_{gd}C_{gd}$. Hence, all the elements can be lumped together, which yields

$$C_{gs,eff} = (1 + M\alpha_{gd})C_{gs}, \tag{6.32}$$

$$L_{s,eff} = \frac{L_s}{1 + M\alpha_{gd}}, \tag{6.33}$$

$$R_{S,eq} = \omega_T L_{s,eff} + \frac{r_{g,NQS}}{1 + M\alpha_{gd}}. \tag{6.34}$$

These equations already reveal part of the story; Since $C_{gs,eff}$ is $1 + M\alpha_{gd}$ times larger, a given input current gives rise to a $1 + M\alpha_{gd}$ times lower v_{gs}, so that the PCC is $(1 + M\alpha_{gd})^2$ times lower. However, before we can draw any further conclusions, the influence of C_{gd} on $R_{S,eq}$ must be known as well. This on his turn requires that L_s (and also L_g) are known. These variables can be eliminated by recognizing that

$$R_{S,eq}(L_g) = \frac{r_{g,NQS}}{1 + M\alpha_{gd}} + \omega_T \underbrace{\left(\frac{1}{\omega_0^2 C_{gs}(1 + M\alpha_{gd})} - L_{g,eq}(L_g)\right)}_{L_{s,eff}(L_s)}, \tag{6.35}$$

which yields the following expression for $R_{S,eq}$:

$$R_{S,eq} \cong \frac{2R_S\left(1+\dfrac{C_{p,fixed}+C_{gd}}{C_{gs}(1+M\alpha_{gd})}\right)^2}{\Psi+\sqrt{\Psi^2-(\Psi-1)^2\left(\dfrac{\omega_T}{\omega_0}\right)^2}}, \tag{6.36}$$

where Ψ is defined as

$$\Psi = 1+\frac{2(C_{p,fixed}+C_{gd})R_S\omega_0^2}{\omega_T}\left(1+\frac{C_{p,fixed}+C_{gd}}{C_{gs}(1+M\alpha_{gd})}\right). \tag{6.37}$$

Just as in Section 6.5, it is only assumed that $\omega_0^2 \ll \omega_T^2$.

As one can see in (6.36), C_{gd} activates two counteracting mechanisms. First, C_{gd} adds to $C_{p,fixed}$ which makes C_p — and hence $R_{S,eq}$ — go up. Secondly, due to the Miller effect, $C_{gs,eff}$ is $1+M\alpha_{gd}$ times larger the C_{gs}, so that the ratio between C_p and $C_{gs,eff}$ becomes lower. This effect makes $R_{S,eq}$ go down again. The net effect thus depends on the magnitude of

$$\frac{1+\alpha_{gd}\dfrac{C_{gs}}{C_{p,fixed}}}{1+M\alpha_{gd}}; \tag{6.38}$$

However, most of the time its value is close to unity so that the impact of C_{gd} on $R_{S,eq}$ is very limited.

Now, how does all the above affect the performance of the LNA? Let us first consider the *PCC*, the main contributor to the gain. It is now given by

$$PCC = \left(\frac{\omega_T}{\omega_0}\right)^2\frac{1}{(1+M\alpha_{gd})^2R_{S,eq}}, \tag{6.39}$$

where $R_{S,eq}$ is defined in (6.36). Clearly, the *PCC* is degraded by a factor of $(1+M\alpha_{gd})^2$. This means that a $1+M\alpha_{gd}$ times larger R_L value is required to achieve the same amount of *PVC* (gain), which on his turn increases the noise contribution of R_L by a factor $1+M\alpha_{gd}$. The noise of the cascode is multiplied by $(1+M\alpha_{gd})^2$, and so is the contribution of the drain noise to the noise figure, F_d-1. The contribution of the non-quasi static gate noise on the other hand remains unaffected.

It should be clear that it is important to keep $M\alpha_{gd}$ sufficiently small in order not to put any burden on the *PCC* and the noise figure. This is especially the case when there is already a significant C_{fixed} at the LNA input; A large $C_{p,fixed}$ results in a large $R_{S,eq}$ value, which makes the drain noise the dominant term in the noise equation.

6.7 LNA Design Equations

In Section 6.4, the expressions have been derived that describe the performance of the LNAs in Fig. 6.2. The impact of several non-idealities (C_p, C_{gd} and M) has been thoroughly examined in Section 6.5 and Section 6.6. In this subsection, all this material is combined and further extended to incorporate the effect of a finite ω_c[5]. This results in the following set of detailed analytical expressions, covering all important LNA specifications:

effective source resistance C_p, C_{gd} and M transform the physical source resistance R_S into an effective source resistance, $R_{S,eq}$, given by

$$R_{S,eq} \cong \frac{2R_S\left(1+\frac{C_{p,fixed}+C_{gd}}{C_{gs}(1+M\alpha_{gd})}\right)^2}{\Psi+\sqrt{\Psi^2-(\Psi-1)^2\left(\frac{\omega_T}{\omega_0}\right)^2}}, \tag{6.40}$$

where the parameter Ψ is defined as

$$\Psi = 1+\frac{2(C_{p,fixed}+C_{gd})R_S\omega_0^2}{\omega_T}\left(1+\frac{C_{p,fixed}+C_{gd}}{C_{gs}(1+M\alpha_{gd})}\right). \tag{6.41}$$

In addition, the value of the physical source resistance R_S is constrained by

$$R_S < \frac{1}{2\omega_0(C_{p,fixed}+C_{gd})\left(1+\frac{C_{p,fixed}+C_{gd}}{C_{gs}(1+M\alpha_{gd})}\right)}. \tag{6.42}$$

load network requirements In order to generate a certain *PVC*, the required R_p of the *RLC* load in the LNA of Fig. 6.2(a) is given by

$$R_p(=R_L) = \frac{\omega_0}{\omega_T}\sqrt{PVC\, R_{S,eq}}\,(1+M\alpha_{gd})\sqrt{1+\left(\frac{\omega_0}{\omega_c}\right)^2}. \tag{6.43}$$

Similarly, to generate a certain G_t, the required R_p of M2 in Fig. 6.2(b) is given by

$$R_p(=2R_L) = 4G_t R_{S,eq}\left(\frac{\omega_0}{\omega_T}\right)^2(1+M\alpha_{gd})^2\left(1+\left(\frac{\omega_0}{\omega_c}\right)^2\right). \tag{6.44}$$

The factors at the end of (6.43) and (6.44) represent the extra amount of R_p that is required to rectify the *PCC* loss due to a finite ω_c.

[5] Until now we implicitly assumed that $\omega_0^2 \ll \omega_c^2$

noise factor The noise factor of both amplifiers in Fig. 6.2 is given by

$$
\begin{aligned}
F \cong 1 &+ \frac{\gamma}{\alpha}\left(\frac{\omega_0}{\omega_T}\right)^2\left(g_m R_{S,eq} + \frac{2}{\kappa}\right)(1 + M\alpha_{gd})^2 \\
&+ \frac{\alpha\delta(1 - |c|^2)}{\kappa g_m R_{S,eq}} \\
&+ 4\gamma\left(\frac{\omega_0}{\omega_c}\right)^2\left(\frac{\omega_0}{\omega_T}\right)^2 g_m R_{S,eq}\frac{(1 + M\alpha_{gd})^2}{M}\left(1 + \left(\frac{\omega_0}{\omega_c}\right)^2\right) \\
&+ \left\{4\frac{\omega_0}{\omega_T}\sqrt{\frac{R_{S,eq}}{PVC}}(1 + M\alpha_{gd})\sqrt{1 + \left(\frac{\omega_0}{\omega_c}\right)^2}\right\} \Big| \left\{\frac{1}{G_t}\right\}.
\end{aligned}
\tag{6.45}
$$

The first three terms represent the noise of the LNA core, i.e. $F_d - 1$, $F_{g,u} - 1$ and $F_c - 1$. The last term is the noise contribution of the load, $F_L - 1$, which is different for both LNAs.

intermodulation performance The IP_3 of both amplifiers is essentially identical:

$$
\begin{aligned}
IP_3(\text{dB}) \cong\ & 11.25 + 10\log_{10}\left(V_{GST}\frac{V_{si-vs}}{1 + \lambda M V_{si-vs}}(2 + \frac{V_{GST}}{V_{si-vs}})(1 + \frac{V_{GST}}{V_{si-vs}})^2\right) \\
& + 20\log_{10}\left(\omega_0 C_{gs}(1 + M\alpha_{gd})\sqrt{50 R_{S,eq}}\right)
\end{aligned}
\tag{6.46}
$$

quality factor In case of the *RLC* resonant LNA of Fig. 6.2(a), the total LNA quality factor approximately equals

$$
Q_{LNA} \triangleq \frac{\omega_0}{3\text{ dB } BW} \cong \underbrace{\frac{1}{2\omega_0 R_{S,eq} C_{gs}(1 + M\alpha_{gd})}}_{Q_{in}} + \underbrace{\omega_0 R_L C_L}_{Q_{out}}.
\tag{6.47}
$$

The quality factor of the output matched LNA of Fig. 6.2(b) highly depends on the exact nature of the matching networks M1 and M2.

Assuming an $R_{S,eq}$ of 200 Ω, a C_{gs} of 370 fF, an $(1 + M\alpha_{gd})$ factor of 1.4, a C_L of 1 pF and an R_L of 150 Ω, and assuming 1.8 GHz operation, Q_{in} and Q_{out} amount to 0.42 and 1.69, respectively.

6.8 The Design of the Cascode Device

Although the amplifying device is the most important device of the LNA, the cascode plays an important role as well. This is clearly illustrated by the fact that virtually all LNA equations in Section 6.7 contain either M, ω_c or both.

In this subsection we will concentrate on optimizing the cascode to decrease the overall noise figure. First, we will calculate what size the cascode device needs to have in order to optimize ω_c; As will be shown later on, the optimum is relatively flat, which will allow us to make abstraction of ω_c and reason — at least for the time being — in terms of M. It is worth noting that this optimization also defines a minimum size for the cascode transistor: a smaller transistor always yields a lower ω_c and a larger M, which increases the noise. In a second step, it will be shown that the noise figure can be further improved by exchanging ω_c against a lower M i.e. by enlarging the size of the cascode.

6.8.1 Optimization of the Cascode Pole

The expression for the pole on the cascode node is given by

$$\omega_c = \frac{g_{m,c}}{\alpha\left((\alpha_{db}+\alpha_{gd})C_{gs}+(1+\alpha_{sb})C_{gs,c}\right)}, \tag{6.48}$$

where $g_{m,c}$ is the transconductance of the cascode, α is defined by (5.62) and α_{xy} is the technology-dependent ratio between C_{xy} and C_{gs} [Stey93]. $g_{m,c}$ can also be formulated in terms of $C_{gs,c}$:

$$g_{m,c} \propto \sqrt{\left(\frac{W}{L}\right)_c I_{DS}} \tag{6.49}$$

$$\propto \sqrt{C_{gs,c}\, I_{DS}}. \tag{6.50}$$

By optimizing the size of the cascode — or in fact its $C_{gs,c}$ —, ω_c can be maximized for a given amplifying device, which yields

$$(\omega_c)_{opt} = \frac{\omega_T}{2\alpha\sqrt{1+\alpha_{sb}}\sqrt{\alpha_{db}+\alpha_{gd}}}. \tag{6.51}$$

The size of the cascode that corresponds to this optimum is defined by

$$(C_{gs,c})_{opt} = \frac{\alpha_{db}+\alpha_{gd}}{1+\alpha_{sb}} C_{gs}, \tag{6.52}$$

or equivalently

$$(M)_{opt} = \alpha\sqrt{\frac{1+\alpha_{sb}}{\alpha_{db}+\alpha_{gd}}}. \tag{6.53}$$

(6.52) clearly shows that ω_c is maximized when the cascode is sized such that the source capacitance of the cascode device matches the total drain capacitance of the amplifying device. Assuming that $\alpha = 0.8$, $\alpha_{gd} = 0.4$, $\alpha_{db} = 0.4$ and $\alpha_{sb} = 0.6$ — the α values of the early 0.25 μm technology used throughout this text — results in:

$$(C_{gs,c})_{opt} \cong 0.5\, C_{gs}, \tag{6.54}$$

$$(\omega_c)_{opt} \cong 0.55\, \omega_T, \tag{6.55}$$

$$M \cong 1.13. \tag{6.56}$$

For the representative LNA input device in Table 5.3 this yields:

$$\left(\frac{\omega_0}{\omega_c}\right)_{opt} \cong 0.15 \tag{6.57}$$

$$M \cong 1.13. \tag{6.58}$$

6.8.2 Optimization of the Overall Noise Figure

Although optimizing ω_c effectively minimizes the noise contribution of the cascode ($F_c - 1$) and the overall effect of the pole, it does not minimize the total noise figure. The reason is that the Miller effect still plays a important role; As shown in (6.45), the drain noise (i.e. $F_d - 1$) is multiplied by a factor $(1 + M\alpha_{gd})^2$, which is as large as 2.1 in case ω_c is maximized. Hence, the noise figure can be reduced further by decreasing M, or in other words, by scaling up the cascode device; The $F_d - 1$ term clearly goes down with decreasing M (due to the increased PCC) while the $F_{g,u} - 1$ term remains unaffected. Also $F_{R_L} - 1$ is reduced by decreasing M because it takes some time before ω_c starts to reverse the picture. The only noise contribution that does not decrease is the contribution of the cascode device itself. There are three reasons for this:

- A larger cascode device features a larger transconductance, and hence a larger noise current.
- The noise current more easily finds its way to the output since the cascode pole frequency is decreased according to the following formula:

$$\omega_c = \frac{(\omega_c)_{opt}}{\frac{M_{opt}}{2M} + \frac{M}{2M_{opt}}} \tag{6.59}$$

- The cascode pole decreases the PCC, which boosts the equivalent input noise.

Consequently, at a certain M value it does no longer pay off to increase the width of the cascode device. Fig. 6.11 summarizes this trade-off for the amplifying device in Table 5.3. The solid line in Fig. 6.11(a) plots the sum of the noise contributions $F_d - 1$ and $F_c - 1$ versus the Miller factor M. The sum is normalized with respect to its value at a Miller factor of 1.13, the M value at $(\omega_c)_{opt}$. As can be seen from the figure, a 20% reduction in noise can be obtained by lowering M from 1.13 to 0.57, i.e. by increasing the size of the cascode from 0.5 C_{gs} to 2 C_{gs}. Fig. 6.11(a) also shows the origin of the optimum: at $M = 0.55$, the increase in $F_c - 1$ (indicated by the dashed line) cancels the improvement in $F_d - 1$ (indicated by the dash-dotted line).

It is worth noting that decreasing the noise figure by 20% comes at the cost of a 4 times larger cascode device, and hence of a significantly larger output capacitance:

$$C_L = C_{L,fixed} + \frac{\alpha^2}{M^2}(\alpha_{gd} + \alpha_{db})C_{gs} \tag{6.60}$$

Even though the excess capacitance can in principle be resonated against an inductor, the achievable output resistance will be lower because the quality factor of the tank inductor is limited (because of physical limitations, process variations,...):

$$R_p = \frac{Q_{inductor}}{\omega_0 C_L}. \tag{6.61}$$

This holds even more for a broadband RC load, where Q_L is by construction equal to one. Therefore, serious consideration should be given whether or not it is worthwhile to reduce the noise figure by such a small amount. Depending on the activated constraint, $C_{gs,c}$ can be set to

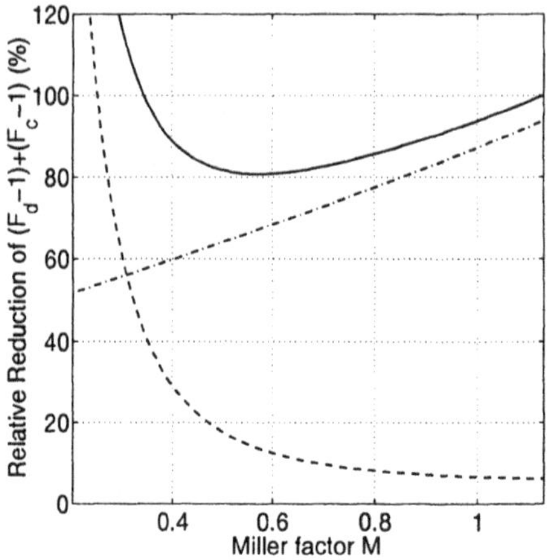

(a) $F_d - 1$ and $F_c - 1$ versus M.

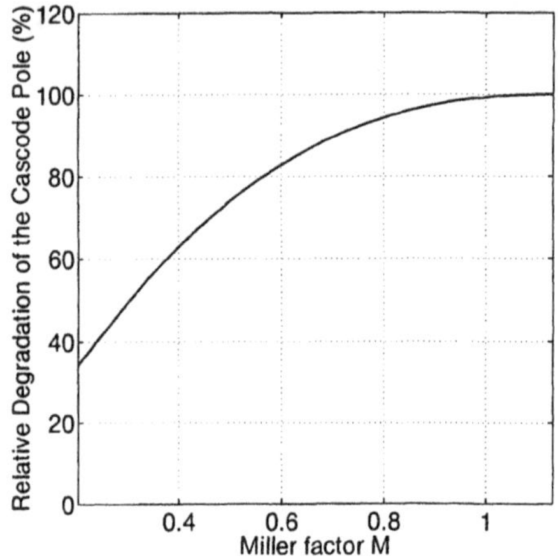

(b) $\omega_c/(\omega_c)_{opt}$ versus M.

Figure 6.11: *The impact of the size of the cascode (M) on the overall noise figure and ω_c. All values are normalized against their corresponding value at a Miller factor of 1.13, the M value at $(\omega_c)_{opt}$.*

$0.5\,C_{gs}$ (when the finite Q of the load is the limiting factor), $2\,C_{gs}$ (in case the noise figure needs the 20% improvement) or some value in between (Belgian compromise).

6.8.3 Unwanted Side-effects Initiated by the Cascode

When connecting the gate of the cascode device to the outside world using a bondwire, some important precautions need to be taken in order not to end up with stability problems or notches in the frequency characteristic. First, if no special countermeasures are taken, the inductance of the bondwire, $L_{g,c}$, and the C_{gd} of the cascode device will enter series resonance at a multiple of ω_0. At this frequency, the output impedance of the LNA — and hence the LNA gain — will inevitably be pulled down to zero (Fig. 6.12(a)). Secondly, the source of the cascode is degenerated by the drain capacitance of the amplifying device. In the same way as inductive degeneration leads to a positive real part in the input impedance, capacitive degeneration leads to a *negative* real part in the input impedance:

$$\Re(Z_{in}) = \frac{g_{m,c}}{g_{m,c}^2 + \omega_0^2 C_{D,tot}^2}\left(1 - \frac{\alpha}{1-\alpha}\frac{C_{D,tot}}{C_{gs,c}}\right) \tag{6.62}$$

$$\cong \frac{1}{g_{m,c}}\left(1 - \frac{\alpha}{1-\alpha}\frac{C_{D,tot}}{C_{gs,c}}\right). \tag{6.63}$$

Again, if no special measures are taken, this negative resistance will give rise to instabilities in conjunction with $L_{g,c}$.

The series resonance can be removed by putting a large decoupling capacitor at the gate of the cascode device (Fig. 6.12(b)). This capacitor shunts $L_{g,c}$ to ground so that the drain impedance

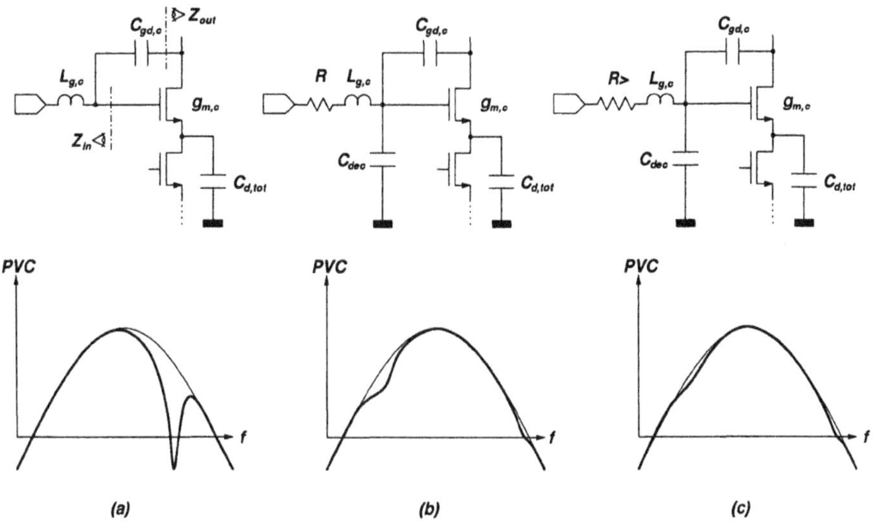

Figure 6.12: *Unwanted side-effects initiated by the cascode: (a) stability problem due to negative Z_{in} as well as a notch in the transfer function, (b) solution to the first problem gives rise to a dip at low frequencies, (c) after insertion of some extra series resistance.*

behaves capacitive again[6]. The instability due to the negative resistance can be canceled by adding an on-chip resistor in series with $L_{g,c}$ (Fig. 6.12(b)). It is worth noting that the decoupling capacitor improves the stability as well, since it lowers the magnitude of the negative input resistance so that less damping resistance is required.

The decoupling capacitor not only solves problems... it also generates a new one; C_{dec} and $L_{g,c}$ will enter parallel resonance at some frequency below ω_0. At this frequency, the cascode device no longer behaves as a cascode because the source admittance of the cascode drops from $g_{m,c} + g_{mb,c}$ to just $g_{mb,c}$, causing another dip in the gain, this time at low frequencies (Fig. 6.12(b)). Hence, some extra series resistance must be inserted in order to kill the quality factor of the parallel resonant tank (Fig. 6.12(c)).

6.9 Systematic LNA Design: A Case Study

Until now, we have focussed on the analysis of the LNA specifications. This has resulted in the bank of expressions in Section 6.7, describing the noise figure, the IP_3 and the gain of an LNA in the design space of the involved devices and in terms of important non-idealities. Now the time has come to take the step to real LNA design.

Designing an LNA is much more than just realizing the specifications; After all, LNA design is also subject to a lot of constraints, e.g. on the input reflection coefficient, on the total quality factor or on the value of external components. Since it is generally not known beforehand which specifications or constraints will be binding in the end, there exists no fixed design trajectory

[6]This is true provided that the ground inductance is low.

which works for an arbitrary set of specs. The design problem may even be over-constrained so that some specs or constraints can never be met simultaneously. In this case, any 'straightforward' design procedure would just leave the designer in the cold without any clue on how to continue.

In this work, a systematic design methodology is proposed which does give feedback to the designer — and not only when it goes wrong. Being based on contour lines, the method allows an efficient identification of un-realizable, inactive and binding constraints and offers a lot of insight in the overall specification dynamics and the design trade-offs involved. In particular, the method reveals the precise impact of a constraint on the optimization goal: minimum power. Since the method does not follow a preprogrammed design trajectory — the design simply follows from the inspection of the design space between the contour lines demarcating the constraints —, the procedure can be used for a broad range of LNA specs. Therefore, the method lends itself perfectly for incorporation in a CAD tool [Giel00].

The method is illustrated by means of a case study on a DCS-1800 power-to-voltage LNA employing the topology of Fig. 6.2(a). A slightly modified version of this LNA is embedded in the 0.25 μm CMOS, 2V DCS-1800 receiver found in Chapter 9. The method can equally be used to design power gain LNAs, — the topology of Fig. 6.2(b) —, of which the 0.8 dB *NF* LNA in Chapter 8 is an example.

Subsection 6.9.1 introduces the design problem, which consists of the boundary conditions, the target specifications, the additional constraints and of course, the optimization goal. Next, the dynamics of all specifications are analyzed in detail as a function of the I_{DS} and the $V_{GS} - V_T$ of the amplifying device. The section concludes with a synthesis of the results: the sizing strategy or "What is the impact of the constraints on the power consumption of an LNA that achieves all the specifications (NF, IP_3 and gain)?".

6.9.1 Target Specifications, Boundary Conditions and Constraints

Table 6.1 shows a representative set of specifications for a DCS-1800 low noise amplifier. With the help of this table, we will be able to explain the trade-offs that are encountered when designing LNA circuits. The table contains four parts: boundary conditions, specifications, additional constraints and the optimization goal.

The table is self-explanatory, except for the two additional constraints, '$Q \leq 2.4$' and '$L_g \leq$ 8 nH'. The first constraint can be explained as follows; The LNA in Fig. 6.2(a) is essentially a bandpass structure. Since the noise figure of a mixer is quite large, any loss with respect to the nominal gain can have a serious influence on the total noise figure budget. Therefore, the -3 dB bandwidth of the LNA needs to be significantly larger than the bandwidth of the application. If a loss of 0.5 dB with respect to the nominal gain can be tolerated, it can be proven that this requires that the -3 dB bandwidth is three times larger than the application bandwidth. However, the maximum loss of 0.5 dB must also be guaranteed in the presence of process variations which may shift the center frequency by an amount $\pm\Delta\omega_0$. It can be shown that this puts an upper limit on the allowable Q:

$$Q \leq \frac{\omega_0}{3\,(\omega_b + 2\Delta\omega_0)}. \qquad (6.64)$$

For the 75 MHz wide DCS-1800 band around 1.84 GHz, an allowable loss of 0.5 dB and a $\Delta\omega_0$

Parameter	Specification	Comments
Technology	TSB 0.25 μm CMOS	an early 2M1P, 0.25 μm CMOS technology
Working Frequency	$\omega_0 = 2\pi \cdot 1.84\,\text{GHz}$	DCS-1800 center frequency
Application Bandwidth	$\omega_b = 2\pi \cdot 75\,\text{MHz}$	DCS-1800 RX band
Source Impedance	$R_S = 50\,\Omega$	typical antenna impedance
Input Impedance	$R_{in} = 50\,\Omega$	a perfect match
Parasitic Capacitance	$C_{p,fixed} \cong 300\,\text{fF}$	bondpad
Load Capacitance	$C_L \cong 1\,\text{pF}$	mixer and inductor parasites
Gain	$PVC \cong 20\,\text{dB}$	target specifications
Noise Figure	$NF \leq 1.8\,\text{dB}$	
Intermodulation	$IP_3 \geq -4\,\text{dBm}$	
Quality factor	$Q \leq 2.4$	additional constraints
Gate Inductance	$L_g \leq 8\,\text{nH}$	
Current Consumption	I_{DS} minimize	optimization goal

Table 6.1: *A representative set of specifications for a DCS-1800 low noise amplifier. Some of the specifications in the table (like e.g. the 1.8* dB *noise figure spec) have been put somewhat tighter than in reality.*

of $\pm 5\%\ \omega_0$ — corresponding to a ΔC of $\pm 10\%$ —, this requires that

$$Q \leq \frac{1840\ \text{MHz}}{3(75\ \text{MHz} + 2 \cdot 92\ \text{MHz})} \cong 2.4. \tag{6.65}$$

Note that if larger process spreads are taken into account, the allowable quality factor can be even lower than 2!

The relevance of the last constraint, '$L_g \leq 8$ nH', depends on the the available technology options. In our ICs, the gate inductor L_g is always implemented as a bondwire (1.2 nH/mm). Whereas a bondwire offers a low cost solution and a high quality factor, it is inherently fragile. Therefore, the bondwire length — and hence the gate inductance — must be limited to a workable value, e.g. 6 to 7 mm (or 8 nH). When either an MCM technology or a low-C_p SMD inductor is available, this constraint will most likely not be activated any longer.

6.9.2 Analysis of the Specification Dynamics

In this section, all LNA specifications are analyzed as a function of the I_{DS} and the $V_{GS} - V_T$ of the amplifying device. The specifications will be visualized using contour lines because they provide us with a general overview at a single glance. In addition, they clearly demarcate the regions where certain specifications/constraints are not met, which allows us to analyze the precise impact of a constraint on the power consumption.

Even though the parameters of the amplifying device are the main design variables, virtually all equations contain the Miller factor M and the cascode pole ω_c, values set by the size of the cascode device. In this design, the cascode transistor is sized for a Miller factor $M = 1$: in each design point — defined by the couple $(I_{DS}, V_{GS} - V_T)$ — W_{casc} is set to $0.57 W_{amp}$.

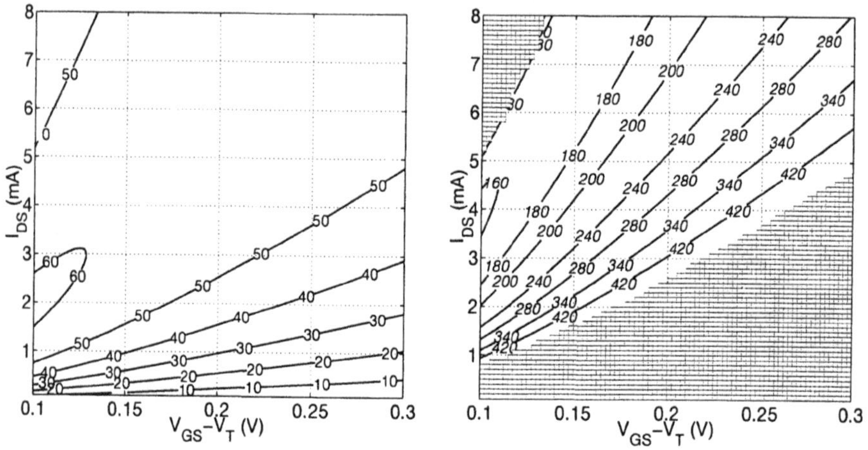

(a) Max. Feasible Source Resistance (Ω). (b) Equivalent Source Resistance (Ω).

Figure 6.13: *Contour plots of the maximum feasible source resistance and the equivalent source resistance.*

It is worth noting that that for any I_{DS} and $V_{GS} - V_T$, the rest of the parameters — like e.g. g_m, ω_T, W/L, C_{xy}, M, ω_c, $R_{S,eq}$, L_g, L_s, R_L and of course NF, IP_3, Q_{LNA} — can be calculated in a single chain-reaction. For each couple (I_{DS}, $V_{GS} - V_T$), the target specification '$PVC \cong 20$ dB' is used to calculate the required R_L, which enables us to calculate Q_{out} and NF.

Fig. 6.13(a) shows a contour plot of the maximum achievable input impedance (i.e. (6.42)) as a function of the I_{DS} and the $V_{GS} - V_T$ of the amplifying device. When the system is required to be directly matched to R_S (i.e. 50 Ω in this case), this is also the maximum allowable value of the physical source resistance $R_{S,max}$. Hence, for all purposes $R_{S,max}$ needs to be larger than 50 Ω. Obviously, there are two 'forbidden' areas on the plot. The area at the bottom right corresponds with transistors that feature a too small C_{gs} compared to the parasitic input capacitance C_p. At the top left of Fig. 6.13, the transistor has become so large that the contribution of its own C_{gd} to C_p limits the achievable input impedance. In the next contour plots, the regions where $R_{S,max} < 50$ Ω will therefore be blanked out.

Before we are able to discuss the specification dynamics, we must first determine the effective source resistance $R_{S,eq}$ when the device is matched to R_S in the presence of a non-zero C_p. For all purposes, the device can be considered being steered by this 'virtual' source resistance. Fig. 6.13(b) shows a contour plot of the effective source resistance $R_{S,eq}$ (i.e. (6.40)). in the design space. A first thing that should be clear from the plot is that — owing to the up-transformation by C_p (discussed in Section 6.5) —, $R_{S,eq}$ is considerably larger than 50 Ω. The smaller the transistor (i.e. small I_{DS} or large $V_{GS} - V_T$), the larger $R_{S,eq}$ becomes. Close to the 'forbidden' region at the bottom right, $R_{S,eq}$ increases even more rapidly. For very large transistors, the dynamics are reversed: the transistor's own C_{gd} starts contributing to the C_p component in the up-transformation network L_g-C_p.

Fig. 6.14(a) shows a contour plot of the LNA noise figure (i.e. (6.45)) in the design space. The noise dynamics as a function of I_{DS} (at any given $V_{GS} - V_T$) can be explained at the aid

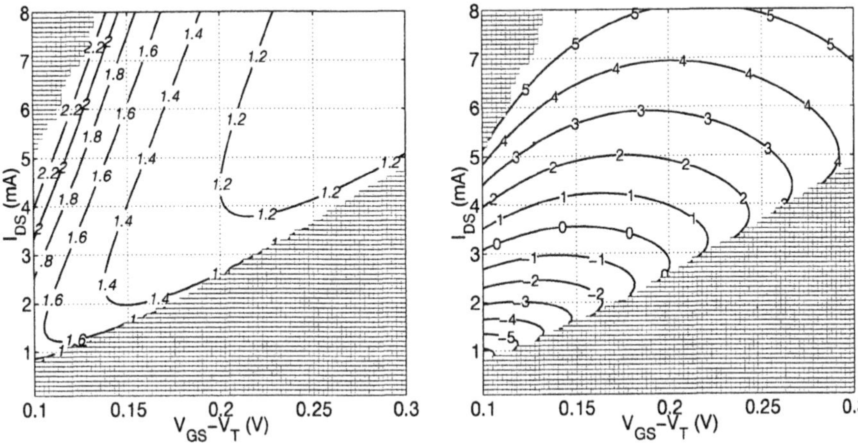

(a) LNA Noise Figure (dB). (b) LNA IP_3 (dBm).

Figure 6.14: *Contour plots of the total LNA noise figure and the input IP_3.*

of Fig. 6.15. At large current levels, the $g_m R_{S,eq}$ product is large because of the high g_m and because the presence of C_p increases the default $R_{S,eq}$ value. Hence, the noise is determined by the classical drain noise. When the current is lowered (while keeping a constant $V_{GS} - V_T$), the g_m of the amplifying device is lowered, and hence the noise figure decreases: the noise transfer function stays largely the same ($TF = f(\omega_0/\omega_T, R_{S,eq})$) while the drain noise current itself goes down. At low current levels, i.e. close to the border of the 'forbidden' region, the lower g_m is counterbalanced by a strong increase in $R_{S,eq}$. Consequently, the $g_m R_{S,eq}$ product starts to grow again, which increases the noise.

In this particular case, the noise is always dominated by the classical drain noise ($F_d - 1$) since $g_m R_{S,eq}$ never becomes lower than $(g_m R_{S,eq})_{opt} = f(\omega_T)$ (see Fig. 6.15(a)). However, when there is a smaller C_p at the input, the $g_m R_{S,eq}$ product generally crosses the $(g_m R_{S,eq})_{opt}$ line, so that the non-quasi static noise component dominates the noise figure in the bottom half of the plot. This is illustrated in Fig. 6.15(b). The same can be observed when the ratio between the operating frequency and the ω_T of the device becomes lower, e.g. when operating at 900 MHz instead of 1.8 GHz or when ω_T is large (e.g. if very low noise figures like 0.8 dB must be achieved).

Fig. 6.14(b) shows a contour plot of the LNA's IP_3 (i.e. (6.46)) as a function of the I_{DS} and the $V_{GS} - V_T$ of the amplifying device. The IP_3 values are considerably better than the ones in Fig. 5.20 since the signal excursions at the gate are much lower due to the larger effective source resistance. When $V_{GS} - V_T$ is kept fixed, the linearity improves with increasing transistor size because of the decreasing quality factor of the input section: the product $C_{gs} R_{S,eq}$ goes up. Only near the 'forbidden' region at the bottom right, the dynamics are reversed because of the steepness of $R_{S,eq}$. The dynamics as a function of $V_{GS} - V_T$ can be understood as follows. When $V_{GS} - V_T$ is decreased — starting from a low $V_{GS} - V_T$ value — while keeping the same current level, the required transistor width increases very rapidly. This causes a very fast drop in the quality factor, which is even able to counteract the decreasing intrinsic linearity. Hence, near

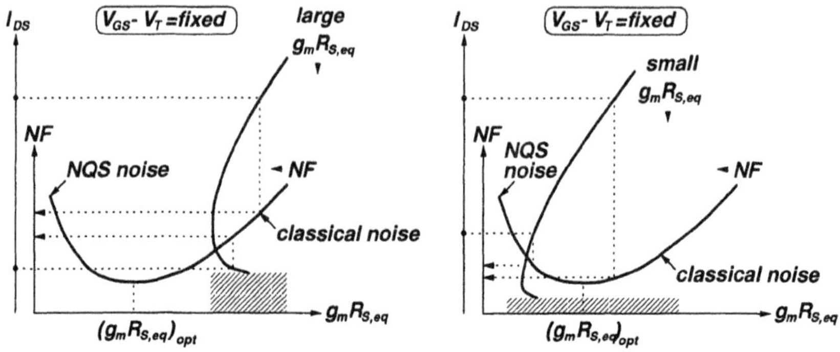

(a) large $g_m R_{S,eq}$ values. (b) small $g_m R_{S,eq}$ values.

Figure 6.15: *Explanation of the noise figure dynamics.*

weak inversion a very good IP_3 can be obtained at extremely low current levels. We will discuss this issue further in the next subsection. At larger $V_{GS} - V_T$ values, the IP_3 again follows the dynamics of the intrinsic linearity, i.e. it increases with increasing $V_{GS} - V_T$.

Fig. 6.16(a) shows a contour plot of the required load resistance — i.e. the R_p of the load inductor — in the design space. At moderate $V_{GS}-V_T$ values (e.g. $V_{GS}-V_T > 0.2$ V), where the ω_T changes relatively slowly, the dynamics are predominantly determined by $R_{S,eq}$: the larger the $V_{GS} - V_T$, the larger $R_{S,eq}$ and hence the lower the PCC and the larger the required R_L. At low $V_{GS} - V_T$ values, the dynamics are controlled by the fast changes in ω_T: a lower $V_{GS} - V_T$ means a lower ω_T and hence the lower the PCC and the larger the required R_L. Fig. 6.16(b) shows the resulting LNA quality factor. The total quality factor consists of two components: the input quality factor and the quality factor of the output section. When there is a significant C_p, the input quality factor (along with the PCC) is generally very low, even significantly lower than one! Consequently, the total quality factor of the LNA almost equals the quality factor of the load, which explains the similarity between Fig. 6.16(a) and Fig. 6.16(b). Again, this issue will be discussed further in the next subsection.

Fig. 6.17(a) plots the contours of the gate inductance L_g as a function of the I_{DS} and the $V_{GS} - V_T$ of the amplifying device. The explanation for the dynamics is self-evident: the smaller the I_{DS} (and the larger the $V_{GS} - V_T$), the lower the C_{gs} of the amplifying device, hence the larger the required L_g. As will be discussed further on, by putting a constraint on the magnitude of L_g, more current is required to achieve a specific noise figure. For the sake of completeness, Fig. 6.17(b) shows the contours of the source inductance L_s.

6.9.3 Contour-Based Sizing

This subsection analyzes the impact of the constraints on the power consumption of an LNA that achieves all the required specifications (NF, IP_3 and gain). The discussion is based on Fig. 6.18, which plots the contour lines of the target NF and IP_3 specifications[7] — 1.8 dB and -4 dBm, respectively — and the constraints on Q and L_g— '$Q \leq 2.4$' and '$L_g \leq 8$ nH', respectively.

[7]The PVC specification is realized by construction.

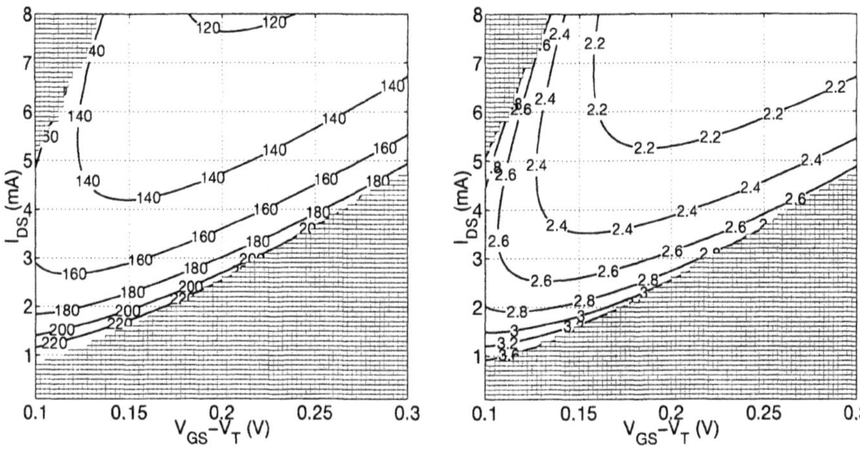

(a) Required Load Resistance R_L (Ω). (b) LNA Quality Factor.

Figure 6.16: *Contour plots of the required load resistance and the resulting LNA quality factor.*

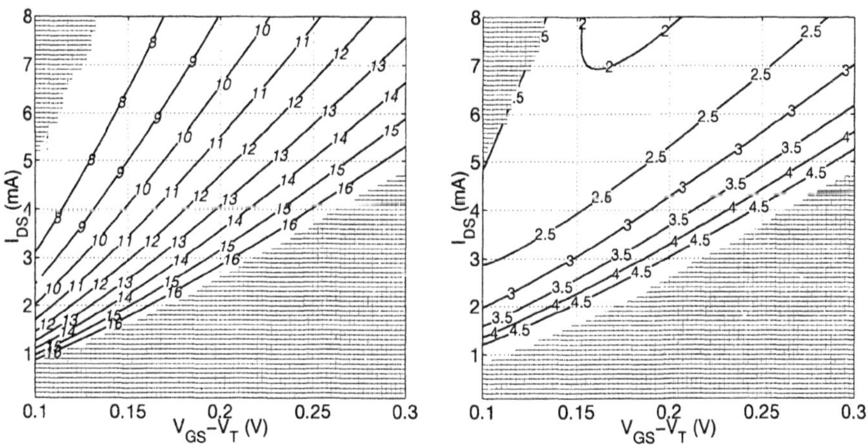

(a) Required Gate Inductance L_g (nH). (b) Required Source Inductance L_s (nH).

Figure 6.17: *Contour plots of the required gate inductance and the required source inductance.*

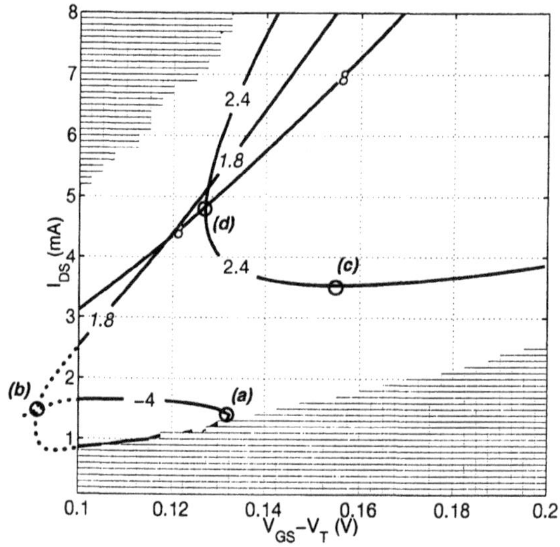

Figure 6.18: *The solutions to* ***(a)&(b)*** *the unconstrained,* ***(c)*** *the partially constrained and* ***(d)*** *the fully constrained design problem. The contour lines of the target NF and IP_3 specifications — 1.8* dB *and -4* dBm, *respectively — are plotted along with the contour lines of the constraints — '$Q \leq 2.4$' and '$L_g \leq 8$ nH'.*

	$NF \leq 1.8$ dB	$IP_3 \geq -4$ dBm	$Q \leq 2.4$	$L_g \leq 8$ nH	I_{DS}	$V_{GS} - V_T$	W/L
(a)	✓ 1.6 dB	✓-4 dBm	✗ 3.4	✗ 16 nH	1.3 mA	0.13 V	100/0.25
(b)	✓ 1.8 dB	✓-4 dBm	✗ 3.2	✗ 11 nH	1.5 mA	< 0.1 V	190/0.25
(c)	✓ 1.4 dB	✓ 0 dBm	✓ 2.4	✗ 11 nH	3.5 mA	0.155 V	210/0.25
(d)	✓ 1.75 dB	✓ 3 dBm	✓ 2.4	✓ 8 nH	4.8 mA	0.125 V	390/0.25

Table 6.2: *The solutions to the unconstrained, the partially constrained and the fully constrained design problem.*

Let us first consider the unconstrained problem, and ask ourselves what is the absolute minimum current that is required to achieve the target *NF* and IP_3 specifications. Apparently, the optimal design — i.e. point (a) in Fig. 6.18 — requires only 1.3 mA at a $V_{GS} - V_T$ of about 0.13 V (see footnote (8) on page 131). It is clear that in this particular case the noise figure specification is not binding: the *NF* is 1.6 dB < 1.8 dB (Fig. 6.14). The Q and the L_g on the other hand clearly out of spec: 3.4 and 16 nH, respectively (Fig. 6.16(b) and Fig. 6.17(a)).

Design (b) in Fig. 6.18 offers a solution at a comparable I_{DS} of 1.5 mA, and $V_{GS} - V_T$ lower than 0.1 V [8].

Since the transistor is then biased in the transition region between strong and weak inversion, its C_{gs} is significantly larger, so that its L_g is 11 nH. This is already much closer to the '$L_g \leq$ 8 nH' constraint. The Q of 3.2 is still out of spec (Fig. 6.16(b)).

The lowest power design that complies to the *NF* and the IP_3 specifications *and* meets the '$Q \leq 2.4$' constraint is design (c) in Fig. 6.18, consuming 3.5 mA at a $V_{GS} - V_T$ of approximately 0.15 V. This is the best solution when L_g is not constrained to a certain maximum size, for instance when an MCM technology is available or when SMDs are allowed. Anyway, the required L_g is still reasonable: 11 nH (Fig. 6.17(a)). Note that that neither the *NF* spec nor the IP_3 spec are binding: the *NF* equals 1.4 dB while the IP_3 is 0 dBm (Fig. 6.14(a) and (b))!

The lowest power solution that complies to all the requirements is design (d) in Fig. 6.18, which consumes 4.8 mA at a $V_{GS} - V_T$ of approximately 0.12 V (see footnote (8) on page 131). Obviously, the lowest power design is fully determined by the constraints, i.e. the couple (L_g, Q). Though this is not a general truth, it is quite typical for 'normal' LNA specifications like the ones in Table 6.1.

Table 6.2 summarizes the solutions of the unconstrained, the partially constrained and the fully constrained design problem.

6.10 Impact of the Source Resistance on Power Consumption

Until now, it has been assumed that the source resistance R_S is set by the boundary conditions (e.g. to a value of 50 Ω). But, what if R_S is a degree of freedom, or, what if an extra matching network is allowed anyway? This subsection examines how this degree of freedom can be exploited to minimize the power consumption for a given set of specs, or, in other words, we will answer the question: "What is the R_S ($R_{S,local}$) value that minimizes the power consumption?".

[8] At the cost of an increased deviation of 10%, the hand calculation model for strong inversion and the model parameters of Table 4.1 on page 52 remain valid for $V_{GS} - V_T$ values between 0.1 V and 0.15 V. At even lower $V_{GS} - V_T$ values, a dedicated weak inversion model needs to be used.

Let us again consider the design problem in Section 6.9. The graphical representation of its solution can be found in Fig. 6.18. During the design of this LNA, we have clearly seen that the input capacitance C_p transforms the 50 Ω source into an equivalent source with a considerably larger impedance, $R_{S,eq}$. Therefore, the *PCC* is much lower than expected based on the 50 Ω R_S value. As a consequence, the noise figure is dominated by the classical drain noise and a lot of pressure is put on the load resistance R_L and hence on the total LNA quality factor[9]. In fact, this is the main reason why the '$Q \leq 2.4$' constraint is binding and puts a lower limit on the power consumption.

If R_S is a degree of freedom, the power consumption can be significantly improved by decreasing R_S. This is illustrated by the following points:

- Because of the larger *PCC*, both the required load resistance R_L and the resulting LNA quality factor are lower at any given $(I_{DS}, V_{GS} - V_T)$ point. Hence, the contour line associated with the (binding) '$Q \leq 2.4$' constraint moves down, which in his turn enables a lower current consumption. This is illustrated by Fig. 6.19(a), which shows the contour lines of the total LNA quality factor at an R_S value of 32 Ω [10]. Compared to Fig. 6.16(b), which shows the contour lines at an R_S value of 50 Ω, the contour lines have been clearly shifted towards lower current values.

- The contour line associated with the '$L_g \leq 8$ nH' constraint moves somewhat down as well, which can be explained by the fact that the capacitive part of the input impedance becomes larger. This also helps reducing the current consumption. This is depicted in Fig. 6.19(b), which shows the contour lines of L_g at an R_S value of 32 Ω. Compared to Fig. 6.17(a), which shows the contour lines at an R_S value of 50 Ω, the contour lines have been clearly shifted towards lower current values.

- Since the classical noise term is proportional to $R_{S,eq}$, the noise figure decreases in every point of the design space, as can be observed from the huge difference between Fig. 6.20(a) and Fig. 6.14(a).

Of course, R_S can not be decreased arbitrarily since a lower R_S also means larger voltage excursions at the gate, which degrades linearity; This makes the contour line associated with the linearity constraint move towards larger current levels, so that the linearity — which is generally not binding (see Fig. 6.18) — can ultimately become binding. Hence, it should come to no surprise that the R_S value that optimizes the current consumption is precisely that value by which the linearity constraint is just binding.

For the design problem in Table 6.1, where R_S ($R_{S,local}$) is a degree of freedom, this yields an optimum R_S value of... 32 Ω. That 32 Ω is indeed the optimum R_S value should be clear from Fig. 6.21. In contrast to solution (d) in Fig. 6.18 — where 4.8 mA is required to meet all specifications and constraints — all requirements are met at a current of only 2.6 mA. It is worth noting that in this solution all three constraints ('$Q \leq 2.4$', '$IP_3 \geq -4$ dBm and '$L_g \leq 8$ nH') are activated at the same time. Also note that the noise figure of 1.3 dB is significantly lower

[9] Due to the large $R_{S,eq}$ value, the quality factor of the input section is generally below 1, so that the total quality factor of the LNA is determined by the quality factor of the RLC tank.

[10] Later on, it will made clear why we have chosen a value of 32 Ω.

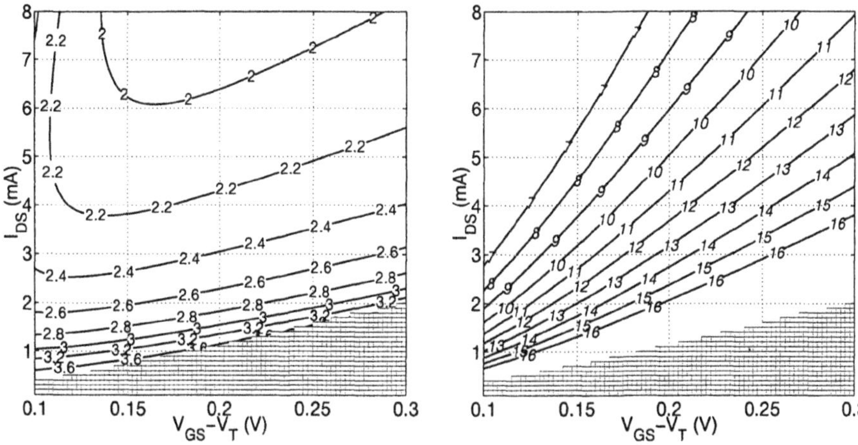

(a) LNA Quality Factor. (b) Required Gate Inductance L_g (nH).

Figure 6.19: *Contour plots of the total LNA quality factor and the required gate inductance at an R_S value of 32 Ω.*

R_S	$NF \leq 1.8$ dB	$IP_3 \geq -4$ dBm	$Q \leq 2.4$	$L_g \leq 8$ nH	I_{DS}	$V_{GS} - V_T$	W/L
50 Ω	✓ 1.75 dB	✓ 3 dBm	✓ 2.4	✓ 8 nH	4.8 mA	0.125 V	390/0.25
32 Ω	✓ 1.3 dB	✓-4 dBm	✓ 2.4	✓ 8 nH	2.6 mA	0.107 V	265/0.25

Table 6.3: *A comparison between the solution for a 50 Ω R_S and the optimum R_S of 32 Ω.*

than the required 1.8 dB. The difference between the solution for a 50 Ω R_S and the solution for the optimum R_S of 32 Ω is summarized in Table 6.3.

To conclude, if the linearity is not binding, the power can be considerably decreased by working with an intermediate reference plane at a lower impedance level ($R_{S,local}$) or using a low-ohmic physical source.

6.11 Fallacies and Pitfalls of LNA Noise Figure Dynamics

6.11.1 The Actual Importance of the Non-Quasi Static Gate Noise

As shown in the previous sections, a non-zero C_p generally makes the equivalent source resistance $R_{S,eq}$ several times larger than the physical source resistance, R_S. Consequently, the classical noise term — which is proportional to $R_{S,eq}$ (see (6.45)) — almost always dominates the noise figure. Hence, it is very tempting to leave the non-quasi static noise for what it seems to be — a purely academic noise term.

Nevertheless, when either the parasites are low, the source resistance is low, or the operating frequency is low compared to the capabilities of the technology, the non-quasi static noise *is* exposed and must be included in the noise calculations:

- When the C_p is low, the up-transformation action of the $L_g - C_p$ network is lower, so that

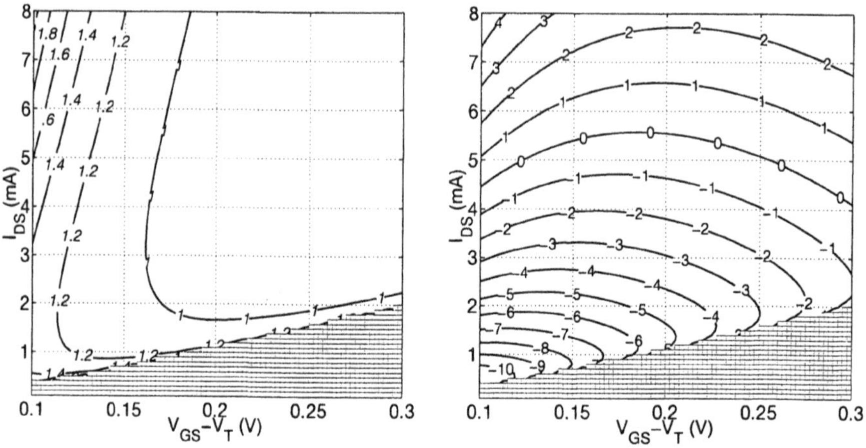

(a) LNA Noise Figure (dB). (b) LNA IP_3 (dBm).

Figure 6.20: *Contour plots of the total LNA noise figure and the input IP_3 at an R_S value of 32 Ω.*

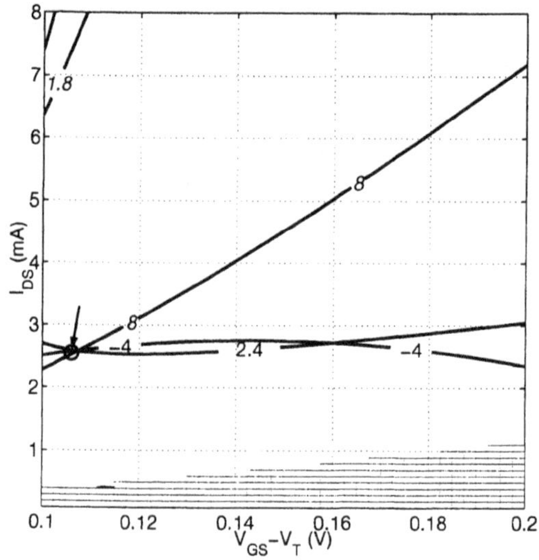

Figure 6.21: *The solution to the design problem in Table 6.1 at an R_S value of 32 Ω. The contour lines of the target NF and IP_3 specifications — 1.8* dB *and -4* dBm, *respectively — are plotted along with the contour lines of the constraints — '$Q \leq 2.4$' and '$L_g \leq 8$ nH'.*

$R_{S,eq}$ — and along with it the drain noise — is less boosted. The same occurs at low R_S values, since $R_{S,eq}$ is directly proportional to the physical source resistance.

- When the ratio between the operating frequency ω_0 and the ω_T of the amplifying device is much lower than say 0.1, the input-referred drain noise can become so low that the non-quasi static noise becomes visible.

The above conditions can also be worded in a single line: the non-quasi static effect does play an important role if the input section generates a significant voltage boost, i.e. if the impedance level of the signal at the gate-source terminal is large at the operating frequency.

At first sight, there is a seeming contradiction in the conclusion that non-quasi static noise is not important at frequencies close to the ω_T of the device; After all, the non-quasi static resistance has precisely been inserted to correctly model the MOS at high frequencies. Nevertheless, this conclusion is correct: owing to the decreased transistor current gain at high frequencies (ω_T/ω), the non-quasi static simply noise drowns in the input-referred drain noise! What is not true, is that the non-quasi static effect is negligible as a whole: $r_{g,NQS}$ may still play a role in the input impedance and still affects the channel charge build-up.

6.11.2 The Tricky Relation between NF, g_m and Current Consumption

If a CMOS analog designer is asked for his opinion on the relation between noise figure and current consumption in common-source based structures, he will most likely say: "The larger the g_m the better the noise figure: if we put some more current in a given transistor, increase its size for a given current, or increase the current while keeping a fixed $V_{GS} - V_T$, etc. the noise figure will always improve." While this reasoning is completely correct in amplifiers with a classical common source input stage, it is no longer unconditionally correct in input matched, common source amplifiers.

In this subsection we will analyze an explain the sometimes counter-intuitive relation between noise figure and power consumption in common-source based, input matched structures. Among others, it will be shown that:

1. the noise figure improvement that occurs when the current is increased is sometimes attributed to an increase in the wrong parameter, in this case g_m (see the example in 6.11.2.1).
2. the noise figure not always improves with an increasing current (see 6.11.2.2); The noise figure only improves if a certain specific condition is met (see 6.11.2.3).

After discussing these two pitfalls in detail, a general rule of thumb is given to always draw the correct conclusions with respect to the noise figure dynamics (see 6.11.2.3). For reasons of clarity, the bulk of the discussion is based on the ideal source-matched input stage, which leads to easily interpretable expressions. The conclusions for the non-ideal input stage — i.e. taking into account C_{gd}, M, ω_c, C_p, etc.— remain largely the same.

6.11.2.1 The First Pitfall: NF versus I_{DS} for a Fixed LNA (Fixed W/L)

Imagine that we are measuring a processed input matched cascode LNA. By construction, its W/L is fixed. Let us now examine increasing the current through the fixed size input device affects the noise figure. Of course, keeping R_{in} at its original value while the current increases requires us to insert a smaller source inductance,

As could have been expected, the noise figure improves with increasing current:

$$F - 1 \propto \frac{1}{\sqrt{I_{DS}}}. \tag{6.66}$$

Many old analog designers instinctively attribute this improvement to the increase in the g_m of the input device. There are two reasons for this. First of all, in many wide-band analog circuits the noise figure is indeed inversely proportional to g_m. Secondly, the observed noise figure dynamics are fully consistent with the dynamics of g_m^{-1}; after all, taking into account the fixed size of the input device, (6.66) can be rewritten in terms of g_m.

$$F - 1 \propto \frac{1}{\sqrt{I_{DS}}} \propto \frac{1}{\sqrt{(W/L)I_{DS}}} \propto \frac{1}{g_m}. \tag{6.67}$$

Yet, the true reason why the noise improves is very different... As shown in (6.10), the noise factor is given by[11]

$$F - 1 \propto n\, g_m R_{in} \left(\frac{\omega_0}{\omega_T}\right)^2. \tag{6.68}$$

Because the size of the device is fixed (remember that we are talking about a processed LNA),

$$g_m \propto \sqrt{I_{DS}} \tag{6.69}$$

$$\omega_T \propto V_{GS} - V_T \propto \sqrt{I_{DS}}, \tag{6.70}$$

so that the noise indeed follows the dynamics of (6.66):

$$F - 1 \propto \frac{\sqrt{I_{DS}}}{I_{DS}} \propto \frac{1}{\sqrt{I_{DS}}}. \tag{6.71}$$

However, as can readily be observed from (6.68) and (6.70), the reason for this is all but the increase in the g_m, as suggested in (6.67). In fact, it is the increasing $V_{GS} - V_T$ that does the trick because it neutralizes the *negative* effect of the increasing g_m (which stands for increasing current noise)! The fact that the noise figure dynamics does follow the dynamics of g_m^{-1} is not only coincidental but very misleading!

6.11.2.2 The Second Pitfall: NF versus I_{DS} at a Fixed $V_{GS} - V_T$

Let us look at another brain-teaser. Consider the device in Fig. 6.22(a), which features a certain W/L and a certain drain current I_{DS}. Using the inductors, the input impedance of the circuit is adjusted to a certain value, R_{in}.

The transistor in Fig. 6.22(b) on the other hand is the same transistor as the one in Fig. 6.22(a) put n times in parallel. Therefore, this transistor features an n times larger W/L, an n times larger I_{DS}, an n times larger g_m,... but exactly the same $V_{GS} - V_T$. The input impedance of this circuit is also adjusted to R_{in}. Note that this requires only that the gate inductance L_g is changed; L_s can remain the same since both devices have an identical $V_{GS} - V_T$ and hence an identical ω_T.

[11] For the sake of simplicity, we limit the discussion to the classical drain noise component; As indicated in Section 6.11.1, this component is almost always the dominant one.

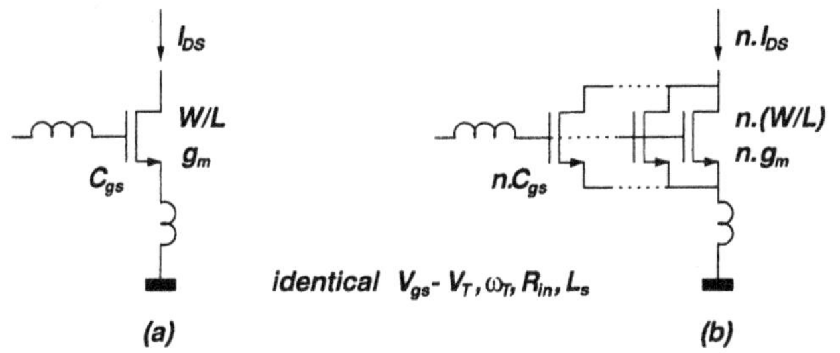

Figure 6.22: *The noise figure does not always improve with increasing current.*

I_{DS}	$V_{GS}-V_T$	g_m	W/L	NF trend	NF	I_{DS} slower than $(V_{GS}-V_T)^3$?
$I_{DS}\nearrow$	**cst.**	$\propto I_{DS}$	$\propto I_{DS}$	$NF\nearrow$	$\propto I_{DS}$	✗
$I_{DS}\nearrow$	$\propto I_{DS}$	**cst.**	$\propto 1/I_{DS}$	$NF\searrow$	$\propto 1/I_{DS}^2$	✓
$I_{DS}\nearrow$	$\propto\sqrt{I_{DS}}$	$\propto\sqrt{I_{DS}}$	**cst.**	$NF\searrow$	$\propto 1/\sqrt{I_{DS}}$	✓

Table 6.4: *The noise figure dynamics along three typical trajectories. A graphical representation of these trajectories can be found in Fig. 6.23.*

Now, since R_{in} and ω_T are the same in both cases, the power-to-current conversion is *identical* as well:

$$PCC_{Fig.\ 6.22(b)} = PCC_{Fig.\ 6.22(a)} = \frac{1}{R_{in}}\left(\frac{\omega_T}{\omega_0}\right)^2. \tag{6.72}$$

However, because of the n times larger g_m, the output noise current of the device in Fig. 6.22(b) is n times *larger* than the noise current of the device in Fig. 6.22(a)! Hence, the noise factor of Fig. 6.22(b) is considerably *worse*:

$$(F-1)_{Fig.\ 6.22(b)} = n\,(F-1)_{Fig.\ 6.22(a)} \tag{6.73}$$

$$\propto n\, g_m R_{in}\left(\frac{\omega_0}{\omega_T}\right)^2. \tag{6.74}$$

This example clearly illustrates that a larger g_m does not always mean a better noise figure.

6.11.2.3 Predicting the NF dynamics along an Arbitrary Trajectory

Since the noise figure dynamics is given by

$$F-1 \propto \frac{g_m}{\omega_T^2} \propto \frac{I_{DS}}{(V_{GS}-V_T)^3}, \tag{6.75}$$

the following rule can be used to predict the noise figure along the $(I_{DS}, V_{GS}-V_T)$ trajectory:
If I_{DS} increases faster than $(V_{GS}-V_T)^3$ along the trajectory, the noise figure deteriorates.
If I_{DS} increases more gradually than $(V_{GS}-V_T)^3$, the noise figure improves.

Table 6.4 illustrates this rule for three different trajectories: increasing the current at 1) a constant $V_{GS}-V_T$, 2) a constant g_m and 3) a constant W/L. The graphical representation is shown

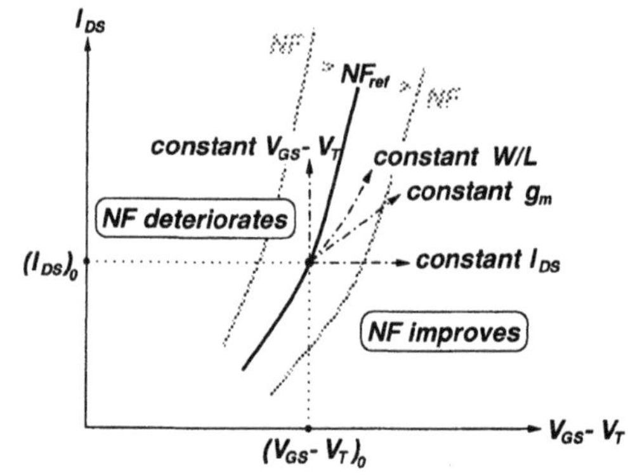

Figure 6.23: *Predicting the noise figure dynamics along an arbitrary trajectory. The three trajectories in Table 6.4 are indicated by the dash-dotted lines.*

in Fig. 6.23. Evidently, the dynamics of (6.75) can also be formulated in terms of other variables, like the W/L and the g_m.

6.12 Impact of a Finite Γ on LNA Performance

As indicated in Subsection 5.2.1, RF filters always need a minimum degree of termination in order to operate reliably, which translates in a specification of the form '$\Gamma < x$ dB'. However, this specification still leaves a lot of design room. The main question that arises now is: "Is it better to design the LNA for a zero Γ, or should one try to keep close to the maximum Γ-value that is allowed by the specification?". This first question is covered in Subsection 6.12.1 and Subsection 6.12.2. Later on, in Section 6.13, the practical implications of a finite reflection coefficient on the LNA performance will be illustrated for the LNA in the case study.

Now, we could go even further... When the system does not require a sharp filter — e.g. because the blocking levels are sufficiently low — there is no real constraint on the value of Γ. This gives rise to a much more radical question: "If we don't really need an input match... should we bother matching then?". The answer to this question is given in Subsection 6.12.3.

6.12.1 Impact of a Finite Γ on PCC and NF

Until now, we have always assumed that the amplifying device was perfectly matched to the source, either using the direct matching scheme (i.e. case (a) in Fig. 6.24: $R_{in} = R_S$) or by means of an intermediate reference plane at $R_{S,local}$ and an extra matching network M (i.e. case (b) in Fig. 6.24: $R_{in} = R'_S = R_{S,local}$). It has been shown in Subsection 5.5.1 that by lowering $R_{S,local}$ with respect to R_S, a substantially larger *PCC* can be obtained than when $R_{S,local}$ equals the source resistance R_S. However, Subsection 5.7.3 has indicated that in case the non-quasi static noise term dominates the noise figure, a lower $R_{S,local}$ value comes at the cost of a larger noise figure.

In this subsection it is examined how the noise figure and the *PCC* are affected by a finite

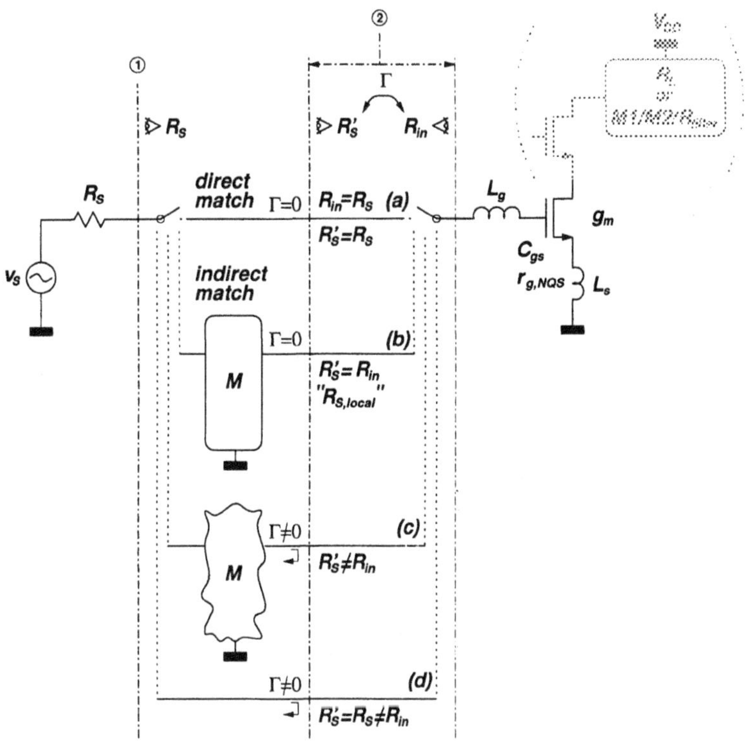

Figure 6.24: *Four different degrees of matching: **(a)** R_{in} directly matched to R_S, **(b)** R_{in} indirectly matched to R_S using an extra matching network M ($R_{in} = R'_S < R_S$), **(c)** a bad indirect match ($R_{in} \neq R'_S$), **(d)** R_S directly connected to $R_{in} \neq R_S$.*

reflection coefficient Γ, or, in other words, when R_{in} (the input impedance of the MOS input section) does not match R'_S (the effective source impedance). The conclusions apply whenever the input network features a finite $\Gamma \in \mathbb{R}$, i.e. when the lossless matching network M does not succeed in transforming the original source impedance R_S into $R'_S = R_{in}$ (case (a) in Fig. 6.24) or when M is simply omitted (case (d) in Fig. 6.24: $R'_S = R_S \neq R_{in}$). Although the expressions are valid at all times, when assessing the quantitative effect on noise and *PCC* we will assume that $R_{in} \leq R'_S \leq R_S$ (e.g. 30 $\Omega[R_{in}] < 50\ \Omega[R_S]$).

In order to emphasize the results and not to get drowned in the calculations, the main conclusions are already written down in 6.12.1.1. People who really want to go to the bottom of it can find the detailed mathematical explanation in 6.12.1.2.

6.12.1.1 Main Conclusions

Subsection 5.5.2 already provides part of the answer to the first question; It states that the extra matching section M to transform R_S into an effective source impedance $R'_S = R_{in}$ is not always required from the point of view of gain; Fig. 5.13 and (5.47) clearly show that, as long as

$$\Gamma \triangleq \frac{R_{in} - R_S}{R_{in} + R_S} \leq 0.25 \quad (-12\ \text{dB}), \tag{6.76}$$

there is hardly any difference between the *PCC* under matched conditions (i.e. when M is present) and the *PCC* when R_S is directly connected to R_{in}; In this case the extra matching network M might as well be omitted (Fig. 6.24(d)). Another conclusion that can be drawn is that — as long as M in Fig. 6.24(c) is lossless — it suffices that

$$\Gamma = \frac{R_{in} - R'_S}{R_{in} + R'_S} \leq 0.25, \tag{6.77}$$

in order to fully benefit from a lower R_{in} value. Hence, from the point of view of gain it is still interesting to let a source with impedance R_S directly drive a system with input impedance $R_{in} \leq R_S$ (Fig. 6.24(d)); Even at a Γ of -12 dB (i.e. $R_{in} \cong 0.6 R_S$), one can still gain about 2 dB compared to when the MOS input section had been directly matched to R_S, i.e. $R_{in} = R'_S$ with $R'_S = R_S$. The 2 dB improvement with respect to the direct match is only 0.3 dB less than if the matching network M had been present (i.e. the indirect match of Fig. 6.24(b): $R'_S = R_{in} \leq R_S$). Consequently, Γ can be traded for a lower cost solution (almost) without any degradation of the *PCC*.

Still, we don't have an idea about the effect of a finite Γ on noise figure. Because the *PCC* stays almost the same, irrespective of the fact that either $R'_S = R_{in}$ or $R'_S \neq R_{in}$ but (6.77) is valid, one could anticipate that both noise figures are nearly identical as well: when directly connecting $R'_S \neq R_{in}$ to R_{in}, the device's noise figure would then be the same as if the source was presenting an effective source impedance $R'_S = R_{in}$ to the device. However, as will be shown in 6.12.1.2, this is *not* the case: instead of being determined by R_{in}, the noise is determined by the R'_S value i.e. *as if* R_{in} *had been set to* R'_S[12].

Though one may be tempted to ask: "So what?", this is a very important result; As mentioned before, R_{in} is often set to a lower value than R_S in order to improve the *PCC*. When the matching network M correctly transforms R_S down to $R'_S = R_{in} < R_S$ (i.e. case (b) in Fig. 6.24) and in

[12]Or, in other words, as if the input reference plane features an $R_{S,local} = R'_S$.

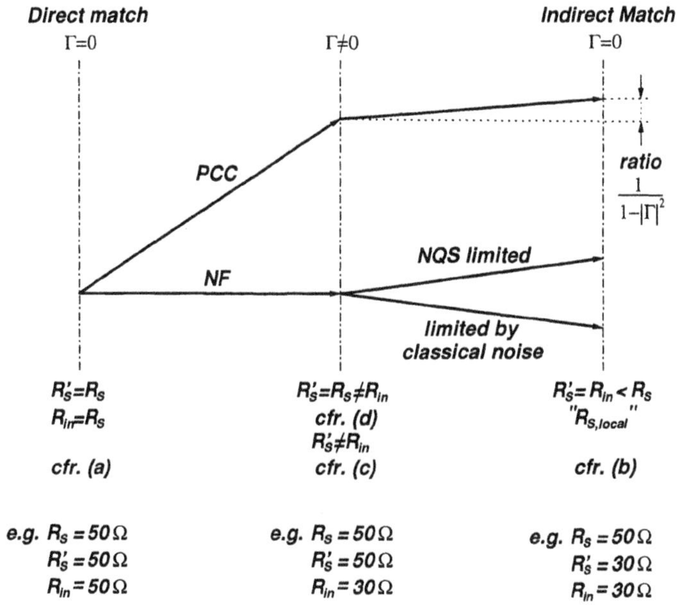

Figure 6.25: *The impact of a finite* Γ *on PCC and noise figure.*

case the noise figure is dominated by non-quasi static noise (see Subsection 5.7.3) this results in a worse noise performance than when R_{in} had been directly matched to the larger R_S value (i.e. case (b) in Fig. 6.24: $R_{in} = R'_S$ with $R'_S = R_S$). However, by actually *designing* for a finite Γ — e.g. by setting R_{in} to a value smaller than R_S and directly connecting R_{in} to R_S like in Fig. 6.24(d) — the same noise figure is obtained as if R_{in} had been set to R_S, which is in fact a lower[13] noise figure!

Fig. 6.25 illustrates the above-mentioned conclusions for exactly the same cases as depicted in Fig. 6.24. The figure clearly demonstrates the effect of lowering R_{in} with respect to R_S and at the same time allowing a finite Γ — e.g. by omitting the extra matching network M and directly connecting R_{in} to R_S:

- A larger gain can be achieved than if R_{in} was simply made equal to R_S; The finite Γ hardly degrades the gain compared to the fully matched case (i.e. $R'_S = R_{in}$), provided that Γ is less than -12 dB. Additionally, at a Γ of -12 dB only 6% of the incident power is being reflected, which means that a filter can still be inserted before the circuit without degradation of the filter characteristic.

[13]When the parasitic input capacitance C_p is low, the noise figure is most likely dominated by non-quasi static noise (see Subsection 5.7.3). In this case, it is indeed advantageous to let the noise figure be determined by the 'larger' R'_S value (because the noise is inversely proportional to the impedance value) instead of the 'lower' R_{in}. When C_p is large ($> 2C_{gs}$), the noise figure is dominated by the classical drain noise term, which is proportional to the impedance value; In this case it is better to actually do the transformation from R_S down to the low R_{in} value, i.e. make $R'_S = R_{in} < R_S$.

R_S	R'_S	R_{in}	Matched?	*PCC* Enhancement	Noise Figure	Comments
50 Ω	50 Ω	50 Ω	Yes	0 dB (ref.)	0.74 dB	Directly matched (no M required)
50 Ω	30 Ω	30 Ω	Yes	2.3 dB	1.1 dB	Indirectly matched via M
50 Ω	50 Ω	30 Ω	$\Gamma \cong -12$ dB	2.0 dB	0.74 dB	Directly connected (no M)

Table 6.5: *A numerical illustration of the impact of a finite Γ on PCC and noise figure. The example is based on the representative LNA input device in Table 5.3 and a classical 50 Ω source (C_p is assumed to be zero). The numbers can be deduced from Fig. 5.16 and Fig. 5.13.*

- Despite the finite Γ, the noise remains determined by the source impedance (R_S) instead of the input impedance (R_{in}): the noise figure is the same as if $R_{in} = R_S$.

  In case the non-quasi static noise is dominant, this means that the finite Γ ensures that the lower R_{in} value is no longer accompanied by a higher noise figure: the noise figure is still determined by the larger R_S value. When the classical noise is dominant, transforming R_S down to the low R_{in} value would have resulted in a lower noise figure: here a finite Γ keeps us from improving the noise. However, the lower noise figure would have come at the cost of an additional matching network M (e.g. a longer bondwire with an external SMD capacitor) which might not have been worth the hassle anyway.

Table 6.5 illustrates these conclusions with a numerical example. The example is based on the representative LNA input device in Table 5.3 and a classical 50 Ω source. Since C_p is assumed to be zero, the noise figure is in this case dominated by the non-quasi static noise term. The numbers in Table 6.5 can be extracted from Fig. 5.16 and Fig. 5.13.

6.12.1.2 Mathematical Explanation

Let's again consider the circuit in Fig. 6.24. The noise factor — neglecting the noise contribution of the load for now — can be written as

$$F \cong 1 + \frac{b(F_d - 1)\big|_{②@R_{in}} + c(F_{g,u} - 1)\big|_{②@R_{in}} + d(F_c - 1)\big|_{②@R_{in}}}{a}, \tag{6.78}$$

where a, b, c and d are defined as:

$$a \triangleq \frac{i^2_{n,out} \text{ owing to } R_S \text{ when } R'_S \neq R_{in}}{i^2_{n,out} \text{ owing to } R_S \text{ when } R'_S = R_{in}}, \tag{6.79}$$

$$b \triangleq \frac{i^2_{n,out} \text{ owing to } i^2_{nd} \text{ when } R'_S \neq R_{in}}{i^2_{n,out} \text{ owing to } i^2_{nd} \text{ when } R'_S = R_{in}}, \tag{6.80}$$

$$c \triangleq \frac{i^2_{n,out} \text{ owing to } i^2_{g,u} \text{ when } R'_S \neq R_{in}}{i^2_{n,out} \text{ owing to } i^2_{g,u} \text{ when } R'_S = R_{in}}, \tag{6.81}$$

$$d \triangleq \frac{i^2_{n,out} \text{ owing to } i^2_{nd,casc} \text{ when } R'_S \neq R_{in}}{i^2_{n,out} \text{ owing to } i^2_{nd,casc} \text{ when } R'_S = R_{in}}. \tag{6.82}$$

The designation "$x\,\big|_{②@R_{in}}$" means that the quantity x is to be evaluated when reference plane ② in Fig. 6.24 is matched to 'R_{in}' at both sides of the reference plane.

The parameter a can be found as follows; Whether the circuit is perfectly matched or not, the available noise power from the source remains the same: $kT\Delta f$ — as is true for all resistive sources. However, when there is a finite Γ, according to (5.48), the *PCC* of the system degrades by a factor $1-|\Gamma|^2$ so that the output noise current that stems from the source is lowered by the same factor. Hence, the parameter a is given by

$$a = 1 - |\Gamma|^2. \tag{6.83}$$

Since the internally generated noise current i_{nd} is the same in both cases, the resulting output noise current $i_{n,out}$ can only change when each i_{nd} propagates differently towards the output. Hence, b is nothing else than the square of the current division factor, which — after some figuring — becomes

$$b = \frac{1\Big/\left(1+\dfrac{R_{in}-r_{g,NQS}}{R'_S+r_{g,NQS}}\right)^2}{1\Big/\left(1+\dfrac{R_{in}-r_{g,NQS}}{R_{in}+r_{g,NQS}}\right)^2} \tag{6.84}$$

$$= \left(\frac{2R_{in}}{R_{in}+R'_S}\right)^2\left(\frac{R'_S+r_{g,NQS}}{R_{in}+r_{g,NQS}}\right)^2 \tag{6.85}$$

$$= (1+\Gamma)^2\left(\frac{R'_S+r_{g,NQS}}{R_{in}+r_{g,NQS}}\right)^2. \tag{6.86}$$

The parameter c can be found by calculating (the square of) the ratio between the gate-source voltages caused by e_{ng} (see (5.69)) in the respective cases. This leads to

$$c = \frac{1/\left(R'_S+R_{in}\right)^2}{1/\left(2R_{in}\right)^2} \tag{6.87}$$

$$= (1+\Gamma)^2. \tag{6.88}$$

The noise current from the cascode transistor is not affected by the quality of the match, so that

$$d = 1 \tag{6.89}$$

Combining (6.78) with (6.83), (6.86), (6.88) and (6.89), and taking into account that

$$\frac{(1+\Gamma)^2}{1-|\Gamma|^2} = \frac{1+\Gamma}{1-\Gamma} \quad (\Gamma \in \mathbb{R}), \tag{6.90}$$

yields the noise factor as a function of Γ:

$$F = 1 + \frac{1+\Gamma}{1-\Gamma}\left(\frac{R'_S + r_{g.NQS}}{R_{in} + r_{g.NQS}}\right)^2 (F_d - 1)\big|_{@R_{in}} + \frac{1+\Gamma}{1-\Gamma}(F_{g.u} - 1)\big|_{@R_{in}} + \frac{1}{1-|\Gamma|^2}(F_c - 1)\big|_{@R_{in}} \tag{6.91}$$

$$= 1 + \frac{R_{in}}{R'_S}\left(\frac{R'_S + r_{g.NQS}}{R_{in} + r_{g.NQS}}\right)^2 (F_d - 1)\big|_{@R_{in}} + \frac{R_{in}}{R'_S}(F_{g.u} - 1)\big|_{@R_{in}} + \frac{R'_S}{R_{in}}\frac{1}{(1-\Gamma)^2}(F_c - 1)\big|_{@R_{in}}. \tag{6.92}$$

If one combines the preceding equation with (5.84) — i.e. $(F_d - 1)\big|_{@R_{S,local}}$ —, (5.85) — i.e. $(F_{g.u} - 1)\big|_{@R_{S,local}}$ — and (6.13) — i.e. $(F_c - 1)\big|_{@R_S}$ — one can clearly see that the terms in R_{in} are systematically factored out, whereupon they are substituted by the corresponding term in R'_S, so that the noise factor becomes

$$F \cong 1 + (F_d - 1)\big|_{@R'_S} + (F_{g.u} - 1)\big|_{@R'_S} + \frac{1}{(1-\Gamma)^2}(F_c - 1)\big|_{@R'_S} \tag{6.93}$$

$$= 1 + \frac{\gamma}{\alpha}\left(\frac{\omega_0}{\omega_T}\right)^2\left(\sqrt{g_m R'_S} + \frac{1}{\kappa\sqrt{g_m R'_S}}\right)^2 + \frac{\alpha\delta(1-|c|^2)}{\kappa g_m R'_S} + \frac{4\gamma}{(1-\Gamma)^2}\left(\frac{\omega_0}{\omega_T}\right)^2\left(\frac{\omega_0}{\omega_c}\right)^2\frac{g_m R'_S}{M}. \tag{6.94}$$

The transition from (6.93) to (6.94) is made by recognizing that $F_d - 1\big|_{@R'_S}$, $F_{g.u} - 1\big|_{@R'_S}$ and $(F_c - 1)\big|_{@R'_S}$ are obtained by replacing the $R_{S.local}$ variable in (5.84), (5.85) and the R_S variable in (6.13) by R'_S.

The only thing that remains to be done is the calculation of the $F_L - 1$ term. If one assumes that the load is always adjusted to generate the same *PVC*, it can be shown that

$$F_{R_L} - 1 = \frac{1}{\sqrt{1-|\Gamma|^2}}(F_{R_L} - 1)\big|_{@R_{in}} \tag{6.95}$$

$$= \frac{1}{1-\Gamma}(F_{R_L} - 1)\big|_{@R'_S}. \tag{6.96}$$

In case of an output matched LNA, where G_t is the parameter of interest, $F_L - 1$ remains the same:

$$F_L - 1 = \frac{1}{G_t}. \tag{6.97}$$

Consequently, it can be concluded that whenever $R'_S \neq R_{in}$, $F_d - 1$ and $F_{g.u} - 1$ have exactly the same value as when R_{in} had been made equal to R'_S (the effective source impedance). The second-order noise contributions $F_c - 1$ and $F_{R_L} - 1$ are larger than when R'_S had been set to R_{in}, but they are still lower than when R_{in} had been set to R'_S. Hence, we can conclude that the noise figure in the unmatched case is almost the same noise figure as when R_{in} had been matched — or set — to the effective source impedance, R'_S.

6.12.2 Impact of a Finite Γ on IP$_3$ and Q$_{in}$

For the sake of completeness, this subsection states the equations for IP_3 and Q_{in} in case R_{in} (the input impedance of the MOS input section) does not match R'_S (the effective source impedance). The conclusions apply whenever the input network features a finite $\Gamma \in \mathbb{R}$.

The IP_3 can be calculated using the same technique as in Subsection 5.8.2. It can easily be shown that the IP_3 is related to the original IP_3 as follows:

$$IP_3(\mathrm{dB}) \cong IP_3\big|_{②@R_{in}} + 20\log_{10}\left(\frac{R'_S + R_{in}}{2\sqrt{R_{in}R'_S}}\right) \tag{6.98}$$

$$= IP_3\big|_{②@R_{in}} + 10\log_{10}\left(\frac{1}{1-|\Gamma|^2}\right). \tag{6.99}$$

The designation "$IP_3\big|_{②@R_{in}}$" indicates that the IP_3 is to be evaluated when reference plane ② in Fig. 6.24 is matched to 'R_{in}' at both sides of the reference plane. As can be derived from (6.99), the IP_3 improves with increasing reflection. This should come to no surprise, since the signal swing at the transistor's input is precisely determined by the power that effectively flows to the transistor.

The quality factor of the input section is now given by

$$Q_{in} \cong Q_{in}|_{②@R_{in}} \frac{2R_{in}}{R'_S + R_{in}} \tag{6.100}$$

$$= Q_{in}|_{②@R_{in}}(1+\Gamma). \tag{6.101}$$

Hence, it follows that by allowing a small amount of negative input reflection ($R_{in} < R'_S$), the quality factor at the input can be reduced hardly without degrading the PCC ($\propto 1 - |\Gamma|^2$), and even with a small improvement in the IP_3 ($\propto 1/1 - |\Gamma|^2$)! This property is very welcome, since the constraint on the total LNA quality factor is often a binding constraint.

6.12.3 Intrinsically Unmatched Input Structures

In this subsection, the performance of two intrinsically unmatched structures are examined: the series resonant and the parallel resonant input section. It will be shown that a good performance does not at all require that a considerable part of the available source power is transferred to the system.

6.12.3.1 Series Resonant Input Structure

Let us consider Fig. 6.26. Fig. 6.26(a) represents a circuit whose input is perfectly matched to R_S. As shown in Subsection 5.3.2, its gain is given by

$$PCC = \frac{1}{R_S}\left(\frac{\omega_T}{\omega_0}\right)^2. \tag{6.102}$$

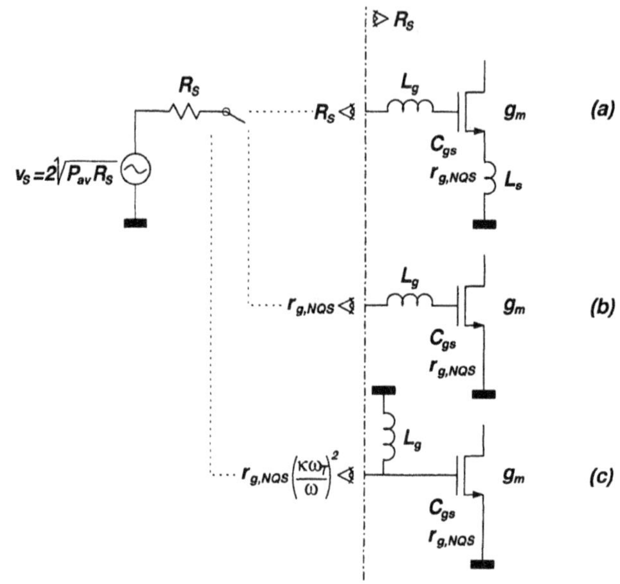

Figure 6.26: *Three different input structures: (a) input matched to R_S, (b) series resonant input structure, (c) parallel resonant input structure.*

When L_s is set to zero and L_g is adjusted to yield series resonance at ω_0, as shown in Fig. 6.26(b), the input impedance is set to its lowest possible level, being the non-quasi static gate resistance $r_{g.NQS}$. The gain of this configuration is then given by

$$PCC_{SR} = \frac{4R_S}{(R_S + r_{g.NQS})^2}\left(\frac{\omega_T}{\omega_0}\right)^2. \tag{6.103}$$

Fig. 6.27 plot equations (6.102) and (6.103) as a function of R_S for the device in Table 5.3.

Obviously, the series resonant structure offers a larger *PCC* than does matching the input to R_S. At normal R_S values, the improvement in gain is even as large as 5 to 6 dB [14], i.e. the gain is virtually the same as if the device was matched to $R_S/2$. One might argue that the gain of a matched structure can always be improved by transforming R_S down to a lower value, but irrespective of the R_S value, a series resonant input structure still offers a few dB more. Of course, when R_S approaches $r_{g.NQS}$, the law of diminishing return starts to play.

Apart from the *PCC*, the noise figure is important as well. The influence of a finite Γ (or omitting M) on noise figure has already been calculated in Subsection 6.12.1. Since this calculation makes no assumption concerning the magnitude of Γ, and since '$L_s = 0$' is a special case of the indirect matching technique, (6.94) equally applies in this case[15]. Hence, a series resonant input structure (Fig. 6.26(b)) features *the same* noise factor as a structure which is matched to

[14] The same information is also contained in Fig. 5.8.

[15] In order not to complicate things, we confine ourselves to the dominant noise terms, $F_d - 1$ and $F_{g,u} - 1$. The effect on the second-order noise terms, $F_c - 1$ and $F_L - 1$, can readily be deduced from (6.94), (6.96) and (6.97).

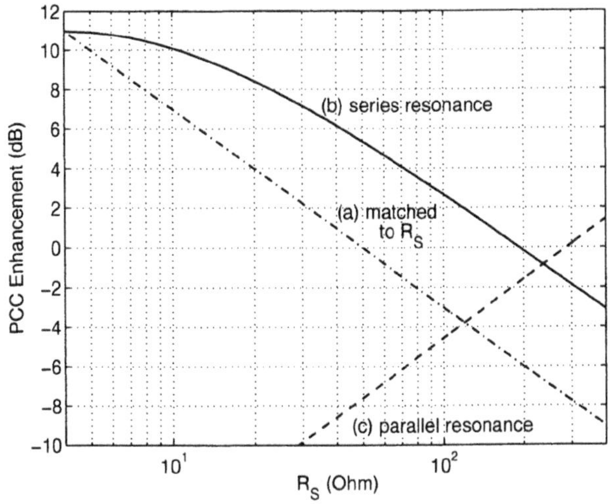

Figure 6.27: *Comparing the PCC of three different input structures: **(a)** input matched to R_S, **(b)** series resonant input structure, **(c)** parallel resonant input structure.*

R_S (Fig. 6.26(a)), i.e. (5.89):

$$F_{SR} \cong 1 + \frac{\gamma}{\alpha}\left(\frac{\omega_0}{\omega_T}\right)^2 \left(\sqrt{g_m R_S} + \frac{1}{\kappa\sqrt{g_m R_S}}\right)^2 + \frac{\alpha\delta(1-|c|^2)}{\kappa g_m R_S}. \tag{6.104}$$

This is a very important conclusion, because it means that the series resonant input structure can offer a much higher gain at an identical noise figure. If this gain had been generated with a structure which was matched to $R_S/2$, the noise figure would probably have been worse (Section 5.7.3). This clearly shows that power matching is *not* a requirement to achieve a good performance; Noise figure as well as gain derive *no* benefit from an power match.

6.12.3.2 Parallel Resonant Input Structure

The parallel resonant input structure is shown in Fig. 6.26(c). When

$$R_S \ll r_{g.NQS}\left(\frac{\kappa\omega_T}{\omega_0}\right)^2, \tag{6.105}$$

the *PCC* and the noise factor are given by

$$PCC_{PR} = 4g_m^2 R_S \tag{6.106}$$

$$F_{PR} \cong 1 + \frac{\gamma}{\alpha}\frac{1}{g_m R_S} + \frac{\delta\alpha}{\kappa}\left(\frac{\omega_0}{\omega_T}\right)^2 (1-|c|^2)\, g_m R_S. \tag{6.107}$$

The *PCC* is plotted by the dashed line in Fig. 6.27. Obviously, at low to moderate R_S values, a parallel resonant structure performs very badly. However, when R_S is large, the parallel resonant structure outperforms both the matched and the series resonant input structure. The same applies

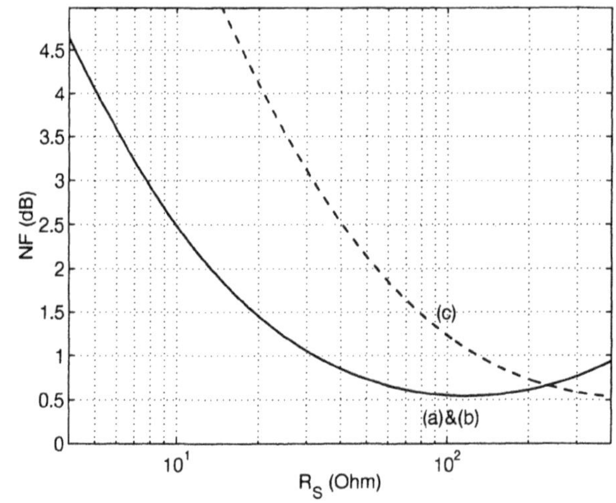

Figure 6.28: *Comparing the noise figure of three different input structures: (a) input matched to R_S, (b) series resonant input structure, (c) parallel resonant input structure.*

to the noise figure; As shown in Fig. 6.28(c), the noise figure is excessive at low R_S values, but at large R_S values the circuit features a lower noise figure than both the matched and the series resonant input structure. Hence, the same conclusion as in 6.12.3.1 applies: unmatched structures sometimes offer a better performance than fully matched structures.

6.13 Systematic LNA Design: The Case Study Revisited

In this section, the case study of Section 6.9 and Section 6.10 is resumed one last time to examine the effect of a finite reflection coefficient on the performance of that particular LNA. This material should help to digest the theoretical treatment in Section 6.12.

6.13.1 From a 50 Ω Match to a 32 Ω Match

The systematic design procedure in Section 6.9 has resulted in a clean 50 Ω matched LNA which achieves the stringent specifications of Table 6.1 at a current consumption of 4.8 mA (see Table 6.6, first row).

Still, Section 6.10 demonstrates that the direct 50 Ω match is not the optimum solution: when an extra (external) matching network is allowed, the power consumption can still be improved upon. In this particular case, the optimum source resistance turns out to be 32 Ω. In short, the reasoning that led to this value is as follows:

- Inspection of the contours in Fig. 6.18 reveals that the 50 Ω matched LNA — design (d) — is fully determined by the constraints on Q and L_g. The IP_3 spec, on the other hand, is far from binding. Strictly speaking, also the noise figure spec isn't. Consequently, any

$R_{S,local}$ Ω	R_{in} Ω	NF dB	IP_3 dBm	Q	L_g nH	I_{DS} mA	$V_{GS}-V_T$ V	W/L	g_m mS	C_{gs} fF
50	50	✓ 1.75	✓ 3	✓ 2.4	✓ 8	4.8	0.125	390/0.25	50	490
32	32	✓ 1.3	✓-4	✓ 2.4	✓ 8	2.6	0.107	265/0.25	30	325
50	32	✓ 1.5	✓-3.7	✓ 2.3	✓ 8	2.6	0.107	265/0.25	30	325

Table 6.6: *A comparison between the optimal solution for a 50 Ω R_S (and 50 Ω R_{in}), the optimal solution for the optimum R_S of 32 Ω (and 32 Ω R_{in}), and the combination of a 50 Ω R_S with a 32 Ω R_{in}.*

improvement in the current consumption requires that the 'Q = 2.4' and 'L_g = 8 nH' contours are shifted down.

- Since the quality factor of the LNA is dominated by the Q of the output section — the Q and the conversion efficiency of the input section is low due to the pad capacitance — a possible solution consists in decreasing the effective source impedance by transforming the 50 Ω source down to a *lower* impedance value, and subsequently matching the input device to this local impedance level. Due to the improved conversion efficiency of the input stage, the Q of the output section can be reduced, lowering the overall Q. Also the required L_g becomes smaller because the input impedance becomes more capacitive.

- The effective source impedance can not be reduced arbitrarily because this would result in a too low IP_3. Hence, at the optimal impedance, the IP_3 specification must necessarily be binding. Fig. 6.21 depicts the contours for the optimal intermediate impedance level — which is 32 Ω for this set of specs. In this case, only 2.6 mA is required to fulfill all the specs of Table 6.1! Observe that the IP_3 constraint has now become binding as well. The noise figure, on the other hand, is better than ever since $R_{S,eq}$ has decreased from 185 Ω— where the classical noise term ($\propto R_{S,eq}$) is dominant — to 97 Ω— which is close to the optimum source resistance.

The properties of this LNA are summarized in the second row of Table 6.6.

6.13.2 Direct Connection between the 50 Ω Source and the 32 Ω Input

In a highly integrated solution, external components are preferably avoided; So, we should at least try to get rid of the extra matching network which transforms the 50 Ω source down to the 32 Ω LNA input. In this regard, two questions must be answered: "Is it allowed to just omit the matching network?" and "What is the impact on the LNA performance? Won't it be destructive?". In this case, the answer to the first question is positive: the 50 Ω source may — technically speaking — be directly connected to the 32 Ω LNA input; A direct connection still yields an S_{11} of -12 dB, which complies to the -10 dB blocking filter spec. Concerning the performance, there is no problem either: omitting the matching network has only a small — and even positive — influence on the contour lines of the binding constraints. Qualitatively[16], the performance of the LNA can be calculated as follows...

[16] Deriving analytical equations that *exactly* describe what happens when there is a finite Γ is very difficult due to the non-zero C_p.

- Because of the -12 dB S_{11}, still 94% of the incident power is flowing to the transistor. As a consequence, the *PCC* of the amplifying device is only 0.28 dB lower than when the matching network had been present At the output this translates into a 3% larger load resistance, and hence in a 3% larger output Q. At the input, things are somewhat more complicated; Using (6.19), and assuming that the resonance between the reactive components is more or less kept, it can be calculated that the effective source resistance is now approximately 152 Ω instead of 97 Ω. Substituting these values in (6.100) reveals that the input Q is roughly 23% lower. The above considerations result in a net Q of approximately 2.3.

- According to (6.99), the IP_3 improves by 0.28 dB to -3.7 dBm.

- The effective source resistance increases from 97 Ω to 152 Ω, so that the net noise figure value increases to about 1.5 dB.

- By construction, the L_g value remains exactly the same: the matching network of the device itself is still designed towards the original 32 Ω source match.

Consequently, the compact solution with the 32 Ω input directly connected to the 50 Ω source performs comparable to the 'optimal' system with an intermediate reference plane matched to 32 Ω! Moreover, the power efficiency of this solution is nearly two times better than the solution with a clean 50 Ω match.

6.14 Conclusion

In this chapter, a systematic design methodology has been developed for the design of power-to-voltage and power gain LNAs. The method is based on a bank of equations describing the performance of the LNA in the whole design space — including all non-idealities. Contour lines are used to efficiently identify un-realizable, inactive and binding constraints. In this way, a lot of insight can be obtained in the overall specification dynamics and the trade-offs involved. In particular, the method reveals the precise impact of a constraint on the optimization goal: power minimization. Since the method does not follow a preprogrammed design trajectory — the design simply follows from the inspection of the design space between the contour lines demarcating the constraints — the method can be used for a broad range of LNA specs. Therefore, the procedure lends itself perfectly for incorporation in a CAD tool.

The strategy is illustrated by means of the design of a power-to-voltage LNA for the DCS-1800 cellular system. The silicon version of this LNA is embedded in the 0.25 μm CMOS, 2V DCS-1800 receiver found in Chapter 9. The method can equally be used to design power gain LNAs, of which the 0.8 dB *NF* LNA in Chapter 8 is an example.

The most important observations in this chapter are as follows:

- The performance of an LNA is strongly affected by the parasitic capacitance C_p originating from the bonding pad, the protection diodes, the C_{gd} of the amplifying device etc.:

  - C_p significantly lowers the attainable input impedance of the LNA, especially at high frequencies. Therefore, it may no longer be possible to inductively match a device to a given source impedance R_S without using an extra matching section. At 1.8 GHz, a 350 fF input capacitance is already enough to make a classical match impossible.

- The presence of C_p decreases the power-to-current conversion (PCC) of the amplifying device. In order to obtain a certain gain, the lower PCC must be compensated by a larger load impedance, making the circuit more sensitive to process variations.
- C_p radically changes the balance between the classical noise and the non-quasi static noise; Non-quasi static noise is pushed to extremely low R_S values so that the noise at normal R_S values is almost certainly determined by the classical drain noise.

  As long as R_{opt} — the optimum source resistance for minimum noise figure — is larger than R_S, C_p can in principle be used to transform R_S into R_{opt}, yielding a quasi perfect noise match! Unfortunately, often the default C_p already transforms R_S into an impedance beyond R_{opt}, so that any additional C_p comes at the cost of extra noise figure.

- In practice, inductively matching an LNA towards a low impedance source (e.g. a 30 Ω source) yields a better overall performance than matching to the typical 50 Ω source:
  - Because of C_p, the classical noise is usually dominant for a 50 Ω source, so that a lower source impedance indeed improves the noise figure. In principle, the lowest noise figure for the device is obtained by lowering R_S until we meet R_{opt}.
  - By lowering the impedance of the input reference plane, the same gain can usually be obtained at a lower overall LNA quality factor.
- As stated previously, the best solution for an LNA is to match its input towards a low impedance source (e.g. 30 Ω). Since the physical source impedance is usually larger (e.g. 50 Ω), this requires an extra matching network M, transforming R_S down to $R'_S = R_{in} =$ 30 Ω. However, even when omitting the extra matching network and directly connecting R_S (50 Ω) to R_{in} (30 Ω) still a better performance is obtained than if the device had been directly matched to R_S (50 Ω).
  - A larger gain can be achieved than if R_{in} was simply made equal to R_S; The finite Γ hardly degrades the gain compared to the fully matched case (i.e. $R'_S = R_{in}$), provided that Γ is less than -12 dB. Additionally, at a Γ of -12 dB only 6% of the incident power is being reflected, which means that a filter can still be inserted before the circuit without degradation of the filter characteristic.
  - The noise is determined by the value of the effective source impedance $R_{S,eq}$. Since $R_{S,eq}$ in the unmatched case is smaller than or equal to the $R_{S,eq}$ when the device is actually matched to R_S, and assuming that the classical noise is dominant, the noise figure is the same or even slightly lower than if the device was directly matched to R_S. A low impedance match would have been better but comes at the cost of an additional matching network M.
- Reflecting structures sometimes offer a better performance at a given source resistance. Input matching is therefore not required to achieve a good performance.

- When either the parasites are low, the source resistance is low, or the operating frequency is low compared to the capabilities of the technology, the input-referred drain noise may become so low that the non-quasi static noise *is* exposed.

Chapter 7

A 0.25 μm CMOS Receiver Prototype for DCS–1800 Cellular Communications

7.1 Introduction

In this chapter, a first attempt is made to combine different RF building blocks to build a low-IF DCS-1800 receiver front-end that is embedded in a highly-integrated CMOS cellular transceiver prototype. During the design, we will get acquainted with the practical problems and the constraints that are encountered when putting these building blocks together.

First, Section 7.2 describes the circuit topologies of the low noise amplifier and the down-converter embedded in the receiver prototype. Next, Section 7.3 explains how these building blocks have been designed. The actual realization along with its experimental results is covered in Section 7.4 and Section 7.5. Finally, some limitations of the current circuit topology are discussed, limitations which will largely be solved in the full-fledged DCS-1800 receiver of Chapter 9.

7.2 Receiver Topology

The low-IF receive path consists of a 1.8 GHz low noise amplifier which is connected to a quadrature down-conversion mixer. Behind the mixer some low-pass filtering is implemented as well. Fig. 7.1 shows the circuit schematic of the receiver front-end.

7.2.1 Low Noise Amplifier

The LNA circuit (Fig. 7.1, top left) consists of a single-stage resistor-loaded cascode amplifier with a low-Q voltage-boosting input section based on the resonator formed with the input inductor. This inductor can be realized as a bondwire. The low quality-factor of the resonance is ensured by the source resistance and makes the transfer function relatively broadband, and hence, insensitive to variations in the bondwire inductance L_g. Provided that the second-order input section is in series resonance at the operating frequency, a gain boost (*PCC* boost) of up to 6 dB can be obtained with respect to a perfectly matched system. at the cost of a worse reflection coefficient (see (6.103)). The LNA is resistor-loaded to avoid miscentering of the gain.

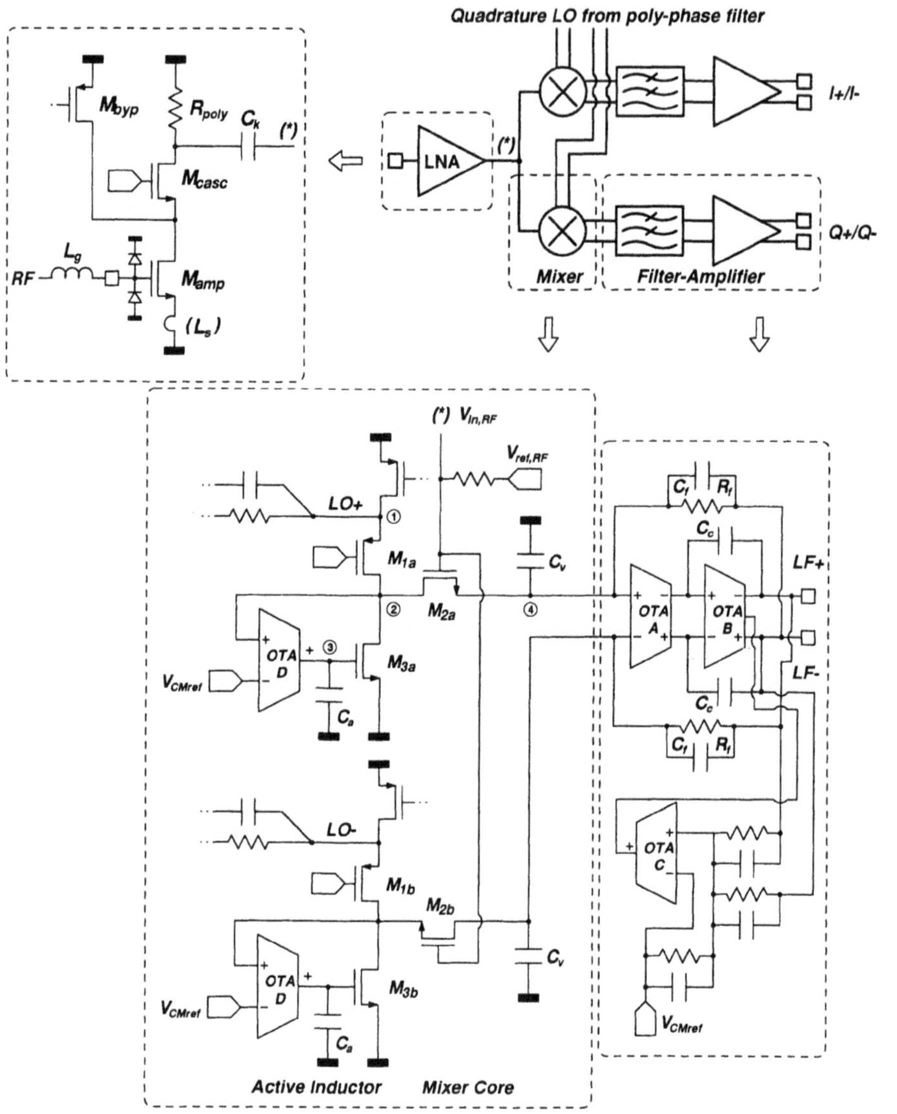

Figure 7.1: *Circuit schematic of the receiver front-end. The schematics of the OTAs can be found in Fig. 7.2.*

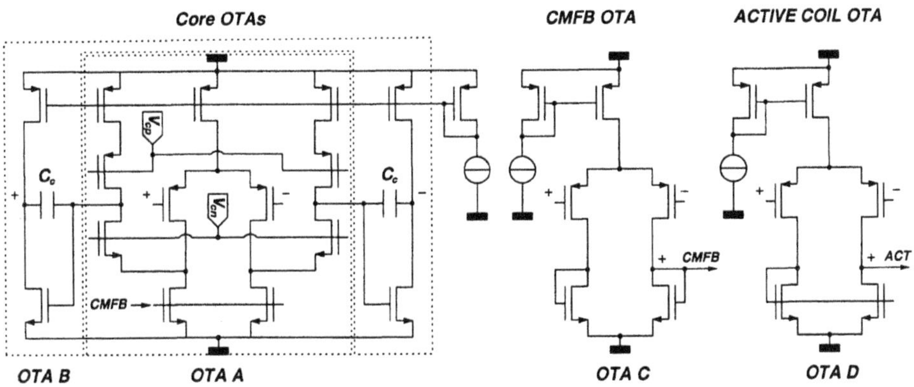

Figure 7.2: *Circuit schematics of the mixer OTAs.*

The output node of the circuit is screened from the drain of amplifying transistor M_{amp} by cascode transistor M_{casc}. This significantly decreases the capacitance on the output node, enabling a higher impedance level at this node and thus a lower power consumption. On the other hand, the drain capacitance of M_{amp} is brought on a low impedance level $(1/(g_m + g_{mb})_{casc})$, which pushes the pole on the intermediate node towards higher frequencies. The cascode transistor is also essential to reduce the Miller effect on the input node and to provide enough reverse isolation, preventing the LO from leaking to the antenna.

Since resistor-loaded amplifiers tend to require a lot of current — the output impedance is restricted by bandwidth considerations — the voltage drop over the load resistor can become considerable. To prevent this voltage drop from determining the minimum power supply voltage, a substantial part of the current is steered through bypass transistor M_{byp}. Since the $V_{GS} - V_T$ of M_{byp} can be made very large, adding this transistor has practically no influence on the parasitic capacitance at the drain node of M_{amp}. Evidently, the current fraction that is steered through M_{byp} can not be made arbitrarily high as it lowers the pole on the cascode node and limits the output swing.

7.2.2 Down-converter with Active Inductor LO Interface

Fig. 7.1 shows the schematic of one of the phases of the quadrature down-conversion mixer and its local oscillator interface. The detailed schematics of the OTAs can be found in Fig. 7.2.

The mixer core is based on the linear mixer presented in [Crol95a]. Instead of the four-transistor cross-coupled structure, only two nMOS transistors (M_{2a} and M_{2b}) — modulated in the linear (triode) region — are used per down-converter. Observe that the gates of the mixer transistors are all connected to the same RF phase; the single-ended to differential conversion of the RF signal is implicitly performed during the mixing process through multiplication with the differential LO phases. Hence, the mixer does neither require a differential LNA nor high-frequency single-ended to differential converters in the signal path.

The demodulated current is fed to a fully differential trans-impedance amplifier which converts the current into an output voltage. At the same time, this trans-impedance amplifier provides a low-frequency virtual ground to the mixer core so as to minimally affect the linearity and the conversion efficiency of the mixing process. At high frequencies the virtual ground capacitors C_v take over. The parasitic mixing product of the LO signal with the DC biasing voltage of the mixer is suppressed by the virtual ground capacitors C_v and the feedback capacitors C_f; This time it is not canceled by cross-coupling since each mixer core consists of only two transistors. The same applies to the square of the LO: the signal at $2f_{LO}$ is filtered by C_v and C_f, and the DC term is rejected by the common mode feedback of the trans-impedance amplifier. The feedback elements, R_f and C_f, combined with C_v, perform second-order low-pass filtering of the blocking levels, preventing saturation at the output of the mixers. This filter also functions as anti-aliasing filter for the oversampled data-converters. In this schematic, a fixed amplification is implemented in the low-pass filter. An AGC function could be implemented by replacing resistor R_f by a bank of switched resistors.

For quadrature down-conversion, the mixer needs to be driven by a quadrature local oscillator signal, e.g. coming from an RC poly-phase filter. Directly connecting the mixer transistors to the RC array of the quadrature generator is not a good idea as the injection of down-converted current might distort the balance in the quadrature generator. Therefore, a uni-directional LO interface is required: one that passes signals in one direction and isolates in the other direction. In this design, the interface is implemented as follows; Transistor M_1 presents a low impedance node (①) to the RC array of the poly- phase filter and level-shifts (folds) the LO current to the drains of the mixing transistors. The folding operation allows the common mode level of the down-converter to be equal to half the supply voltage, maximizing the available output swing. This property is difficult to realize with a source follower. Moreover, an important drawback of using followers is that the baseband currents resulting from the down-conversion process can induce parasitic currents at the double frequency because they are injected in a predominantly quadratic transistor. This gives rise to parasitic baseband terms.

Apart from providing enough LO drive, the interface also needs to suppress the low frequency voltages that are generated at the drain nodes of the mixing transistors due to the injection of down-converted current; Since the circuit acts as an inverting amplifier, any voltage on these nodes is amplified towards the output and counteracts the down-converted signal. In addition, these voltages may generate extra unwanted mixing terms. Therefore, a conducting path must be provided to the low-frequency current so that this current can flow unhindered to ground (and hence to the trans-impedance amplifier). Note that this problem does not pose itself in cross-coupled mixers, as the low-frequency current components are ideally canceled out in that topology.

In order to lead the injected currents to ground, an active inductor circuit (gyrator) is added to the buffer between the poly-phase filter and the drains of the down-conversion mixers. The active coil interface works as follows. An OTA senses the low-frequency voltage excursions of the drain nodes of the mixing transistors and feeds back a low-pass filtered, amplified version of the signal to the same node via M_3. In this way, a low-ohmic path is created for the low-frequency injected currents. Due to the internal pole caused by C_a at node ③, the impedance on node ② behaves inductively for higher frequencies; For all purposes, this circuit acts as a coil with a small series resistance. At radio frequencies, the feedback mechanism is disabled, so that the LO current is fully available for the down-conversion mixers.

7.3 Design Considerations

7.3.1 Down-converter and Active Inductor LO Interface

The multiplication process in the mixer transistors is based on the fact that a MOS device in the linear region (triode region) behaves as a voltage-controlled resistor:

$$i_{DS}(t) = \beta v_{DS}(t)\left(v_{GS}(t) - V_T - n\frac{v_{DS}(t)}{2}\right). \tag{7.1}$$

The best way to employ this mixing principle is to force the LO voltage at the drain by an ideal voltage source, and apply the RF signal at the gate. This leads to the most linear conversion. Yet, generating a low-impedance LO at the drains is far from obvious.

Fortunately, the mixer can also operate by forcing an LO current through the mixer transistors. This mechanism is employed here. Though it may seem weird at first sight, the linearity of the conversion process can be good as well. The reason is that any parasitic high-frequency current which is temporarily generated through non-linearities can not flow into high impedance node ②, so that it gives rise to a counteracting voltage. This reasoning holds as long as the impedance of the mixer transistors is sufficiently small compared to the rest of the impedance on node ②.

In the current-driven mixer, the saturation current of the mixer transistors needs to be larger than the LO current in order to remain in the triode region:

$$\frac{\beta}{2n}(V_{GS} - V_T)_2^2 \ll I_{LO}, \tag{7.2}$$

where I_{LO} is the amplitude of the LO current that is injected into the mixer transistors. Otherwise the linearity of the mixing process is degraded. Under these conditions, the conversion gain is calculated as

$$CVG \cong \frac{R_f I_{LO}}{(V_{GS} - V_T)_2}, \tag{7.3}$$

Observe that the gain can be controlled by adjusting the resistor R_f.

By dividing the white output noise power of this structure by the square of the conversion gain, the noise figure due to white noise can be calculated as

$$NF \cong 10\log_{10}\left(\frac{\beta(V_{GS} - V_T)_2^3}{25 I_{LO}^2}\right). \tag{7.4}$$

At first sight, the noise figure drastically improves with decreasing $V_{GS} - V_T$. However, the β must increase aggressively to still fulfill the linearity constraint in (7.2), largely counteracting the effect of the decreasing $V_{GS} - V_T$. It is therefore much more realistic to look at the noise figure dynamics while keeping the relative degree of saturation, given by

$$\rho_{sat} \triangleq \frac{\beta(V_{GS} - V_T)_2^2}{2n I_{LO}}, \tag{7.5}$$

constant. This yields

$$NF \cong 10\log_{10}\left(\frac{2n\rho_{sat}(V_{GS} - V_T)_2}{25 I_{LO}}\right). \tag{7.6}$$

Consequently, for a given linearity and for a given I_{LO}[1] it is best to use a relatively low $V_{GS}-V_T$ and hence a large β. In our design, the LO current that flows through the mixer transistors is 1.3 mA, 3 dB [2] less than the current injected into M_1. The $V_{GS}-V_T$ of transistor M_2 has been set to 0.35 V and its (W/L) to 55/0.25. As a result, the transistor features an I_{DSsat} of more than 3 mA and a gate capacitance of 110 fF. This yields a white noise figure of about 18 dB for each of the mixers. Further lowering the noise figure by decreasing the $V_{GS}-V_T$ is virtually impossible as the required $(W/L)_2$ increases quadratically with decreasing $V_{GS}-V_T$ in order to comply to (7.2).

The sizing of the active inductor needs to be done very carefully for $1/f$ noise reasons; In contrast to the externally injected noise components that are actively drawn away from the mixer input nodes, the internally generated low-frequency noise of the gyrator loop propagates to the output. Mainly the area of the differential pair in the gyrator's OTA (OTA D in Fig. 7.2) needs to be large enough in order not to generate too much $1/f$ noise and DC offset. However, one can not go arbitrarily far in increasing their WL; If the differential pair in the gyrator has a too large WL, the injected LO current leaks away through the parasitic gate-source capacitance, decreasing the conversion gain of the mixer and therefore increasing the total noise figure. Indeed, the expression for the equivalent in-band input noise power of the mixer due to $1/f$ noise is approximately given by

$$dv_{in}^2 \cong \frac{KF_p}{WLC_{ox}} \log\left(\frac{200\text{ kHz}}{1\text{ kHz}}\right) \frac{A_0^2}{CVG_0{}^2}\left(1+\frac{(WL)^2}{(WL)_p^2}\right), \tag{7.7}$$

where the first factor denotes the equivalent $1/f$ input noise power of the active coil (KF_p = 7.9e-30 C^2/cm^2), A_0 is the gain of the mixer when it is used as a differential amplifier, CVG_0 is the conversion gain of the mixer when the capacitive load due to the active coil is zero and $(WL)_p$ is the critical area of the OTA input transistor that generates a pole at 1.8 GHz on node ②. The optimum WL can readily be derived and is given by

$$(WL)_{opt} = (WL)_p. \tag{7.8}$$

In this design, the optimum WL is given by $160 \times 0.25\ \mu m^2$, corresponding to a load capacitance of about 210 fF. It is important to note that for a given WL, the maximum W/L ratio must be selected, because this leads to the lowest white noise contribution for a given bias current. Note that this is the basic reason why the current coming from the poly-phase is attenuated by 3 dB. In the end, the active inductor provides a resistive impedance of 5 Ω at low frequencies. Due to the internal pole at 2 MHz— caused by a 20 pF C_a — the node impedance behaves inductively for higher frequencies. The complete circuit acts as a 0.4 μH coil with a series resistance of 5 Ω.

The OTAs in the trans-impedance amplifier consist of folded-cascode, miller compensated two-stage OTAs (see Fig. 7.2), yielding a high quality virtual ground. Each two-stage OTA draws 8 mA from a 2.5 V supply. For the detailed equations regarding the design of these OTAs, the reader is referred to 9.3.1.5, 9.3.1.6 and page 203 in Subsection 9.3.2.

---

[1] I_{LO} is limited by the impedance of the poly-phase filter and the transconductance of M_1.

[2] The 3 dB comes from the loss at node ②.

7.3.2 Low Noise Amplifier

The limiting factor in the design of this kind of LNA is often the gain. Since the load is a simple resistor and the capacitive parasites are not canceled by a parallel inductance, achieving a sufficiently large LNA gain requires a large *PCC* from the amplifying device. To minimize the power consumption, the value of the load resistor must be maximized subject to the output bandwidth constraints. Since the total input capacitance of the I and Q mixers sum to 600 fF, the load resistor is limited to 92 Ω. This value sets the 3 dB bandwidth of the load impedance to about 3 GHz, leading to a relative loss of about 1.5 dB at 1.8 GHz.

At series resonance, the *PVC* of the LNA core is given by

$$PVC_{SR} = \frac{4R_S R_{L,eff}^2}{(R_S + r_{g,NQS})^2}\left(\frac{\omega_{T,eff}}{\omega_0}\right)^2, \tag{7.9}$$

where $\omega_{T,eff}$ is given by

$$\omega_{T,eff} = \frac{g_m}{C_{gs}(1 + M\alpha_{gd}) + C_p}, \tag{7.10}$$

with C_p the additional parasitic capacitance on the input node. Since $r_{g,NQS}$ can safely be neglected — g_m will turn out to be large anyway — realizing gain at a minimum power consumption means achieving the $\omega_{T,eff}$ at minimum current.

In principle, there exists an optimal size for the input device which minimizes the current consumption; Analogously to the derivation in Subsection 6.8.1, it can be shown that the optimum device size is defined by

$$(C_{gs})_{opt} = \frac{C_p}{1 + M\alpha_{gd}}. \tag{7.11}$$

In this design, C_p equals 400 fF (270 fF from the bondpad, and 130 fF from the protection diodes and the biasing resistor) and $M\alpha_{gd}$ is about 0.4. The total input capacitance for minimum power thus equals 800 fF. However, series resonance at 1.8 GHz requires an inductor of 10 nH which is quite large. Therefore, the C_{gs} of M_{amp} has been set to 500 fF [3]. Together with a 5.5 nH inductor — which is a more reasonable value — this leads to a flat gain at 1.84 GHz. To achieve the specified gain, a g_m of 85 mS is required, which for a C_{gs} of 500 fF leads to a current of 14.5 mA at a $V_{GS} - V_T$ of 0.25 V. This $V_{GS} - V_T$ value is sufficiently large to obtain an IP_3 of -1 dBm in this configuration. In addition, the large g_m makes that non-quasi static noise is negligible while the large $\omega_{T,eff}$ keeps the classical noise contribution low (despite the large g_m):

$$F_{SR} \cong 1 + \frac{\gamma}{\alpha}\left(\frac{\omega_0}{\omega_{T,eff}}\right)^2 g_m R_S. \tag{7.12}$$

As a result, the total noise figure of the LNA — including bondpad parasites — is less than 2.5 dB.

If the 14.5 mA current would flow through the load resistor, a voltage headroom of 1.3 V would be required. This problem is overcome by steering part of the current trough M_{byp}. In this design the current is equally divided between the load resistor and pMOS transistor M_{byp}. The

[3] increasing C_p also lowers the required inductance, but would boost the current requirements.

$(I_{DS})_{amp}$	14.5 mA	$(I_{DS})_{casc}$	7.5 mA
$(V_{GS}-V_T)_{amp}$	0.25 V	$(V_{GS}-V_T)_{casc}$	0.25 V
$(W/L)_{amp}$	400/0.25	$(W/L)_{casc}$	200/0.25
$(g_m)_{amp}$	85 mS	$(g_m)_{casc}$	44 mS
$(C_{gs})_{amp}$	500 fF	$(C_{gs})_{casc}$	260 fF
$(W/L)_{byp}$	90/0.25	$C_{bp}+C_{protect}$	400 fF
$(V_{GS}-V_T)_{byp}$	0.6 V	L_g	5.5 nH
R_{poly}	92 Ω	C_k	2 pF

Table 7.1: *Transistor sizes and component values of the LNA.*

width of transistor M_{casc} is chosen so that the cascode pole and the output pole are optimally positioned. This leads to a slightly smaller aspect ratio than the one which follows from (6.54):

$$(W/L)_{casc} = 0.36(W/L)_{amp}. \tag{7.13}$$

The resulting transistor sizes are summarized in Table 7.1.

7.4 Realization

The receive path discussed in this chapter has been embedded in a full-CMOS DCS-1800 transceiver prototype [Stey98, Stey00b]. Fig. 7.3 shows the micro-photograph of the transceiver IC, realized in a standard 2M1P 0.25 μm CMOS process. The transceiver integrates LNA, down-converters, VCO, quadrature generator, up-converter and pre-amplifier on a single 8.6 mm^2 die. A high level of integration is achieved by using the low-IF topology for reception, a direct quadrature up-conversion topology for transmission and an oscillator with on-chip integrated inductor. The final objective of this design is to develop a highly-integrated solution for the DCS-1800 cellular system that can be built by flanking a CMOS transceiver chip by a minimum of surrounding components: only an antenna, a duplexer, a power amplifier and a baseband signal processing chip.

The receive path, comprised of the LNA, the down-conversion mixers and the active coils, can be discerned at the left. A large part of the area is occupied by the capacitance of the active inductors and the capacitors on the virtual ground. The rest of the capacitors (e.g. near the bonding pads of the down-converter and in the LNA) function as decoupling capacitance. The RF input bonding pad of the LNA has a grounded n-well underneath it to prevent capacitive injection of parasitic high frequency signals on this node. This is important for reducing LO feed-through. In addition, the bonding pad capacitance has been reduced somewhat by using an octogonal layout.

7.5 Experimental Results

To measure the transceiver prototype in receive mode, the IC is mounted die-on-board on a thick film alumina substrate and placed in a CuBe shielding (see Fig. 7.4). The RF input signal is supplied through SMA connectors and a 50 Ω stripline. Unless stated otherwise, the supply voltage of the IC is set to 2.5V.

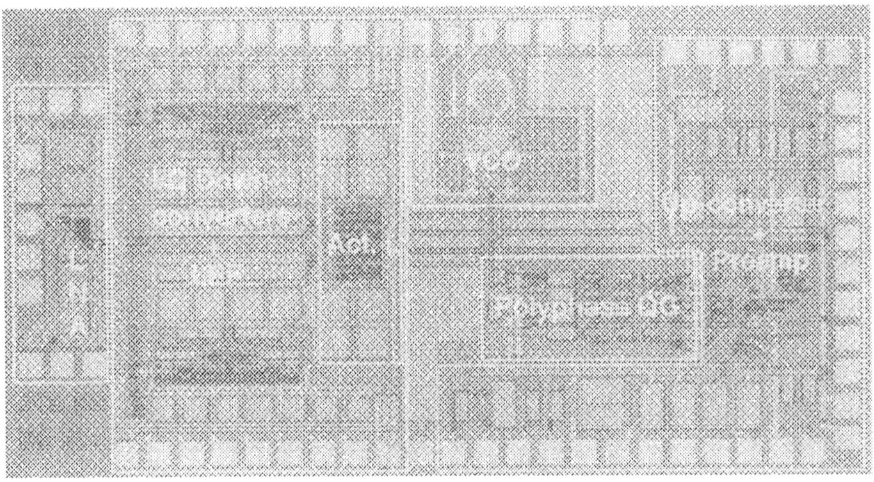

Figure 7.3: *Micro-photograph of the complete transceiver IC.*

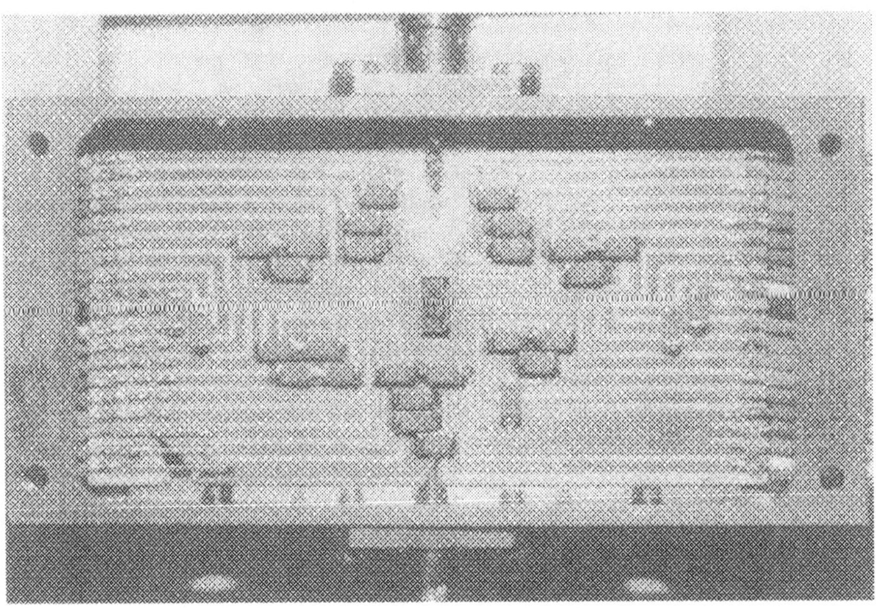

Figure 7.4: *The transceiver mounted on an alumina substrate in a CuBe box.*

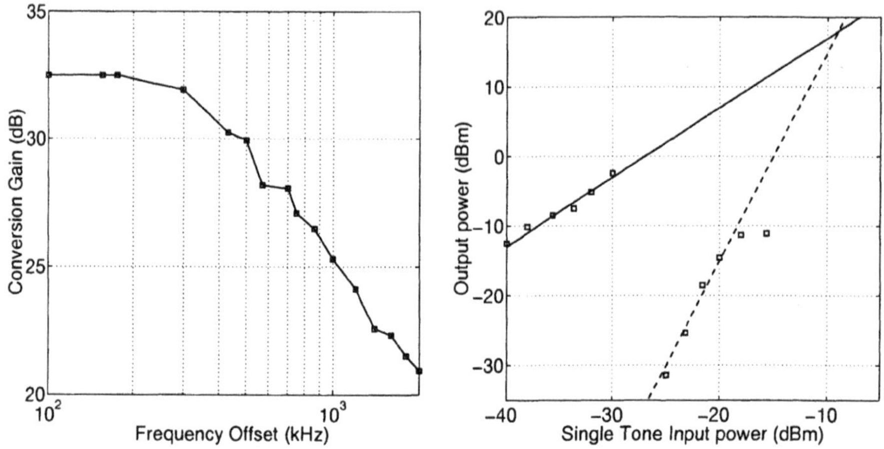

(a) Conversion Gain. (b) Third-Order Intercept Point.

Figure 7.5: *Gain and intermodulation characteristics.*

To measure the conversion gain characteristic of the receiver, the VCO reference voltage is adjusted to produce a 1.88 GHz LO signal. A 1.88 GHz $- x\ kHz$ signal is applied at the input terminal and is converted to baseband. Fig. 7.5(a) shows the measured conversion gain versus carrier frequency offset. The conversion gain is 32.5 dB in a $2 \cdot 530$ kHz band around the carrier.

The IP_3 of the receive path is extracted by applying two large out-of-channel interferers, positioned in the DCS band but out of the 1 MHz wide conversion channel, at a frequency offset of respectively 5 MHz and 9.9 MHz from the 1.88 GHz carrier. The generated third-order intermodulation product is down-converted to an in-channel low frequency signal which can be plotted versus the available power of a single input tone. Fig. 7.5(b) shows the results for the receive path at 2.0V. This is the worst case situation for the transceiver chip with respect to linearity; The reason for this is that at 2.0V the mixer transistors are biased at a very low overdrive voltage. Consequently part of the conversion is realized in chopping mode instead of in linear mode, because the LO drives the mixer transistors into the saturation region. Extrapolation of the linear output and the intermodulation product still yields an input IP_3 of -9 dBm.

The phase and amplitude errors between the I and Q channels are measured by connecting both channels to a sampling scope (see Fig. 7.6(a)). During the transceiver measurements, strong AM-band broadcast signals (mainly around 320 and 621 kHz) were perturbing the reference voltage of the open-loop VCO. This gave rise to parasitic AM modulation of the VCO amplitude, making reliable measurements of the quadrature accuracy for frequencies below 1 MHz very difficult. For that reason, the I and Q channel quadrature accuracy measurements had to be done at 1.3 MHz (i.e. $> 2 \cdot 621$ kHz). The measured amplitude and phase mismatch is respectively 0.41 dB and 0.2 degrees. This is sufficient to ensure the required mirror signal suppression (image rejection) of 32 dB (see Table 3.2).

The DC offset at the output of the receiver is about 100 mV and mainly originates from the amplified offset of the active coils.

The noise figure is measured by applying a 50 Ω noise source at the receiver input, measuring

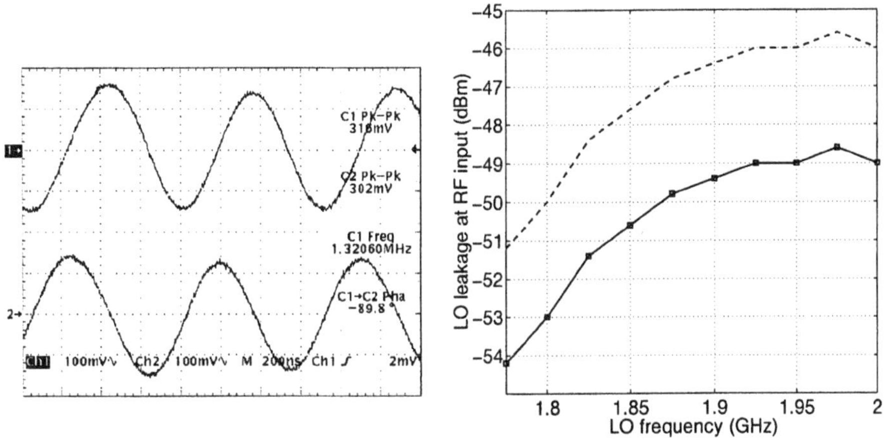

(a) LF Quadrature Accuracy. (b) LO leakage.

Figure 7.6: *Quadrature accuracy and spurious LO emission.*

the output noise power in a spectrum analyzer and referring it back to the receiver input. At 2.5V, the measured noise figure of the total receiver is 4.9 dB, which confirms that the front-end meets the sensitivity required by the DCS-1800 system (see Table 3.2).

To measure the LO leakage towards the antenna, the LNA input is directly connected to a spectrum analyzer. The local oscillator frequency is swept from 1.775 GHz to 2.000 GHz by adjusting the VCO reference voltage. Fig. 7.6(b) plots the LO power which is measured in the spectrum analyzer (dashed line). Because in the full system a duplex/blocking filter is inserted between the antenna and the LNA input, the net LO leakage will be lower by an amount equal to the insertion loss of the duplexer, which is nominally 3 dB. The net LO leakage is shown in Fig. 7.6(b) (solid line). As can be derived from the figure, the net LO power that is injected into the antenna is always lower than -48.5 dBm, which is within the DCS-1800 specifications (see Table 3.2).

The measured S_{11} of -7 dB can be improved by inserting some extra source inductance, decreasing the S_{11} while maintaining a good LNA noise figure. The overall noise figure will increase somewhat because the decrease in LNA gain.

The experimental results of the transceiver in receive mode along with the results of the rest of the transceiver building blocks [Stey98, Stey00b] are summarized in Table 7.2. The experimental results demonstrate that the key specifications of the DCS-1800 can indeed be met in CMOS.

7.6 Discussion

During the design of the receiver, several important issues have come up that limit the performance one way or the other. These aspects deserve some extra attention:

- In any receiver, there are nodes where the RF signals, the LO signals and the down-converted signals meet. Since there is necessarily a low-frequency conducting path be-

Receive Path	Type	Low-IF
	Conversion gain	32.5 dB
	Third order intercept point (IP_3)	-9 dBm
	Image rejection ratio (*IMRR*)	33 dB
	Amplitude accuracy	0.4 dB
	Phase accuracy	0.2 deg
	Noise figure (*NF*)	4.9 dB
	LO leakage	-48.5 dBm
	Reflection coefficient (S_{11})	-7 dB
	DC offset	100 mV
	Power consumption	76 mW
Transmit path	Type	Direct Up
	Output power	-10 dBm
	Spurious suppression	35 dBc
	Image suppression	35 dBc
	LO feedthrough	-42 dBc
	Power consumption	50 mW
VCO	Tuning range	26.9 %
	Phase noise	
	at 600 kHz offset	-123.6 dBc/Hz
	at 3 MHz offset	-136 dBc/Hz
	Power consumption	15 mW
Quadr. gen. & LO Interface	Type	Polyphase
	Power consumption	100 mW
	Supply voltage	2.5 V
	Total power consumption	240 mW
	Die area	8.3 mm^2

Table 7.2: *Summary of the measured transceiver specifications. Along with the experimental results of the receive path described in this work, the results of the VCO [Demu02] and the transmit path [Borr01] are shown as well.*

tween this node and the output, the DC offset and the $1/f$ noise generated at that node can propagate unhindered to the output. Since a low DC offset and low $1/f$ noise require large transistors, and therefore cause a large capacitive load, there is almost always a clash between low-frequency performance and RF performance. Yet, not much can be done about this as the optimum is very flat. Therefore, $1/f$ noise often dominates the noise figure.

- The high-frequency poly-phase filter that is used to generate the quadrature LO is low-ohmic. This makes the interface circuits quite power-hungry. In fact, in this transceiver the poly-phase dominates the total power consumption.
- The input impedance of the trans-impedance amplifier must not only be low to take up all the generated current, but also to make sure that the mixing process is not influenced. When going towards lower voltages, the gain of the OTAs will tend to drop, which may complicate the implementation of a VGA function.
- By not using a cross-coupled mixer structure, there is only a small margin between the measured LO leakage and the maximum allowable LO leakage.
- The load resistor of the LNA must be replaced by an on-chip inductor — be it a low-Q one; Its power consumption is too much determined by the gain.

These issues will largely be solved in the transceiver of Chapter 9.

7.7 Conclusion

In this chapter, the design and measurement of a first prototype of a full-CMOS DCS-1800 receiver has been presented. The receiver integrates the low noise amplifier, I/Q mixers with active coils and IF trans-impedance amplifiers in a standard 2M1P 0.25 μm CMOS process. The front-end achieves a conversion gain of 32.5 dB, a total noise figure of 5 dB, an IP_3 of -9 dBm and an image rejection ratio of 33 dB. The receiver (including its LO interface) draws 46.5 mA from a 2.5V supply.

The receive path has been embedded in a single-chip CMOS transceiver, which is — to the author's knowledge — the first transceiver demonstrating the feasibility of achieving cellular specifications in CMOS. A high level of integration is achieved by using the low-IF topology for reception, a direct quadrature up-conversion topology for transmission and an oscillator with on-chip integrated inductor. The final objective of this design is to develop a complete transceiver system for DCS-1800 wireless communications at 1.8 GHz that can be built with a minimum of surrounding components: only an antenna, a duplexer, a power amplifier and a baseband signal processing chip.

In Chapter 9, the design of a completely new receive path will be covered, this time for a full-fledged CMOS DCS-1800 transceiver.

Chapter 8

A 0.8 dB NF, ESD-protected CMOS LNA

8.1 Introduction

Apart from cellular standards there exist other challenging applications that might be a candidate for CMOS integration. An example of such an application is the Global Positioning System [GPS96]; This system will be an obligatory feature in the upcoming 3G terminals.

Yet, the performance requirements of a high-end GPS receiver are quite tough; To keep such a receiver from giving up in urban canyons or in woods, it needs to be extremely sensitive, which, in his turn, imposes severe demands on the noise figure and the gain of the LNA. To cope with these tough requirements, often high-performance GaAs MESFET low noise amplifiers are used, since they are capable of offering excellent noise figures in the order of 1 dB at large power gains of 20 dB. Using a GPS LNA as test vehicle, this chapter will prove that even in a standard submicron CMOS technology, an extremely low noise figure ($\leq$ 1 dB) and a high gain (20 dB) can be achieved at the same power consumption as commercially available GaAs LNA solutions.

The 0.25 μm CMOS LNA described in this chapter [LeJa01] offers a noise figure as low as 0.8 dB at a power gain of 20 dB while consuming only 9 mW! The presented LNA outperforms all previously published CMOS LNA's with respect to noise figure, gain and power consumption; It features the lowest noise figure, the second highest gain and a power consumption which is among the lowest published to date in CMOS. In addition, the IC contains an ESD protection on the RF input pad which is capable of protecting the LNA against -1.4 kV to 0.6 kV HBM pulses. This demonstrates that an excellent performance can still be achieved while at the same time providing > $\pm$0.5 kV ESD protection.

First, Section 8.2 and Section 8.3 review the frequency plan and the power levels of the GPS application, and of course, the typical requirements for a GPS LNA. The topology of the realized LNA is presented in Section 8.4. The actual design trade-offs and the layout optimizations are covered in Section 8.5 and Section 8.6. The experimental results and the ESD performance are discussed in Section 8.7 and Section 8.8, respectively. To conclude, Section 8.9 compares the realized LNA with existing state-of-the-art CMOS low noise amplifiers.

	P code	C/A code
L1	-133 dBm	-130 dBm
L2	-136 dBm	-136 dBm

Table 8.1: *Minimum GPS receive power levels.*

8.2 The GPS Frequency Plan in a Nutshell

The GPS signal is broadcast at two frequencies: a primary signal at 1.575 GHz (L1 band) and a secondary broadcast at 1.2276 GHz (L2 band). The transmitted information in these bands consists of a continuous 50 bps stream, containing data like e.g. the satellite location, the satellite time and the necessary clock corrections. Before transmission, this data is spread to a much larger bandwidth by multiplication by a wide-bandwidth pseudo-random (PRN) code, commonly known as direct-sequence spread spectrum modulation (DS–SS). In the receiver, the signal is de-spread by correlating it with an identical PRN sequence. Combining the received satellite data with the computed time of arrival then yields the position information.

At each GPS frequency, two different DS–SS modulations can potentially be present at the same time, each using its own spreading code [GPS96, p. 12]:

- The C/A code or coarse acquisition code, which is the code for civil use, providing the standard positioning service. It is a short, un-encrypted PRN code broadcast at a chipping rate of 1.023 MHz.

- The P code or the protected code, which is a code for military use only providing the so-called precise positioning service. This is a much longer, encrypted code at a chipping rate of 10.23 MHz.

Currently, the L1 band carries both the civil C/A code and the military P code whereas the L2 band only carries the military P code. Hence, at this moment civilian users can only determine their position using the L1 band. However, as requested by many civil users, by 2003 also the C/A code will be broadcasted on the L2 band. The presence of such an extra civil GPS signal offers some advantages; First of all, the signals in the L1 and the L2 band are generated synchronously. Hence, a user who is able to receive and decode the signals in both bands can in principle correct for the ionospheric group delay, improving the position accuracy [GPS96, p. 144]. In addition, the L2 band can serve as a backup to the conventional GPS L1 link.

8.3 GPS Power Levels and LNA Requirements

Even though the GPS satellites transmit about 50 W, due to the large distance between the GPS satellites and the earth's surface, the received signal power is extremely low. Table 8.1 shows the minimum specified received signal strength for a user receiver employing a 0 dBIC antenna [GPS96, p. 84]. The satellite is assumed to be at an elevation angle above 5 degrees. For civil GPS, the second column (the C/A code) is the relevant one. In the L1 band the minimum received power is -130 dBm (-160 dBW or 0.1 fW), which is — assuming a realistic antenna noise

Noise figure (*NF*)	1 dB	1.4 dB
Power gain (S_{21})	21.5 dB	18.5 dB
Input reflection coefficient(S_{11})	(-10 dB)	(-10 dB)
Output reflection coefficient (S_{22})	(-10 dB)	(-10 dB)
Reverse isolation ($-S_{12}$)	(30 dB)	(30 dB)
Third order intercept point (IIP_3)	(-8 dBm)	(-8 dBm)
Supply voltage	5 V	3 V
Current consumption	6 mA	3.3 mA
Power consumption	30 mW	10 mW

Table 8.2: *Data sheet of the Motorola MRFIC1501 GaAs GPS LNA at 1.2276* GHz *(GPS L2). The data between brackets were only specified at 1.575* GHz *(GPS L1) and 5* V *supply.*

temperature of 100 K[1] — 14.6 dB lower than the integrated antenna noise power of -115.4 dBm. In the L2 band, the minimum received power is even 6 dB lower, so that the signal power is 20.6 dB below the antenna noise. However, due to the processing gain in the digital back-end — basically the ratio 1.023 MHz/50 bps, which is 43.1 dB— the signal comes out of the background noise with a the theoretical SNR_{max} — i.e. assuming a perfect, noiseless receiver — of 28.5 dB and 22.5 dB in the L1 and the L2 band, respectively.

In practice, the *SNR* of the received signal is much worse; In urban canyons or when tree foliage shadows the user, the minimum received power often is much lower than the specified -130 dBm; The attenuation can be as high as 10 to 20 dB [GPS96, p. 576]. This highly degrades the *SNR* of the received signal, especially the *SNR* of the L2 signal since its power is 6 dB lower. In addition, the physical temperature of the receiver chip-set is in the order of 290 K, which generally is much larger than the antenna noise temperature. And, as if it weren't enough, the receiver features a finite noise figure, even with respect to a source at 290 K. As a result, the *SNR* of the signal after de-spreading becomes

$$SNR = SNR_{max} - attenuation - 10\log_{10}\left(1 + (F_{receiver} - 1)\frac{290\ \mathrm{K}}{T_a}\right), \tag{8.1}$$

where $F_{receiver}$ is the noise factor of the receiver — evaluated at 290 K — and T_a is the antenna noise temperature.

Hence, to keep the receiver from giving up early, the receiver noise figure must be very low, which in his turn poses severe demands on both the noise figure and the gain of the RF input amplifier. Since we are in a no-compromise situation, often high-performance GaAs MESFET low noise amplifiers are used, which are capable of offering noise figures in the order of 1 dB at gains of 20 dB. Today's GPS receivers generally contain the MRFIC1501, a GaAs LNA from Motorola [MRFIC98], whose performance is listed in Table 8.2. This IC is especially designed for the GPS application; Its maximum gain (21.5 dB) and its minimum noise figure (1 dB) is achieved near the L2 band. In practice, it is mostly used in the L1 band — the only civil GPS band to date —, where it features a gain of 18 dB and a noise figure of 1.1 dB. Hence, to prove the suitability of CMOS for building extremely sensitive receivers, one must demonstrate the

[1] An antenna temperature of 75–100 K is typical, depending upon the antenna pattern and the amount of ground temperature [GPS96, p. 344].

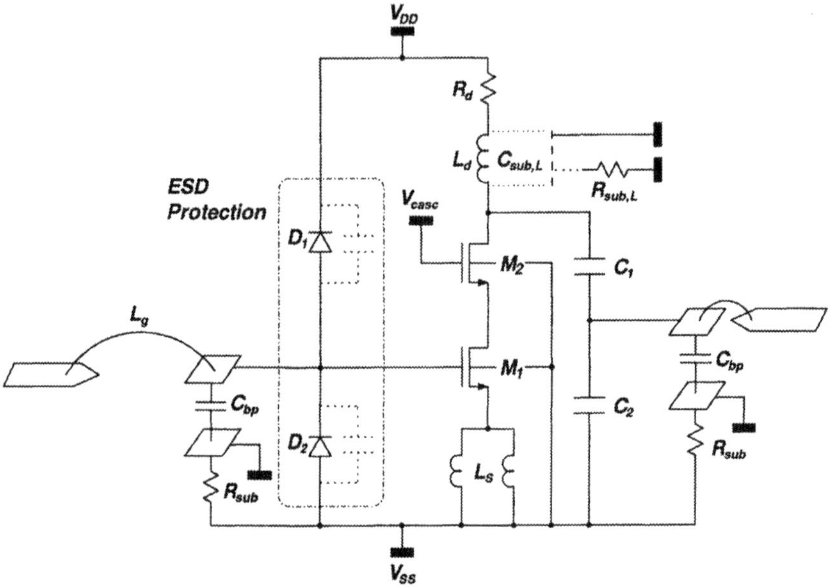

Figure 8.1: *LNA topology.*

feasibility of achieving very low noise figures ($\leq$ 1 dB) and large gains (18 to 20 dB) at a power consumption comparable to GaAs solutions. In this chapter, a CMOS LNA is developed which consumes less than 10 mW— the power consumption of the MRFIC1501 when it is biased at its lowest voltage — while offering a performance comparable to the MRFIC1501 at 30 mW!

8.4 Topology

Fig. 8.1 shows the proposed schematic of the CMOS GPS LNA. The topology is an instance of the template in Fig. 6.2(b) on page 103. Essentially, the schematic has already been discussed in detail in Section 6.3. Nevertheless, for the sake of completeness the most important properties are restated here. Where possible, we refer to the relevant sections for the mathematical explanation.

The LNA consists of a cascoded, inductively degenerated gain stage (M_1) that converts the antenna power into a current while presenting a 50 Ω input impedance to the antenna. The properties of this input stage (*NF*, *PCC*, IP_3) along with the influence of all sorts of parasitic effects have been the subject of Chapter 5 and Chapter 6. The LNA input is protected against ESD by two reverse-biased diodes. In case of a positive ESD pulse, the upper diode D1 leaks the ESD charge to the power supply while in case of a negative ESD pulse, the lower diode D2 leaks the charge to ground.

Cascode transistor M_2 drastically reduces the Miller effect on the C_{gd} by ensuring a low impedance at the drain of the amplifying device. This keeps the Miller effect from degrading the *PCC* as well as from increasing the input referred noise (see Section 6.6). In addition, the cascode improves the reverse isolation, increases the stability, and makes that the output matching network and the input matching network no longer influence each other.

Since the output current is offered at a high impedance level — let us call it Z_{out} — the output power is maximized by transforming Z_{out} down to the 50 Ω load. Evidently, the first element in this transformation network needs to be an inductor because it must carry the DC current from the power supply to the LNA core. Due to the limited quality factor of this inductor — i.e. its finite series resistance R_d —, the effective output impedance of the structure becomes significantly lower than Z_{out}; It is limited by the R_p of the load inductor, given by

$$R_p \triangleq (Q_L^2 + 1)R_d \cong \frac{\omega_0^2 L_d^2}{R_d}, \tag{8.2}$$

where R_d is the series resistance of the inductor. Matching the 50 Ω load towards the original output impedance Z_{out} would have been highly unpractical anyway; The selectivity of the LNA would have been enormous — making it extremely sensitive to process variations — and it is known that the simulated Z_{out} value is very unreliable.

Achieving a maximum power transfer requires that the 50 Ω load is transformed into the complex conjugate of the effective output impedance ($\cong R_p$) at the drain of the cascode; In other words, the matching network must transform the 50 Ω load into a resistive path with impedance R_p and at the same time generate the exact amount of parallel capacitance to cancel out the effective inductance at the drain of the cascode. Therefore, the rest of the matching network must contain two degrees of freedom. In this particular circuit, these degrees of freedom are offered by the quasi-lossless capacitive divider C_1/C_2 [Floy99]; In fact, for each realizable inductor — i.e. each combination of L_d, R_p and $C_{sub,L}$ — there exists a realizable combination of C_1 and C_2 values that provides the correct impedance, provided that

1. The inductor is not self-resonant at frequencies near or below the operating frequency.

2. The required C_2 value is larger than the sum of the parasitic capacitance of the output bondpad and the stray capacitance of C_1 towards the substrate.

For a given *PCC*, the required gain (G_t) can be obtained by scanning the design space for the inductor that offers the required R_p while complying to the above constraints. So far the description of the most important functional components. The next sections focus on the actual design and its optimized layout.

8.5 Design

The maximum performance of an LNA is generally set by the allowable bias current. Therefore, the initial design goal should be to try to achieve the same raw performance as the MRFIC1501 at its best (the left column of Table 8.2), while drawing the same total bias current (6 mA)[2]. Under these conditions, it should still be possible to consume less than 10 mW, the power consumption of the MRFIC1501 when it is biased at its lowest voltage. After all, a power consumption of

[2]Later on, it will be shown that the 6 mA is very useful to keep the size and the quality factor of the involved inductors within certain limits.

nMOS						pMOS					
L_{eff}	K_n	V_{Tn}	Θ_n	Λ_n	err	L_{eff}	K_p	V_{Tp}	Θ_p	Λ_p	err
μm	μA/V^2	V	V^{-1}	V^{-1}	%	μm	μA/V^2	V	V^{-1}	V^{-1}	%
0.2	192	0.52	4.15	0.07	1	0.2	55	0.50	3.87	0.07	2

α	α_{gd}	α_{gb}	α_{db} (0.5 V)	α_{sb} (0.5 V)
0.83	0.23	0.16	0.68	0.65

Table 8.3: *Hand calculation parameters for the 0.25 μm CMOS technology of [KS] (extracted for $V_{GS} - V_T$ values between 0.1 and 0.3 V).*

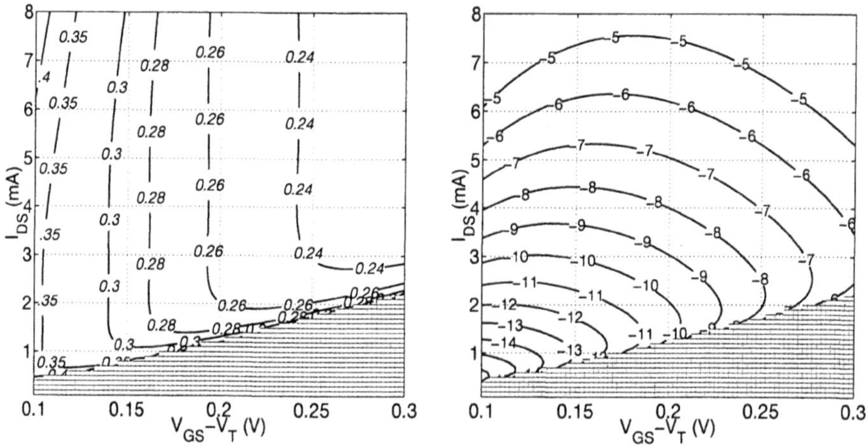

(a) LNA Noise Figure (dB). (b) LNA IP_3 (dBm).

Figure 8.2: *Contour plots of the total LNA noise figure and the input IP_3.*

10 mW at a current of 6 mA requires the power supply voltage to be less than 1.6 V, which is definitely enough to accommodate an LNA of the cascode type.

In order to clarify the design trade-offs, Fig. 8.2, Fig. 8.3 and Fig. 8.4, show contour plots of the most important LNA properties in the design space of the amplifying device. All the values are calculated using the equations in Section 6.7 and the hand calculation parameters in Table 8.3. In these plots it is assumed that the input capacitance C_p is set to 210 fF— 110 fF for the bondpad and 100 fF for the protection diodes. Fig. 8.2(a) depicts the noise figure of the LNA under ideal circumstances (i.e. assuming a lossless L_g, L_s etc.). As can be seen from the plot, the noise figure is extremely low in the whole design space. The LNA doesn't even need the available 6 mA: according to the plot, a noise figure as low as 0.3 dB can already be achieved at a drain current of only 1.1 mA! Fig. 8.2(b) plots the contour lines of the IP_3 of the LNA. Clearly, 6 mA is not even required from a linearity perspective: an IP_3 of −6 dBm can already be obtained at about 3 mA. Hence, from the point of view of noise figure and linearity, 6 mA seems to be quite an overkill. However, there are a few reasons why the 6 mA are very welcome... For instance, Fig. 8.3(a) indicates that the required load inductor R_p — or more accurately, the effective R_p at the drain of the cascode — becomes large when biasing the input stage at low current levels. This can be attributed to the drop in the efficiency of the amplifying

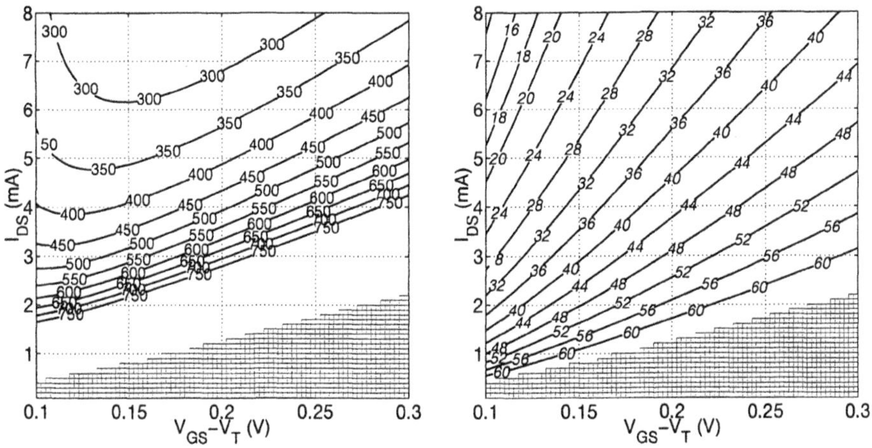

(a) Load Inductor R_p (Ω). (b) Gate Inductance L_g (nH).

Figure 8.3: *Contour plots of the required inductor R_p and the required L_g.*

device due to the increase in the $R_{S,eq}$ (see also Fig. 8.4(a)). In addition, as shown in Fig. 8.3(b), the required gate inductance becomes very large at low current levels and/or moderate $V_{GS} - V_T$ values.

The fact that inductors with a large R_p indeed pose some problems, can be explained by the following reasoning. Any practical system must be able to tolerate process variations; For instance, one of the requirements could be that the operating frequency must lie in the -3dB bandwidth of the amplifier despite a $\pm\Delta\omega_0$ deviation in the center frequency. This requirement translates into the following constraint on the total LNA quality factor:

$$Q_{LNA} \leq \frac{\omega_0}{2\Delta\omega_0}. \tag{8.3}$$

For instance, in case $\Delta\omega_0 = \pm 0.1\omega_0$, Q_{LNA} must be smaller than 5. In his turn, the Q_{LNA} value consists of two contributions: the quality factor of the input — generally close to one because of the bondpad/protection capacitance — and the quality factor of the output section, which is mainly determined by the inductor. Stating R_p in terms of the quality factor of the inductor yields

$$R_p = (Q_L + \frac{1}{Q_L})\omega_0 L_d \cong Q_L \omega_0 L_d. \tag{8.4}$$

Since Q_L needs to be smaller than the value calculated in (8.3) (diminished by the quality factor of the input section), a large R_p inductor must necessarily exhibit a relatively large inductance value. Yet, considering that the resonance frequency must remain the same, this strongly limits C_1 and $C2$, which makes the matching network very sensitive to external parasitics. The self-resonance frequency of the inductor poses no problem since the inductor is implemented in the top metal layer.

In view of the above, the I_{DS} and the $V_{GS} - V_T$ of the amplifying device have been set to 6 mA and 0.14 V, respectively. At this point, the quality factor of the input section equals 0.9 (see Fig. 8.4(b)), leaving a maximum of 4.1 for Q_L. On the other hand, Fig. 8.3(a) indicates that

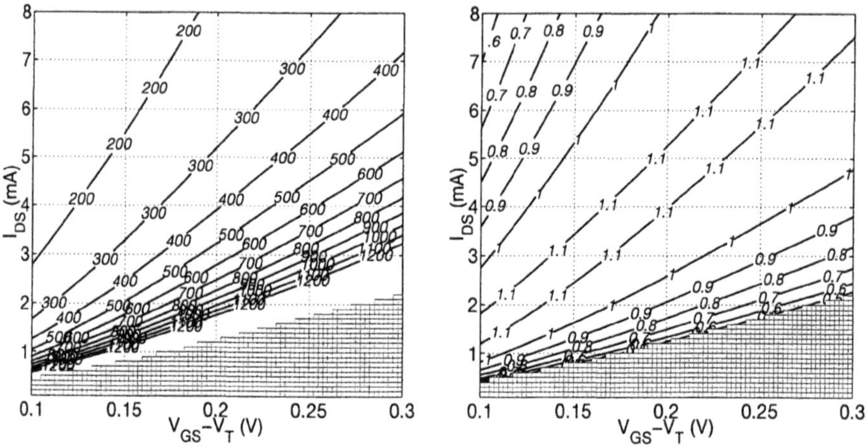

(a) Equivalent Source Resistance $R_{S,eq}$ (Ω). (b) Input Quality Factor.

Figure 8.4: *Contour plots of the equivalent source resistance $R_{S,eq}$ and the input section Q.*

$(I_{DS})_{amp}$	6 mA	$(I_{DS})_{casc}$	6 mA
$(V_{GS}-V_T)_{amp}$	0.14 V	$(V_{GS}-V_T)_{casc}$	0.2 V
$(W/L)_{amp}$	450/0.25	$(W/L)_{casc}$	250/0.25
$(g_m)_{amp}$	64 mS	$(g_m)_{casc}$	46 mS
$(C_{gs})_{amp}$	320 fF	$(C_{gs})_{casc}$	190 fF
L_d	10.5 nH	C_1	0.96 pF
R_d	20 Ω	C_2	1.16 pF
D_1&D_2 area	2x50 μm^2	L_g	24 nH
C_{bp}	110 fF	L_s	1 nH

Table 8.4: *Transistor sizes, component values and biasing information.*

the effective R_p must be around 310 Ω. A few iterations with FastHenry [FastHen, Kamo94] have resulted in a 10.5 nH inductor with a 20 Ω series resistance — equivalent with a Q_L of 4 and an R_p of about 330 Ω. Evidently, all other element values can now be calculated. The resulting transistor sizes and component values are summarized in Table 8.4.

8.6 Layout

The analysis in the previous section assumes that all passive components — except for L_d — are lossless and hence, noiseless. The layout must ensure that this is effectively the case. How this is done is indicated in this section.

transistors Both the amplifying device and the cascode device employ a finger structure. The gate resistance has been reduced as much as possible by using short fingers (5/0.25). Since these

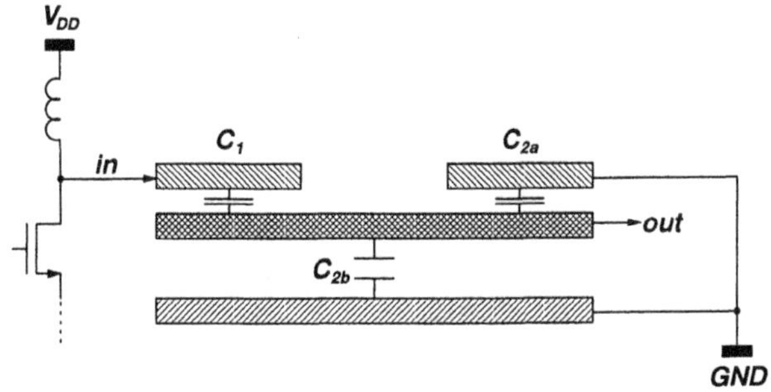

Figure 8.5: *Cross-section of the capacitive divider in the output matching network.*

short fingers allow the distance between the bulk contacts and the middle of the channel to be lower, the effective substrate resistance can be decreased so that less noise is injected through the back gate (g_{mb}). The bulk contacts also shield the g_{mb} from injected output signals, cutting a possibly dangerous feedback loop.

bonding pads To prevent the substrate resistance from injecting noise currents, the input pad is shielded from the substrate by a grounded metal layer underneath the bondpad [Rofo98b, Colv99] (see Fig. 8.1). A detailed explanation of its function can be found in Section 6.5.3. At the same time, this action increases the quality factor of the bondpad structure so that less useful signal is lost. Since the shield is implemented in the first metal layer and the pad itself is implemented in the fourth metal layer, the insertion of the metal shield only has a minor effect on the pad capacitance.

ESD diodes The diffusions connected to the supply lines completely enclose the diffusion regions connected to the RF input terminal to reduce the series resistance of the two-diode protection as much as possible. Evidently, the spacing between the p and n diffusions needs to be minimized since we are dealing with high-ohmic substrate material.

load inductor The load inductor is implemented in the fourth and the third metal layer, minimizing its parasitic capacitance towards the substrate. A patterned ground shield underneath the inductor (Fig. 8.1) shields the inductor from the lossy substrate by providing a low-ohmic path to ground. In this way the series resistance of the parasitic coil capacitance is significantly reduced, yielding both a higher inductor R_p and a larger quality factor [Yue98]. The R_p of the inductor is now solely determined by the inductor's series resistance instead of by the unpredictable substrate resistance. The ground shield is implemented in the first metal layer and is patterned in order to prevent eddy currents from flowing. Note that the patterned ground shield only lowers the loss associated with the electrical field: it does not affect the magnetic losses.

output capacitors The capacitors of the output matching network (C_1 and C_2) are implemented as metal–insulator–metal (MIM) capacitors featuring a very low series resistance. The stray capacitance to the substrate is intercepted by a ground shield and also contributes to C_2

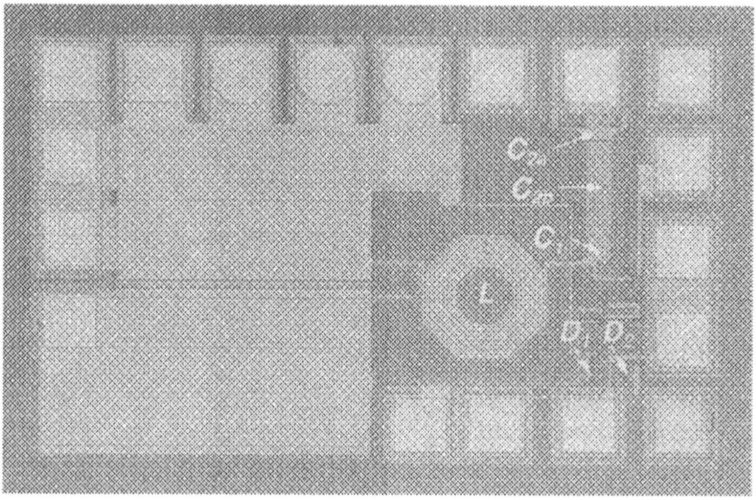

Figure 8.6: *Micro-photograph of the 0.8* dB *NF, ESD-protected CMOS LNA.*

(see Fig. 8.5). The ground shield has originally been inserted to bypass the substrate resistance, improving the quality factor of the output network. The ground shields in this layout — the shields underneath the pads and the output capacitors, the patterned ground shield underneath the inductor — also serve another purpose: increasing the on-chip reverse isolation. Ultimately, the reverse isolation is believed to be limited by the cross-talk between the external bondwires.

high level layout The power supply lines, the gate of the cascode device (and the associated bondwires) are bypassed to ground using 40 pF decoupling capacitors. Special care has been taken to sufficiently damp the parasitic resonances (see Subsection 6.8.3). A photograph of the IC is shown in Fig. 8.6. The IC is implemented in the 0.25 μm 4M1P CMOS process of Table 8.3 and occupies an area of 0.66 mm^2.

8.7 Experimental Results

To measure the LNA, the IC is glued onto a thick film ceramic substrate and all the pads are wire bonded to 50 Ω strip-lines. Apart from providing the connection, some of these bonding wires also serve as inductor: the source inductor L_s is for instance implemented as two short bondwires. Also the gate inductor is implemented as a bondwire because of its low series resistance and its low parasitic capacitance. The substrate is then mounted in a Copper-Beryllium box which shields the LNA from external interference and serves as reference ground. The connection to the external world is provided through two SMA connectors. A large amount of vias connect the local ground plane underneath the IC to the reference ground. Finally, the LNA is biased in its nominal 9 mW regime, i.e. drawing 6 mA from a 1.5 V supply.

First, the complete S-parameter set has been measured using the HP network analyzer. The

Noise figure (NF)	0.79 dB
Power gain (S_{21})	20 dB
Input reflection coefficient(S_{11})	-11 dB
Output reflection coefficient (S_{22})	-11.5 dB
Reverse isolation ($-S_{12}$)	> 31 dB
Third order intercept point (IIP_3)	-10.8 dBm
HBM ESD-immunity	0.6 kV (+) 1.4 kV (−)
Supply voltage	1.5 V
Current consumption	6 mA
Power consumption	9 mW

Table 8.5: *Experimental results in the GPS L2 band (i.e. at 1.2276* GHz*).*

forward gain (transducer power gain, S_{21}) is measured to be a flat 20 dB in a 100 MHz wide band around the GPS L2 frequency of 1.2276 GHz (1.2–1.3 GHz). The -3 dB bandwidth is approximately 400 MHz (1.05–1.45 GHz). At the same time, the reverse isolation ($-S_{12}$) is better than 31 dB over the whole frequency range of the network analyzer (300 kHz–3 GHz). In the L2-band, the input reflection coefficient (S_{11}) and the output reflection coefficient (S_{22}) are -11 dB and -11.5 dB, respectively. Both reflection coefficients are better than -10 dB in a 100 MHz wide band around the GPS L2 frequency of 1.2276 GHz (1.2–1.3 GHz). Due to the increased resistivity of the top metal layer, the R_p of the coil had become 20 percent lower than originally simulated, which resulted in a lower S_{21} and a larger S_{22}. The gain degradation has been compensated for by lowering the input impedance to 30 Ω by decreasing the nominal L_s value. In this way, extra *PCC* (and hence gain) is created in exchange for a larger S_{11} and a lower linearity. According to Section 6.12, the noise figure of the 50 Ω R_S/30 Ohm R_{in}-configuration is approximately the same as in case of a normal 50 Ohm R_S/50 Ω R_{in}-configuration.

The noise figure of the LNA has been measured directly using a noise figure meter. At the GPS L2 frequency, an extremely low noise figure of 0.79 dB is measured (including the noise of the strip-lines!). In addition, the noise figure remains below 1.2 dB in the 200 MHz wide frequency range between 1.1 GHz and 1.3 GHz. Also the sensitivity to nearby interferers has been evaluated; In the L2 band, the input-referred third-order intercept point (IIP_3) and the 1-dB compression point are -10.8 dBm and -24 dBm, respectively.

The IC has been tested for ESD-immunity as well. HBM (Human Body Model) ESD-tests have shown that the LNA is capable of surviving positive ESD pulses up to 0.6 kV (zaps measured with respect to V_{DD}) and negative ESD pulses down to -1.4 kV (zaps measured with respect to ground), exceeding the 0.5 kV specification. The susceptibility to positive zaps with respect to ground and negative zaps with respect to V_{DD} could not be measured owing to the absence of an on-chip clamping circuit. More information on the measured ESD performance and how the ESD performance can be improved is given in Section 8.8.

It is worth noting that all the measurements have been performed from SMA connector to SMA connector, i.e. without de-embedding the substrate parasitics like strip-line resistance, connector non-idealities, etc. The measurement results in the GPS L2 band are summarized in Table 8.5. The detailed measurement plots can be found in [LeJa01].

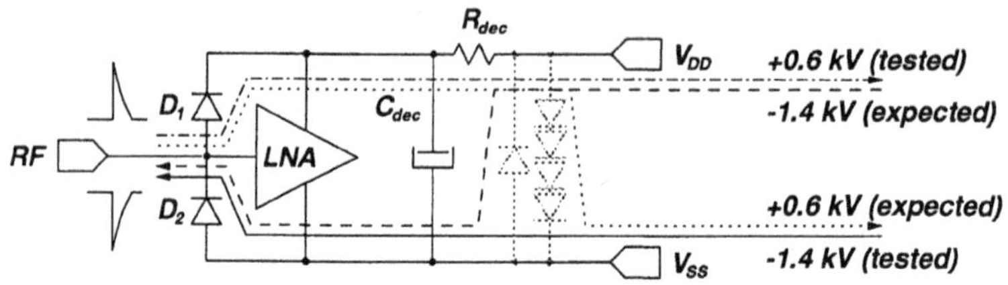

Figure 8.7: *The origin of the measured ESD performance.*

8.8 ESD Performance

8.8.1 Measured ESD Susceptibility

A mentioned before, HBM ESD-tests have shown that the LNA is capable of surviving positive ESD pulses up to 0.6 kV (with respect to V_{DD}) and negative pulses down to -1.4 kV (with respect to ground). The origin of this performance can be clarified using Fig. 8.7. The bottom diode (D2) protects the input against negative zaps with respect to ground through its forward conduction mechanism (solid line). D2 is therefore responsible for the -1.4kV value. Positive zaps with respect to V_{DD} are covered by top diode D1 (dash-dotted line). However, the series resistance R_{dec}, originally inserted in the V_{DD} path to damp any possible resonance between the power supply bondwire and the decoupling capacitors, always lies in the discharge path and therefore limits the positive ESD performance to the lower 0.6 kV value.

8.8.2 Expected ESD Performance with an On-chip Clamp

Diode D2 offers no protection against positive zaps with respect to ground; Even though the diode can in principle clamp the voltage to its breakdown voltage, it will most likely be destroyed during the ESD event. Therefore, in case of a positive zap with respect to ground, the top protection diode must conduct the positive ESD current to the V_{DD} from where it must be directed to ground through a low-resistance power supply clamp (a string of large diodes). However, since this clamp was not implemented on the test chip, we could only test the susceptibility to positive ESD pulses with respect to V_{DD}. For exactly the same reason we could only test the susceptibility to negative ESD pulses with respect to ground and not with respect to the V_{DD}. Nevertheless, since the clamp diodes are very large structures which contribute almost no series resistance to the ESD discharge path, the LNA should definitely be able to withstand 0.6 kV positive zaps with respect to ground (dotted line) and -1.4 kV negative zaps with respect to V_{DD} (dashed line) — that is of course if the clamps are effectively there. So, the first thing which ought to be done is to include the power supply clamp on the IC.

8.9 Discussion and Comparison with Existing CMOS LNAs

To be able to position this work with respect to existing LNAs, Table 8.6 lists the performance of the CMOS power-gain LNAs published to date.

Ref.	Freq. (MHz)	NF (dB)	P_{DC} (mW)	G_t (dB)	IIP_3 (dBm)	ESD (kV)	L (μm)	NF_{sys} (dB)	$((F-1)P_{DC})^{-1}$ (dB)
[Kara96]	900	2.2	20	15.6	-3.2	-	0.5	3.1	18.8
[Shah97]	1575	3.8	12	17	-6	-	0.35	4.3	17.8
[Shae97]	1500	3.5	30	22	-9.3	-	0.6	3.7	14.3
[Zhou98]	900	4.1	18	12.3	-	-	0.6	5.3	15.5
[Stub98]	940	2.1	24	12	+4	±1	0.5	4.1	18.3
[Haya98]	900	1.8	9	14.8	-2.5	-	0.35	3.0	23.4
[Huan98a]	900	1.74	45	14.3	-2.8	-	0.25	3.1	16.4
[Huan98b]	900	1.85	27	16.2	-7.2	-	0.25	2.8	18.4
[Floy99]	900	1.2	30	22	-1	-	0.8	2.7	20.2
[Gram00a]	900	1.75	27	10	+4.7	-	0.35	4.7	18.7
[Gram00b]	900	1.05	9	11.4	-2	-	0.35	3.7	26.0
[Gram00b]	900	0.9	18	13.5	0	-	0.35	2.8	23.8
[Gram00b]	900	1.2	13.5	12.8	-3	-	0.35	3.2	23.7
[LeJa01]	1227	0.79	9	20	-10.8	-1.4/0.6	0.25	1.3	27.5

Table 8.6: *Performance summary of recently published CMOS LNAs.*

The 0.8 dB noise figure offered by this LNA is the lowest noise figure ever reported in CMOS. In addition, the LNA is the first sub-1dB noise figure LNA at such a low power consumption (9 mW); The only other sub-1dB *NF* LNA is the 0.9 dB noise figure LNA in [Gram00b], which consumes twice the power and features a 6 dB lower gain!

The power gain of 20 dB is the second largest gain in the table. In a system, this LNA suppresses the noise of the subsequent stages by a factor of 100, making it very suitable for applications where a low system noise figure is required. Most published CMOS LNAs that attain a low noise figure simply *not* have sufficient gain to yield a low noise receiver. This can be shown by looking at the system noise figure, defined as

$$NF_{sys} = 10\log_{10}\left(F_{LNA} + \frac{F_{mixer} - 1}{G_t}\right), \tag{8.5}$$

which is a good figure-of-merit to evaluate the capability of an LNA when it is inserted in a real system. Table 8.6 states NF_{sys} for all LNAs in case the mixer noise figure amounts to 12 dB. Whereas all CMOS LNAs in the table offer system noise figures between 3 to 5 dB, the presented CMOS LNA enables total receiver noise figures as low as 1.3 dB! And, since the noise figure of CMOS mixers is usually higher than 12 dB, the difference between the presented LNA and the existing LNAs is even more pronounced in reality.

It is important to note that both the input and output reflection coefficient (-11 dB and -11.5 dB, respectively) comply to the -10 dB filter termination requirement. The 0.8 dB noise figure is thus accompanied by an acceptable amount of reflection[3]. Moreover, all the matching networks are integrated on-chip (except for the input bondwire, which can not be avoided in this topology). All other low-power low-noise amplifiers ([Haya98, Gram00a, Gram00b]) use

[3] in contrast to some other LNA's that feature a good noise performance at the cost of a lousy input match, like e.g. [Kara96] and [Floy99].

external input and output matching networks, and exploit these additional degrees of freedom to balance a noise match with an input match; For instance, by using an intermediate reference plane a a lower impedance value ($R_{S,local}$), the noise figure at a given power consumption can be dramatically lowered because $F - 1 \propto R_{S,local}$ (see Section 6.7). In a solution with a single external inductance this degree of freedom is not available since the value of $R_{S,local}$ is a design constraint: it is simply 50 Ω.

The linearity –although somewhat lower than the rest — is more than acceptable; The measured IP_3 value of -10.8 dBm is even better than the linearity required by the GSM cellular system (-14 dBm), and certainly more than enough for the GPS application (less than -20 dBm) because the received signal power is very low[4] at all times.

Apart from its raw performance, one of the most important properties of this LNA is that it features some degree of protection against ESD; The IC is fitted with an ESD protection on the RF input, which is capable of protecting the LNA against -1.4 kV to 0.6 kV HBM zaps. Almost none of the published low noise amplifiers include any ESD protection.

To conclude, Fig. 8.8 shows yet another way of positioning this work with respect to previously published CMOS LNA's. The figure on the vertical axis is a measure of the power efficiency with which the LNA noise figure has been realized, defined as

$$\eta_{NF} = -10\log_{10}(\frac{(F-1)P_{DC}}{1\ \mathrm{W}}), \tag{8.6}$$

while the horizontal axis represents the LNA gain — the LNA's capability of determining the system noise figure. The more the LNA is positioned towards the upper right corner, the better its performance. The presented LNA clearly outperforms all other published CMOS LNAs in this regard. [5]

The reader may ask oneself why the IP_3 has not been included in any of the above performance metrics. The reason is that the only IP_3 value which is relevant in the end is the IP_3 of a complete receiver system. This number includes the linearity of the LNA as well as the linearity of the down-conversion mixers. In sensitive receivers, and especially in CMOS receivers, the gain of the LNA needs to be quite large in order to bring the signal sufficiently above the noise floor of the mixers. As a result, the IP_3 of the down-conversion mixer generally dominates the IP_3 of a receiver system. Therefore, the IP_3 is not so much a spec for an LNA as it is for a mixer; As long as the linearity of the LNA is large enough, there is no problem whatsoever. In this context, "large enough" could mean "larger than -14.5 dBm" i.e. the linearity requirement of a GSM-900 receiver, the toughest linearity spec that exists in today's cellular systems. Only if the IP_3 value is so large that the blocking filter can be omitted, the LNA's IP_3 is worth talking about again.

[4]Contrary to the DCS–1800 system, there are no in-band blocking signals present. The out-of-band blockers are attenuated more than 40 dB by the antenna filter so that they do not cause any problem either.

[5]Until now there is no commonly accepted way of combining IP_3, NF, G_t, ω_0 and P_{DC} in a figure-of-merit which only depends on technology variables (like e.g. the effective channel length). As a result, the performance comparison in this chapter does not take into account the evidently positive effect of a smaller technology on LNA performance; The performance metrics in this chapter only compare the raw performance of the LNA's with respect to each other.

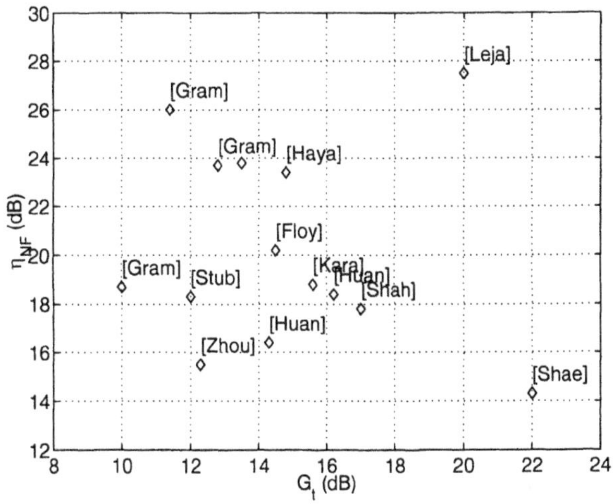

Figure 8.8: *Comparison between the raw performance of the realized LNA and previously published CMOS LNAs. The figure on the vertical axis is a measure of the power efficiency with which the LNA noise figure has been realized while the horizontal axis represents the LNA gain.*

8.10 Conclusion

Even in a standard submicron CMOS technology, an extremely low noise figure (< 1 dB) can be combined with a high gain (20 dB) at the same power consumption as commercially available GaAs LNA solutions. The LNA presented in this chapter outperforms all previously published CMOS power gain LNA's with respect to noise figure, gain and power consumption; It features the lowest noise figure published to date in CMOS (0.8 dB), the second highest gain published to date in CMOS (20 dB), and offers all this at a power consumption which is among the lowest in CMOS (9 mW). In addition, the IC is fitted with an ESD protection on the RF input, which is capable of protecting the LNA against -1.4 kV to 0.6 kV HBM zaps. This demonstrates that an excellent performance can still be achieved while at the same time providing $> \pm 0.5$ kV ESD protection. The performance has been optimized by taking special precautions during layout.

Chapter 9

A 2V CMOS DCS–1800 Receiver Front-End

9.1 Introduction

CMOS IC technology is scaling down at a very fast pace, enabling an ever increasing speed in the analog as well as in the digital part. On the other hand, the created speed margin is consumed very rapidly by the ever growing demand for bandwidth, which exerts a strong push on the center frequency of the applications. Therefore, RF CMOS circuits are predestined to go to even deeper submicron technologies.

There is however one problem which comes up; The lower feature sizes are accompanied by a decreasing supply voltage. The allowable V_{DD} roughly scales proportionally with gate length to keep the electrical field in the channel and in the gate oxide constant (constant field scaling):

$$V_{DD}(\mathrm{V}) \cong 10L_{min}(\mu\mathrm{m}). \tag{9.1}$$

At the same time, the V_T does not decrease as fast, so that the maximum gate over-drive tends to become lower.

While the net effect on the performance of digital circuits is undoubtedly positive — both a lower power consumption and a higher switching speed —, the effect of the lower V_{DD} on the power consumption (or even the realizability) of the analog circuits greatly depends on the class and the topology of the building blocks; As long as the supply voltage is not a binding constraint for the proper working, the performance or the current consumption of a given analog block, its power consumption does improve with decreasing V_{DD}. A typical example of such a circuit is an LNA. The story is completely different when the supply voltage is a binding constraint, which is generally the case in high-accuracy oversampling data converters. There, a lower supply voltage leads to lower signal levels, so that a given *SNR* requires more current. While this might be remedied by foreseeing a dedicated MOS transistor with a thicker oxide (enabling a larger V_{DD}), this is not an option for the RF part; To achieve the raw speed, the RF circuits really need the full capability of the process. Hence, in order for an RF circuit topology to be portable towards lower gate lengths, the V_{DD} must be able to scale down.

The 0.25 μm receiver front-end in Chapter 7 already operated at a low supply voltage of 2.5V, the nominal supply voltage for a 0.25 μm CMOS digital IC. Yet, even though the receiver IC is capable of functioning at supply voltages below 2.5V, it becomes increasingly more difficult to achieve the specifications required by DCS-1800 at these low voltages. We would meet exactly the same problems when future applications force us to go to deeper submicron technologies.

The limitations largely originate from the linear mixer; If the supply voltage is lowered, and the common-mode output level is kept at $V_{DD}/2$, the maximum over-drive voltage of the mixer transistors decreases. For example, if we put the IC of the previous chapter at 2V—corresponding to a common-mode level of 1V—, the maximum over-drive voltage of the mixing transistors would only be

$$1\text{V} - V_T(@V_{SB} = 1\text{V}) \approx 0.2\text{V}. \tag{9.2}$$

While this is still sufficient to build a switch, it does not suffice for the linear mixer; At low over-drive voltages the local oscillator drives the mixer transistors out of the linear region, into saturation, degrading the linearity of the mixing process. This effect is further aggravated when going to smaller gate lengths since the V_T doesn't scale proportionally with V_{DD}.

Another reason why the mixer topology is not so suited for low voltage operation, is that a low input impedance needs to be provided to the mixer transistors in order not to degrade the conversion gain and the conversion linearity. The OTAs in the transimpedance amplifier (see Fig. 7.2) must therefore feature a large gain-bandwidth and a large DC gain, not in the least because of the low impedance of the mixer transistors themselves. This issue is even more important when implementing some AGC in the loop of the transimpedance amplifier. High gains can only be achieved at a low power consumption by using cascoded OTAs or gain-boost circuits, configurations which run into problems at low voltages.

In this chapter, we will build a 0.25 μm CMOS RF receiver front-end which is capable of operating at a supply voltage of 2V. Designing the front-end towards even lower voltages (like e.g. 1.8 V) has little sense since the data converters require sufficiently large input signals anyway in order to keep their power consumption low. When lower voltages are required the only solution is to use a separate V_{DD} for the low-frequency IF part of the IC. The threshold voltages of the available 0.25 μm technology are quite large as well: an V_T of 0.65V for the nMOS and a V_T of 0.75V for the pMOS. At 2.0V, the common-mode level is only 1V, which is about the minimum common-mode level at which a conventional nMOS differential pair can work:

$$0.15\text{V} + 0.15\text{V} + V_T(@V_{SB} = 0.15\text{V}) \approx 1\text{V}. \tag{9.3}$$

The structure of this chapter is as follows. First, Section 9.2 introduces the topology of the realized 2V receiver front-end. The design trajectory of the down-conversion mixers, the variable-gain amplifier-filters and the low noise amplifier is discussed in Section 9.3 and Section 9.4, respectively. The sizing procedure is based on a set of analytical equations describing the performance parameters of the building blocks in terms of the operating point of the involved transistors. The realized receiver front-end forms the receive path of a highly integrated 2V DCS-1800 transceiver. Its layout is covered in Section 9.5. Finally, Section 9.6 describes the experimental results of the receive path, which demonstrate that the raw analog performance required for the DCS-1800 cellular system can indeed be met in a standard 0.25μm 2M1P CMOS process.

9.2 Receiver Topology

Fig. 9.1 shows the topology of the realized 2V receiver front-end. The receiver employs the low-IF architecture described in Section 3.3: the wanted DCS-1800 channel is converted into a differential quadrature signal, centered at the 100 kHz IF. The receive path consists of an input-matched single-ended LNA which feeds its output signal into two (I&Q) down-conversion mixers. Each of these mixers is then followed by a differential variable-gain amplifier-filter.

The low noise amplifier is an inductively matched power-to-voltage LNA with an RLC resonant load — the topology of Fig. 6.2(a), repeated at the top left of Fig. 9.1. The properties of this LNA have been discussed in detail in Section 6.3. The LNA directly drives the down-conversion mixer without intermediate image filter, avoiding an LNA output matching circuit and a costly filter component.

The active quadrature mixers embedded in the front-end employ a novel low-voltage current-folding switching mixer topology, shown at the bottom of Fig. 9.1. It is especially designed to be able to operate at supply voltages as low as 2.0V. In short, its operation is as follows. First, the single-ended RF input signal from the LNA is fed to a highly linear cascoded transconductor $M_1 - M_2$ which converts the signal into an RF current. The current which enters node ② is periodically folded to the left and to the right by two commutating pMOS switches (the differential pair M_{4a} and M_{4b}), carrying out the multiplication between the LO and the RF signal and converting the signal from single-ended to differential. Note that the mixer features a single-ended input, enabling the use of a single-ended LNA.

The resulting low-frequency output current flows into a differential transimpedance amplifier which converts the current into a voltage. Since the value of the transimpedance elements can be controlled and the elements are frequency-dependent, the transimpedance amplifier realizes both the variable-gain function and the filter function. It is worth noting that the mixer and the variable-gain amplifier-filter are seamlessly integrated together; No extra interface circuits are required to connect the different building blocks or to do extra signal conversions. In the following paragraphs, the raison d'être and the exact function of all elements in the schematic is discussed in detail.

Cascode transistor M_2 is inserted to isolate the drain of M_1 (node ①) from the switch node (node ②). In this way, the signal at $2\omega_{LO}$ which is generated at node ② through rectification of the LO signal is attenuated before it appears on node ①. The linearity of the voltage to current conversion is thus minimally affected by the switching process. M_2 also lowers the LO feed-through to the mixer input and reduces the Miller effect on the C_{gd} of M_1.

A large part of the drain current of M_1 is provided through pMOS bleeder M_6. In this way, the current through M_2 and M_3 is greatly reduced so that the current noise of M_3 and the contribution of M_2 and M_3 to the total capacitance on the switch node is minimized. Since the pMOS bleeder transistor M_6 can have a very large $V_{GS} - V_T$, its current noise contribution can be made much smaller than when M_3 had to provide the current. The minimum current that still needs to flow through M_2 is set by three factors: the Miller effect on node ①, the pole on node ① and the maximum RF signal current.

The requirement for the switch pair M_{4a} and M_{4b} to be a *folding* switch is a direct consequence of the limited supply voltage; Otherwise, the stack of transistors would not have fitted between the rails without omitting the cascode device. Another advantage of using pMOS switches is that the residual $1/f$ noise is much less than when nMOS transistors are used for the switches. It is worth noting that the pMOS switches are always biased in the saturation regime: from the perspective of an RF signal entering node ②, each side works just like a folded cascode.

The differential output current coming out of the switches is fed to a transimpedance VGA which converts it into a voltage by forcing it through a linear network using high gain OTAs.

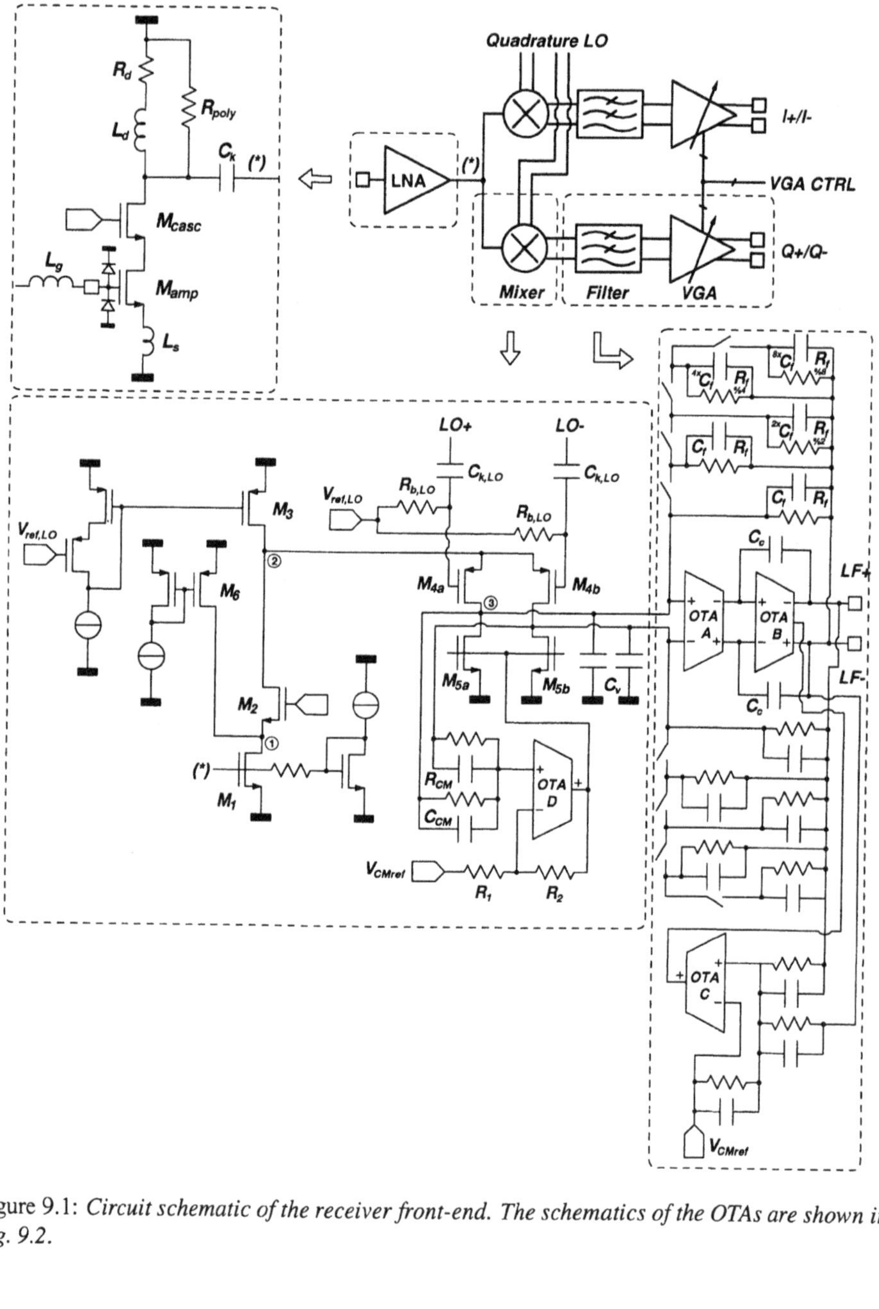

Figure 9.1: *Circuit schematic of the receiver front-end. The schematics of the OTAs are shown in Fig. 9.2.*

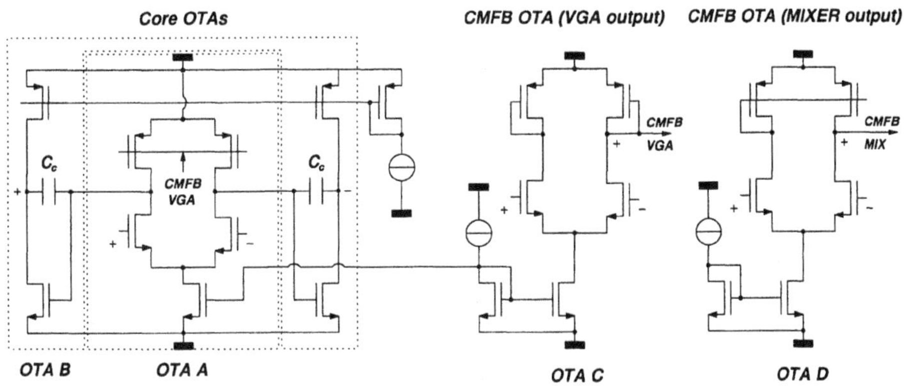

Figure 9.2: *Circuit schematics of the OTAs.*

At low frequencies, the VGA presents a low input impedance to the mixer core (node ③), so as to maximally take in the down-converted current. At high frequencies, the capacitors C_v take over from the VGA loop to ensure that injected LO currents and the generated high frequency mixing products are shorted to ground. Since the trans-impedance amplifier does not interact with the mixing process itself, the low-frequency input impedance of the VGA is allowed to be larger than in case of the linear mixer in the previous chapter; There is a better isolation between the mixing function and the efficiency with which the current is processed. Therefore, the OTAs in the trans-impedance amplifier do not need to contain cascodes, which facilitates low voltage operation. In addition, the larger input impedance makes that the VGA function can be implemented by increasing the value of the trans-impedance elements, and this without requiring more gain-bandwidth in the OTAs.

The VGA itself consists of a two-stage fully-differential Miller-compensated OTA — the schematics of OTA A and OTA B are depicted in Fig. 9.2 — with a bank of five matched RC elements in the feedback:

$$(R_f, C_f), (R_f, C_f), (\frac{R_f}{2}, 2C_f), (\frac{R_f}{4}, 4C_f) \text{ and } (\frac{R_f}{8}, 8C_f). \tag{9.4}$$

Apart from providing the actual current-to-voltage conversion, the RC elements also perform first order filtering of the blocking levels and anti-aliasing for the oversampling A/D converters. The global feedback, the large loop gain and the intrinsic linearity of the RC feedback elements make the current-to-voltage conversion in the VGA very linear. The circuit is therefore capable of providing a rail-to-rail output signal with minimum distortion and intermodulation. The gain of the VGA can progressively be decreased by 6 dB by switching extra RC elements in parallel. As a result, the ratio between the largest gain and the lowest gain is 24 dB. The switches — large nMOS transistors controlled by a digital word — are inserted at the virtual ground node of the amplifier to ensure that they experience a minimum signal swing and always stay biased in the linear region. At a supply voltage of 2V, the over-drive voltage of the switches is still 0.2V.

Both the differential VGA and the differential mixer core are fitted with a common-mode feedback. The detailed schematics of the internals of OTA C and OTA D can be found in Fig. 9.2. The feedback at the output of the mixer is absolutely necessary since the common-mode feedback on the output of the transimpedance VGA does not control the common-mode level of the mixer core. Of course, both reference voltages must be identical in order to prevent DC current from flowing through the transimpedance elements. The common-mode feedback of the mixer core consists of a single stage OTA which senses the common-mode output voltage and feeds it back via M_{5a} and M_{5b}. At high frequencies, the resistors sensing the common-mode output voltage of the mixer core are bypassed by capacitors C_{CM} to stabilize the loop. The low-frequency loop gain and the non-dominant pole of the loop are controlled by the resistive feedback R_1/R_2.

9.3 The Down-Conversion Mixer and the Filter/VGA

This section covers the design of the down-conversion mixer and the variable-gain amplifier-filter, the two intimately coupled blocks at the bottom of Fig. 9.1. First, in Subsection 9.3.1, approximate expressions are derived for the most important performance parameters of the mixer core and the VGA. The actual design flow is described in Subsection 9.3.2.

9.3.1 Design Equations

9.3.1.1 Conversion Gain

The task of the mixer core consists in converting the RF voltages into IF currents. Its conversion transconductance can be calculated as

$$g_{m,mixer} \triangleq \frac{i_{LF,left} - i_{LF,right}}{v_{in}} \tag{9.5}$$

$$\cong g_{m,1} \frac{\frac{2}{\pi}\left(1 - \frac{\sqrt{2}(V_{GS} - V_T)_4}{\pi V_{LO}}\right)}{\sqrt{1 + \left(\frac{\omega_0}{\omega_{p,1}}\right)^2}\sqrt{1 + \left(\frac{\omega_0}{\omega_{p,2}}\right)^2}}, \tag{9.6}$$

where V_{LO} is the amplitude of the differential local oscillator and $\omega_{p,1}$ and $\omega_{p,2}$ are the poles on node ① (the cascode pole) and node ② (the switch pole), respectively.

The numerator of (9.6) is the conversion efficiency of the switches themselves [Rude97], defined as the ratio between the generated low-frequency current and the high-frequency current injected into the switches. Its expression has been derived with the aid of Fig. 9.3(a)–(c) and makes sense as long as $\sqrt{2}(V_{GS} - V_T)_4 \leq V_{LO}$, i.e. as long as the local oscillator is able to switch M_{4a} and M_{4b} on and off completely. The term in $V_{GS} - V_T$ represents the conversion gain degradation owing to the fact that the local oscillator signal is a sine wave instead of a square wave. After all, the smaller the LO amplitude or the larger the $V_{GS} - V_T$ of the switches, the longer it takes for a switch to move from the on-state to the off-state, and hence, the longer both switches conduct signal current at the same time. Since any signal current which is common to both sides does not contribute to the differential output current, the conversion efficiency is necessarily lower. The numerator of (9.6) has been derived assuming that the instantaneous current division factor, given by

$$\frac{i_{left}(t) - i_{right}(t)}{i_{injected}(t)} = \frac{g_{m,4a}(t) - g_{m,4b}(t)}{g_{m,4a}(t) + g_{m,4b}(t)}, \tag{9.7}$$

increases linearly from zero to one when going from the balanced state ($V_{LO}(t) = 0$) to the unbalanced state ($V_{LO}(t) = \sqrt{2}(V_{GS} - V_T)_4$). This approximation is shown in Fig. 9.3(c). Note that the transition time from balanced to unbalanced, Δt, can be approximated by [Rude97]:

$$\Delta t \cong \frac{\sqrt{2}(V_{GS} - V_T)_4}{\omega_0 V_{LO}}. \tag{9.8}$$

The factors in the denominator of (9.6) represent the signal loss associated with the poles on the internal nodes before the actual down-conversion takes place. The expressions for the poles are given by

$$\omega_{p,1} = \frac{g_{m,2} + g_{mb,2}}{C_{n,1}} \tag{9.9}$$

$$\omega_{p,2} \cong \sqrt{2}\frac{g_{m,4} + g_{mb,4}}{C_{n,2}}. \tag{9.10}$$

(9.10) is the expression for the pole when either M_{4a} or M_{4b} is switched off. After all, as mentioned before, the down-conversion is predominantly realized in the unbalanced state.

When the mixer is connected to the transimpedance amplifier, the total conversion gain is given by

$$CVG \triangleq \frac{v_{out,diff}}{v_{in}} \tag{9.11}$$

$$= g_{m,mixer}\frac{R_f}{2^i}, \tag{9.12}$$

where R_f is the feedback resistance in the transimpedance amplifier when all the switches are open (the resistance at maximum VGA gain) and i denotes the number of closed switches.

9.3.1.2 Noise Figure

In the following paragraphs the different contributors to the eigen[1] noise figure — the intrinsic noise of the mixer weighed against the background noise power coming from the source — of the combination mixer-VGA are discussed.

input transconductor The noise figure contribution of the input transconductor — consisting of M_1, M_2, M_3 and M_6 — is given by

$$F_{(M_1,M_2,M_3,M_6)} \cong \frac{\gamma}{\alpha 50}\left(\frac{2}{g_{m,1}} + \frac{2g_{m,6}}{g_{m,1}^2} + \frac{2g_{m,3}}{g_{m,1}^2}\left(1 + \left(\frac{\omega_0}{\omega_{p,1}}\right)^2\right) + \frac{2g_{m,2}}{g_{m,1}^2}\frac{1}{1 + \left(\frac{\omega_{p,1}}{\omega_0}\right)^2}\right). \tag{9.13}$$

[1]The exact definition for the eigen noise figure can be found in Subsection A.3.1.

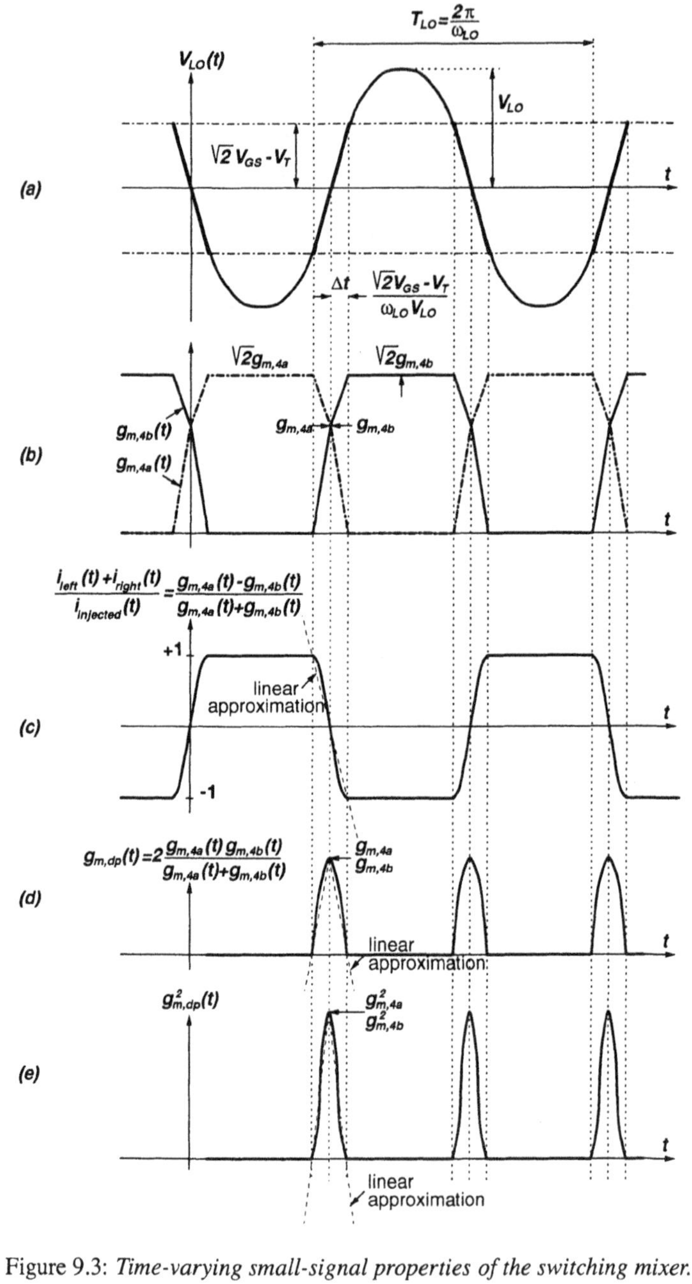

Figure 9.3: *Time-varying small-signal properties of the switching mixer.*

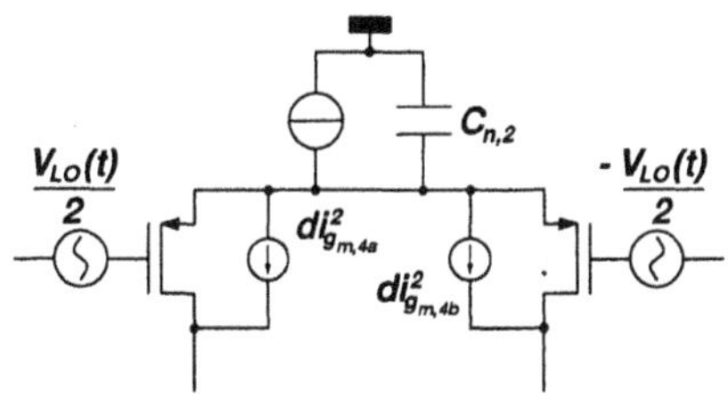

Figure 9.4: *The white noise of the switches can be modeled by white cyclostationary noise sources with a time-varying propagation to the differential output.*

The factors two in each noise term express that apart from the noise at the RF frequency, also the internally generated noise on the image frequency is converted to the IF. The factor containing $\omega_{p,1}$ in the third term takes into account the loss due to the cascode pole. The factor containing $\omega_{p,1}$ in the last term indicates that the noise current generated within cascode transistor M_2 can only propagate to the output if it finds a way through the parasitic capacitances on the cascode node. Note that M_1, M_2, M_3 and M_6 contribute only white noise; Their $1/f$ noise is up-converted by the commutating switches.

switches M_{4a} and M_{4b}... and their biasing resistors To calculate the contribution of the white noise coming from the switches, it is important to take into account that the operating point of the switch transistors, and along with it the noise and its propagation to the output, changes periodically with time. Such a random process, whose statistics are periodic functions of time, is called cyclostationary. A complete description of this class of processes requires the concept of a time-varying power spectral density (PSD) [Terr99].

The derivation employs two basic laws [Hull93, Phil00]:

1. Cyclostationary noise sources can be modeled as modulated stationary noise sources i.e. their time-varying PSD can be obtained by incorporating the time-varying aspect in the variables that express the magnitude of the noise and its propagation towards the output in terms of the operating point of the device.

2. When cyclostationary noise is passed through a filter with a bandwidth much lower than the frequency of the LO, as is typically the case in a down-converter, the noise again becomes stationary. The PSD of the resulting signal is then obtained by averaging the time-varying PSD over one period of the LO.

Consider the switches M_{4a} and M_{4b} when they are being modulated by the local oscillator (see Fig. 9.4). Since the virtual ground capacitors C_v and the common-mode feedback ensure that the switch devices remain in saturation, the instantaneous PSD of each white noise current is given by

$$S_{w,4a|4b}(f,t) = 4\frac{\gamma}{\alpha}kT g_{m,4a|4b}(t). \tag{9.14}$$

The time-varying PSD of the differential output noise current is then given by

$$S_{w,out}(f,t) = 4\frac{\gamma}{\alpha}kT\left(g_{m,4a}(t)p_{4a}^2(t) + g_{m,4b}(t)p_{4b}^2(t)\right), \tag{9.15}$$

where the propagation factors $p_{4a}(t)$ and $p_{4b}(t)$ are defined as

$$p_{4a|4b}(t) = \frac{i_{left}(t) - i_{right}(t)}{i_{ds,4a|4b}(t)}. \tag{9.16}$$

When calculating the propagation factors, it is generally assumed that $\omega_{p,2}$ is much larger than the operating frequency, or equivalently, that $C_{n,2}$ is negligible. In that case, $p_{4a}(t)$ and $p_{4b}(t)$ are given by

$$p_{4a}(t) = \frac{2g_{m,4b}(t)}{g_{m,4a}(t) + g_{m,4b}(t)} \tag{9.17}$$

$$p_{4b}(t) = \frac{2g_{m,4a}(t)}{g_{m,4a}(t) + g_{m,4b}(t)}, \tag{9.18}$$

indicating that the noise current of a given switch transistor can only contribute to the output noise in the small time window when also the other transistor is conducting i.e. near the zero crossings of the local oscillator (two times per period). The factors two stem from the fact that the differential noise current is considered. Substituting (9.17) and (9.18) into (9.15) yields:

$$S_{w,out}(f,t) = 16\frac{\gamma}{\alpha}kT\frac{g_{m,4a}(t)g_{m,4b}(t)}{g_{m,4a}(t) + g_{m,4b}(t)} \tag{9.19}$$

$$= 8\frac{\gamma}{\alpha}kT g_{dp}(t), \tag{9.20}$$

where $g_{dp}(t)$ is the time-varying transconductance of the switches considered as differential pair. Its time-domain waveform is depicted in Fig. 9.3(d). The time-average PSD then amounts to [Terr99]:

$$S_{w,out}(f) = 8\frac{\gamma}{\alpha}kT\frac{1}{T_{LO}}\int_0^{T_{LO}} g_{dp}(t)\,dt. \tag{9.21}$$

Using the linear approximation of $g_{dp}(t)$ in Fig. 9.3(d) and employing (9.8), it can be shown that (9.21) boils down to

$$S_{w,out}(f) = 8\frac{\gamma}{\alpha}kT g_{m,4a}\frac{\sqrt{2}}{\pi}\frac{(V_{GS} - V_T)_4}{V_{LO}}. \tag{9.22}$$

In the high-frequency mixer implemented on the IC, the above reasoning no longer holds. Here, the noise current generated in each device can propagate to the differential output not only in the small time window when both transistors are conducting, but also when the switches are in the unbalanced state: at that moment the noise can still flow through $C_{n,2}$. The output noise can be approximated as follows...

Let us assume for the moment that $\omega_{p,2}^2 \ll \omega_0^2$. In that case, any high-frequency noise current injected into the switch node flows entirely through $C_{n,2}$. As a result, $p_{4a}(t)$ and $p_{4b}(t)$ approximately amount to one. The resulting time-varying PSD of the output noise then becomes:

$$S_{w,out}(f,t) \cong 4\frac{\gamma}{\alpha}kT\left(g_{m,4a}(t) + g_{m,4b}(t)\right). \tag{9.23}$$

Using the linear approximation of $g_{m,4a}(t)$ and $g_{m,4b}(t)$ in Fig. 9.3(b) and averaging (9.23) over the period of the LO eventually yields the following expression for the time-average PSD:

$$S_{w,out}(f) \cong 8\frac{\gamma}{\alpha}kTg_{m,4}\left(\frac{\sqrt{2}}{2}+\frac{\sqrt{2}-1}{\pi}\frac{(V_{GS}-V_T)_4}{V_{LO}}\right). \tag{9.24}$$

Finally, it can be taken into account that in reality not all the noise current flows through $C_{n,2}$. In first order, this yields the following approximate expression for the PSD:

$$S_{w,out}(f) \cong 8\frac{\gamma}{\alpha}kTg_{m,4}\left(\frac{\sqrt{2}}{2}+\frac{\sqrt{2}-1}{\pi}\frac{(V_{GS}-V_T)_4}{V_{LO}}\right)\frac{1}{1+\left(\frac{\omega_{p,2}}{\omega_0}\right)^2}. \tag{9.25}$$

Referring this noise to the input of the mixer and normalizing against the reference voltage noise density $4kT50$ leads to the noise figure contribution:

$$F_{(M_4)_w} \cong \frac{\gamma}{\alpha 50}\frac{2g_{m,4}}{g_{m,mixer}^2}\left(\frac{\sqrt{2}}{2}+\frac{\sqrt{2}-1}{\pi}\frac{(V_{GS}-V_T)_4}{V_{LO}}\right)\frac{1}{1+\left(\frac{\omega_{p,2}}{\omega_0}\right)^2}. \tag{9.26}$$

The contribution of the $1/f$ noise of the switches and the white noise of the biasing resistors can be calculated in a similar way. These noise contributions can be modeled by stationary[2] noise voltages at the gates of the switches (see Fig. 9.5) which propagate towards the differential output using the square of the transconductance of the differential pair (depicted in Fig. 9.3(e)). This gives rise to the following expression for the time-varying PSD:

$$S(f,t) = \left(2dv_{1/f}^2 + 2dv_{R_b}^2\right)g_{dp}^2(t). \tag{9.27}$$

Using the linear approximation of $g_{dp}^2(t)$ depicted in Fig. 9.3(e), (9.27) can be averaged over one period of the LO, yielding

$$S(f) = \left(2dv_{1/f}^2 + 2dv_{R_{b,LO}}^2\right)g_{m,4}^2\frac{\sqrt{2}}{\pi}\frac{(V_{GS}-V_T)_4}{V_{LO}}. \tag{9.28}$$

Eventually, it can be shown that the contribution to the noise figure becomes

$$\begin{aligned} F_{(M_4)_{1/f}} + F_{(R_{b,LO})} = {} & \frac{2g_{m,4}^2}{C_{ox,p}W_4L_4g_{m,mixer}^2}N(KF_p, AF_p, f_2, f_1)\frac{\sqrt{2}(V_{GS}-V_T)_4}{\pi V_{LO}} \\ & + \frac{2g_{m,4}^2R_{b,LO}}{50g_{m,mixer}^2}\frac{\sqrt{2}(V_{GS}-V_T)_4}{\pi V_{LO}} \end{aligned} \tag{9.29}$$

where N is given by

$$N(KF, AF, f_2, f_1) = \frac{KF}{-AF+1}\left(\frac{f_2^{-AF+1}-f_1^{-AF+1}}{4kT50\cdot(f_2-f_1)}\right). \tag{9.30}$$

[2]because they are independent of the operating point of the switches and do not depend on the value of the local oscillator signal either.

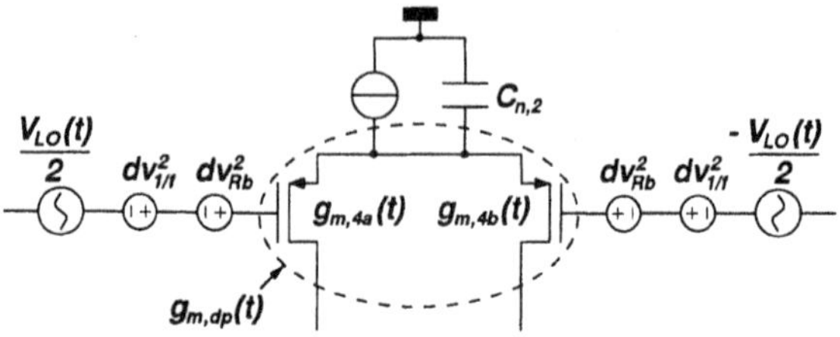

Figure 9.5: *The $1/f$ noise of the switches and the white noise of the biasing resistors can be modeled by stationary noise voltages at the gates of the switches which propagate towards the output with the square of the transconductance of the differential pair.*

It is important to note that the $1/f$ noise of the switches is *not* reduced due to the switching process in the mixer; Unlike in [Gier99], the transistors are switched until V_{GS} dips just below V_T, where they are still in (weak) inversion and thus far away from accumulation. The long term correlation of the $1/f$ noise is thus not destroyed.

CMFB transistors M_{5a} and M_{5b} nMOS transistors M_{5a} and M_{5b} inject both white noise and $1/f$ noise currents into the VGA. Their noise contribution is calculated as

$$F_{E,SSB(M_{5a},M_{5b})} \cong \frac{\gamma}{\alpha 50} \frac{2g_{m,5}}{g_{m,mixer}^2} + \frac{2g_{m,5}^2}{C_{ox,n} W_5 L_5 g_{m,mixer}^2} N(KF_n, AF_n, f_2, f_1), \tag{9.31}$$

with N given in (9.30).

transimpedance VGA The contribution of the transimpedance amplifier (VGA) to the mixer noise figure can be calculated to be

$$F_{(VGA)} \cong \frac{\gamma}{\alpha 50} \frac{2}{g_{m,mixer} CVG} + \frac{dv_{in,OTA}^2}{4kT 50\, CVG^2}. \tag{9.32}$$

The first term denotes the noise of the feedback resistors, written in terms of the conversion gain of the combination mixer-VGA and the conversion transconductance of the mixer. The second term represents the white noise of the OTA embedded in the VGA; It reflects that at low frequencies, the OTA's output noise nearly equals its input noise since the OTA is practically in unity feedback. Observe that the noise contribution is minimum when it is most needed: at low input signals, or equivalently, at high conversion gains.

total noise figure The total eigen noise figure can be broken up in two parts:

$$F_{E,SSB} = F_{E,SSB(white)} + F_{E,SSB(1/f)}, \tag{9.33}$$

where the noise figure contribution due to white noise, $F_{E,SSB(white)}$, is given by

$$F_{E,SSB(white)} \cong \frac{\gamma}{\alpha 50}\left(\frac{2}{g_{m,1}} + \frac{2g_{m,6}}{g_{m,1}^2} + \frac{2g_{m,3}}{g_{m,1}^2}\left(1+\left(\frac{\omega_0}{\omega_{p,1}}\right)^2\right) + \frac{2g_{m,2}}{g_{m,1}^2}\frac{1}{1+\left(\frac{\omega_{p,1}}{\omega_0}\right)^2} + \frac{2g_{m,4}}{g_{m,mixer}^2}\left(\frac{\sqrt{2}}{2}+\frac{\sqrt{2}-1}{\pi}\frac{(V_{GS}-V_T)_4}{V_{LO}}\right)\frac{1}{1+\left(\frac{\omega_{p,2}}{\omega_0}\right)^2} + \frac{2g_{m,5}}{g_{m,mixer}^2} + \frac{2}{g_{m,mixer}CVG}\right) + \frac{dv_{in,OTA}^2}{4kT50\cdot CVG^2} + \frac{2g_{m,4}^2 R_{b,LO}}{50g_{m,mixer}^2}\frac{\sqrt{2}(V_{GS}-V_T)_4}{\pi V_{LO}} \tag{9.34}$$

and the noise figure due to $1/f$ noise, $F_{E,SSB(1/f)}$, is given by

$$F_{E,SSB(1/f)} \cong \frac{2g_{m,5}^2}{C_{ox,n}W_5L_5g_{m,mixer}^2}N(KF_n, AF_n, f_2, f_1) + \frac{\sqrt{2}(V_{GS}-V_T)_4}{\pi V_{LO}}\frac{2g_{m,4}^2}{C_{ox,p}W_4L_4g_{m,mixer}^2}N(KF_p, AF_p, f_2, f_1). \tag{9.35}$$

The expression for N is given in (9.30).

9.3.1.3 Linearity

The linearity can be calculated assuming that the intermodulation products generated in the mixer are solely due to the non-ideal voltage-to-current conversion of the input transconductor. This results in the following estimate for the IP_3 (see also (4.32)):

$$IP_3[\text{dBmv}] \cong 11.25 + 10\log_{10}\left(\frac{(V_{GS}-V_T)_1}{\Theta+\Lambda M}(2+r_{vs})(1+r_{vs})^2\right) \tag{9.36}$$

where M is given by

$$M = \frac{g_{m,1}}{g_{m,2}+g_{mb,2}} \tag{9.37}$$

and r_{vs} denotes the relative amount of velocity saturation, defined as

$$r_{vs} = \Theta(V_{GS}-V_T)_1 = \frac{(V_{GS}-V_T)_1}{V_{si-vs}}. \tag{9.38}$$

Observe that the linearity is a function of the $V_{GS} - V_T$ of the input device only.

9.3.1.4 DC offset

The DC offset at the output of the differential VGA is given by

$$\sigma^2(v_{os}) \cong \left(g_{m,4}\frac{R_f}{2^i}\right)^2 \sigma^2(v_{os,dp}) \tag{9.39}$$

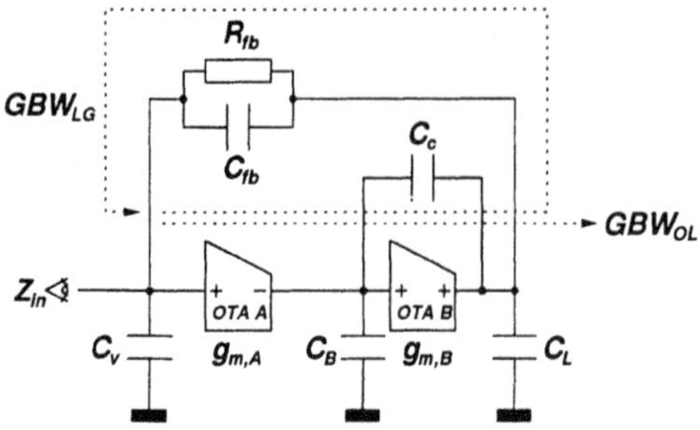

Figure 9.6: *Small signal equivalent circuit of one half of the VGA. The actual implementation of the OTAs can be found in Fig. 9.2.*

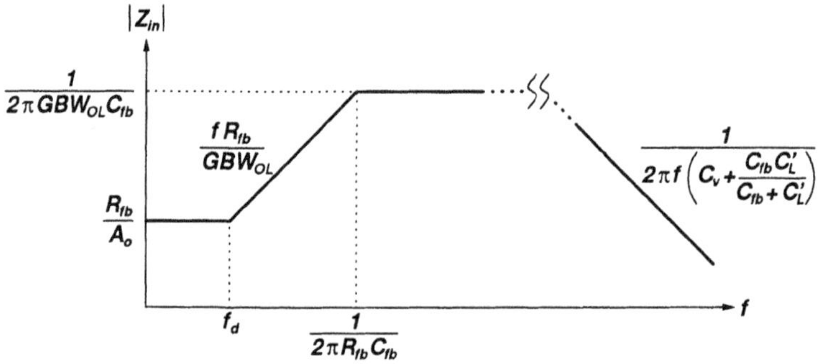

Figure 9.7: *The input impedance at each VGA input terminal.*

where $\sigma^2\left(v_{os,dp}\right)$ is approximately given by [Pelg89]

$$\sigma^2\left(v_{os,dp}\right) \cong \sigma^2\left(\Delta V_T\right) + \frac{\left(V_{GS}-V_T\right)_4^2}{4}\sigma^2\left(\frac{\Delta\beta}{\beta}\right). \tag{9.40}$$

This leads to

$$\sigma^2\left(v_{os}\right) \cong \left(\frac{R_f}{2^i}\right)^2\left(g_{m,4}^2\sigma^2\left(\Delta V_T\right) + I_4^2\sigma^2\left(\frac{\Delta\beta}{\beta}\right)\right). \tag{9.41}$$

9.3.1.5 VGA input impedance

In order to be able to calculate the impedance level of each VGA input node, Fig. 9.6 depicts the small signal equivalent circuit of one half of the differential VGA. In this figure, R_{fb} and C_{fb} represent the effective resistance and capacitance in the feedback, C_v is the capacitance on the virtual ground (including the parasitic capacitance of the input stage,i.e. OTA A in Fig. 9.2), C_B is the parasitic input capacitance of the output stage (OTA B in Fig. 9.2), C_c is the Miller compensation capacitance, C_L is the external load capacitance, and $g_{m,A}$ and $g_{m,B}$ are the transconductances of the respective OTAs.

The result of the calculation is shown in Fig. 9.7. At low frequencies, the input impedance simply equals the feedback resistance divided by the open loop gain of the OTA. As soon as the dominant pole f_d (the internal Miller pole) of the OTA occurs, the input impedance increases proportionally with frequency until the cut-off frequency of the coarse channel select filter is reached. From then on, the impedance becomes flat for a few decades. Its value is given by

$$|Z_{in}| = \frac{1}{2\pi GBW_{OL} C_{fb}}, \tag{9.42}$$

where GBW_{OL} is the open loop gain-bandwidth of the two-stage OTA (see Fig. 9.6), given by

$$GBW_{OL} = \frac{g_{m,A}}{2\pi C_c}. \tag{9.43}$$

This impedance level is relevant for the wanted signal and its adjacent channels.

At radio frequencies, far beyond the gain bandwidth, the OTAs are no longer active and the input impedance behaves capacitive again:

$$|Z_{in}| = \frac{1}{\omega\left(C_v + \dfrac{C_{fb}C'_L}{C_{fb} + C'_L}\right)} \tag{9.44}$$

where C'_L is given by

$$C'_L = C_L + \frac{C_B C_c}{C_B + C_c}. \tag{9.45}$$

This expression can be used to calculate the attenuation of the LO signal.

9.3.1.6 VGA stability

The gain-bandwidth of the loop (see Fig. 9.6) is given by

$$GBW_{LG} = GBW_{OL}\frac{C_{fb}}{C_v + C_{fb}} \tag{9.46}$$

$$= \frac{g_{m,A}}{2\pi C_c}\frac{C_{fb}}{C_v + C_{fb}}. \tag{9.47}$$

The non-dominant pole (the pole of the output stage) is then

$$f_{nd} = \frac{g_{m,B}}{2\pi\left(C_{L,eff}\left(1 + \dfrac{C_B}{C_c}\right) + C_B\right)} \tag{9.48}$$

where $C_{L,eff}$ is the effective load capacitance that must be driven by the output stage, given by

$$C_{L,eff} = C_L + \frac{C_{fb}C_v}{C_v + C_{fb}}, \tag{9.49}$$

The zero associated with the Miller compensation capacitance is given by

$$f_z = \frac{g_{m,B}}{2\pi C_c}. \tag{9.50}$$

Both frequencies must lie sufficiently far above GBW_{LG} to guarantee the phase margin and gain margin. Observe that the stability constraint is most stringent at the lowest gain level; At that time, the feedback capacitance is the largest so that gain-bandwidth of the loop — given in (9.47) — is maximum and the frequency of the non-dominant pole — given in (9.48) — is the lowest.

9.3.2 Design Procedure

The following paragraphs discuss the actual design procedure of the mixer and the VGA/filter i.e. the sizing of the mixer core and the VGA elements and the high level design of the OTAs. In principle, the design procedure described in this section can be automated. Yet, one must be aware of the fact that during the design many decisions have been taken that depend on the values of the specifications. Therefore, a different set of specifications generally changes the path which needs to be followed.

input transistor M_1 The input transistor is the most critical transistor of the mixer; Its $V_{GS} - V_T$ determines the linearity, its g_m controls the noise figure, and its area controls the gain mismatch between the quadrature paths. Let us first look at the linearity requirement.

Since the required mixer IP_3 is very high — in the order of +17 dBm$_V$— the over-drive voltage of input transistor M_1 needs to be large. As can be seen from (9.36) or Fig. 4.5, depending on the channel length a $V_{GS} - V_T$ value of 0.5 to 0.6V is required, which inevitably leads to a low transconductance efficiency. Even though minimum length transistors indeed offer the required linearity at a lower $V_{GS} - V_T$ value compared to non-minimum length transistors, the use of minimum length transistors does not lead to the most power-efficient solution; By choosing a larger length transistor, a better transconductance efficiency can be obtained despite the fact that a larger $V_{GS} - V_T$ is required for the same linearity. As the required $V_{GS} - V_T$ is determined by the channel length — which is set (later on) by mismatch considerations — it is necessary to start the design procedure with a reasonable guess on the required $V_{GS} - V_T$. The channel length and the $V_{GS} - V_T$ that result from the iterative design process are 0.5 μm and 0.6V, respectively.

The g_m of M_1 (and hence the current through M_1) is determined by noise considerations; To minimize the power consumption, the lowest g_m is to be chosen that still satisfies the noise specification. An initial value for the required g_m can be obtained by making an estimate for the so-called excess white noise figure, which weighs the mixer noise factor due to white noise against the eigen noise factor of the input transconductor:

$$\epsilon \triangleq \frac{F_{E,SSB(white)}}{\frac{\gamma}{\alpha 50 g_{m,1}}}. \tag{9.51}$$

The reason why the white noise is taken as a reference is simply that it is easier to make an educated guess for ϵ when $1/f$ noise is not included. The minimum g_m is then given by

$$g_{m,1} \geq \frac{\epsilon\gamma}{\alpha 50 F_{E,SSB(white)}}. \tag{9.52}$$

After some iterations, $F_{E,SSB}$ can then be set to $(F_{E,SSB})_{spec}$ while ϵ can be made consistent with (9.51). In this design, the $g_{m,1}$ has converged to 13 mS (5.1 mA at $V_{GS} - V_T$ of 0.6V). Note that it is not recommended to use larger values for $g_{m,1}$ than required from the point of view of noise since this not only increases the current of the main branch, but also the minimum current through the switches.

Finally, the area of M_1 needs to be set. M_1 plays a dominant[3] role in the gain mismatch between the I and Q paths and hence in the image rejection ratio. Using [Pelg89], it can be shown that the contribution to the gain mismatch is given by

$$\sigma^2\left(\frac{\Delta g_{m,1}}{g_{m,1}}\right)=\frac{\sigma^2(\Delta V_T)}{(V_{GS}-V_T)^2}+\sigma^2\left(\frac{\Delta\beta}{\beta}\right). \tag{9.53}$$

For the 75 μm × 0.5 μm (225 fF) transistor implemented on the IC, $\sigma(\Delta V_T)$ and $\sigma\left(\frac{\Delta\beta}{\beta}\right)$ are 3.5 mV and 0.6%, respectively, so that the spread on the gain error amounts to 0.85%. At 3σ, the gain error is 2.5% or 0.22dB, which still complies to the image rejection requirement.

cascode M_2, bleeder M_6 and current source M_3 As mentioned in Section 9.2, the current through the cascode transistor (I_2) needs to be minimized in order to maximally reduce both the parasitic drain capacitance and the current noise of $M3$ ($\propto g_{m,3}$). This is very important since the $V_{GS}-V_T$ of $M3$ is generally quite small (0.3V), and hence both its transconductance per unit current and its W/L per unit current is large. The rest of the current, $I_6 = I_1 - I_2$, needs to be provided by pMOS bleeder transistor M_6. As this device can have a large $V_{GS}-V_T$ (maximally $V_{DD}-s(V_{GS}-V_T)_1$ with $s \geq 1$ a safety factor), its current noise can be made much smaller than the corresponding noise contribution of M_3 when M_3 had to provide this current.

The minimum I_2 value is determined by the Miller effect, the pole on node ① and the maximum signal current injected into node ①. In the mixer implemented on the IC, I_2 is set by this last constraint:

$$I_2 \geq s g_{m,1} v_{in,max}, \tag{9.54}$$

where $v_{in,max}$ is the amplitude of the maximum input signal of the mixer, and s is a safety factor. The reason for the existence of this specification is of course that it can not be tolerated that M_2 is switched off by the RF signal current. It is good design practice to make $s \geq \sqrt{2}$[4]. In this design, $v_{in,max}$ is -10 dBmv (a sine with an amplitude of 0.1V) and s is taken 2[5], setting $I_{2,min}$ to 2.5 mA.

For several reasons, M_2 must be biased at a low $V_{GS}-V_T$ value, at the transition between weak and strong inversion. First of all, a low $V_{GS}-V_T$ value enables the use of a larger $V_{GS}-V_T$ for pMOS current source M_3. This reduces both its noise contribution and capacitive loading of the switch node. Even at this low $V_{GS}-V_T$, M_2 itself does not contribute much capacitance to node ②, partly because it's an nMOS device carrying less current, partly because its drain junction features a large reverse voltage. Secondly, the V_{DSsat} of M_2 — and hence its $V_{GS}-V_T$— must be low to accommodate the signal swing on node ②. A small $V_{GS}-V_T$ also efficiently reduces the signal swing on node ① and therefore improves isolation; A simple small-signal

[3]The mismatch between the RC transimpedance elements in the VGA, is less important because their area can be much larger.

[4]$\sqrt{2}$ is the ratio between the peak amplitude of an AM signal of x dBm and a sine wave of x dBm.

[5]s is taken somewhat larger than $\sqrt{2}$ to ensure that M_2 still operates as a cascode, even at large signal amplitudes.

analysis reveals that

$$v_{①} \cong v_{②} \frac{1}{r_{o,2}} \frac{1}{\omega C_{n,②}} \sqrt{1 + \left(\frac{\omega_{p,1}}{\omega}\right)^2} \tag{9.55}$$

$$\cong v_{②} \frac{1}{n g_{m,2} r_{o,2}} \frac{\omega_{p,1}}{\omega} \sqrt{1 + \left(\frac{\omega_{p,1}}{\omega}\right)^2} \tag{9.56}$$

$$\sim (V_{GS} - V_T)_2^3. \tag{9.57}$$

The last reason why a small $V_{GS} - V_T$ is required is the reduction of the Miller effect. In the end, the pole on node ① sets a minimum on the $V_{GS} - V_T$ (since $\omega_{p,1} \propto \omega_{T,2} \propto (V_{GS} - V_T)_2$). The cascode implemented on the IC features a $V_{GS} - V_T$ of 0.2V, corresponding to an $\omega_{p,1}$ of 10 GHz.

Current source transistor M_3 must supply the required current at the maximum possible $V_{GS} - V_T$. In this way, the transconductance of M_3 — and hence its noise current — is reduced. In this design, the $V_{GS} - V_T$ of M_3 is set to 0.32V.

switches M_{4a} and M_{4b} The sizing of the switches is difficult and easy at the same time; On one hand, the $V_{GS} - V_T$ of the switches controls in a complicate way the balance between the $1/f$ noise and the white noise of both M_4 and M_5. We will come back to this later. On the other hand, the design room for $V_{GS} - V_T$ is so small that little can be optimized. Let us first examine this last aspect.

The $V_{GS} - V_T$ of the switch pair must be low enough in order to make sure that the local oscillator signal can completely switch off the transistors. Beyond this point, the performance of the mixer drops very rapidly. Therefore, the $V_{GS} - V_T$ must obey the following inequality:

$$s\sqrt{2}(V_{GS} - V_T)_4 \leq V_{LO}, \tag{9.58}$$

where s is again a safety factor. For a 0 dBm$_V$ LO signal, the absolute maximum for $V_{GS} - V_T$ is thus 0.21V. Of course, it is recommended to take some more margin.

There is also a minimum feasible value for the $V_{GS} - V_T$ of the switch pair. This can be understood as follows. The DC current through each of the switches must be large enough to ensure that the RF signal current always finds its way to the output:

$$I_4 \geq \frac{s g_{m,1} v_{in,max}}{2\sqrt{1 + \left(\frac{\omega_0}{\omega_{p,2}}\right)^2}}. \tag{9.59}$$

The term in $\omega_{p,2}$ represents the current loss on the switch node and indicates that a lower pole frequency reduces the minimum required current for the switches. This is already the first part of the story. A second important aspect is that the LO section which drives the gates of the mixer switches can only cope with a limited capacitive load, limiting the area of the switches to

$$W_4 L_4 \leq \frac{C_{LO,max}}{C_{ox}}. \tag{9.60}$$

Combining (9.59) and (9.60) with the fact that I_4 is determined by the W/L and the $V_{GS} - V_T$

of the switches, yields the following minimum value for the $V_{GS} - V_T$:

$$(V_{GS} - V_T)_{4,min} \cong \sqrt{\frac{I_{4,min} L_4^2}{K_p W_4 L_4}}. \tag{9.61}$$

Even when choosing the minimum length for M_4, the minimum feasible over-drive voltage can be considerable. The reason is that a low noise figure generally requires a larger $g_{m,1}$, and hence also a larger $I_{4,min}$. In our design, the minimum $V_{GS} - V_T$ is about 0.15V.

Now that we know that the $V_{GS} - V_T$ of the switch pair can only reside in a small window — in our case, between 0.15V (maximum allowable capacitance of the switches attained) and 0.21V (end of the switching mixer regime with a 0 dBm LO) — it should be indicated how the actual $V_{GS} - V_T$ value must be determined. Depending on which noise mechanism in M_4 and M_5 plays the dominant role, the $V_{GS} - V_T$ must be chosen at the low end or at the high end of the window:

- By choosing a $V_{GS} - V_T$ at the lower end, the conversion efficiency of the switches can be increased so that the input-referred $1/f$ noise of both M_4 and M_5 is reduced. It is worth noting that the reduction of $\omega_{p,2}$ that is associated with the decrease in $V_{GS} - V_T$ has only a small effect on the $1/f$ contributions; It can be shown that the boost in the input-referred $1/f$ noise introduced by the pole is compensated for both M_4 and M_5:
  - in case of M_4, by the shorter time that both switches are on, and
  - in case of M_5, by a corresponding decrease in $I_{4,min}$ and therefore also in the $1/f$ noise current itself.
- By choosing a $V_{GS} - V_T$ at the high end of the window, the white noise of M_4 and the white noise of $R_{b,LO}$ can be reduced; A larger $V_{GS} - V_T$ decreases the transconductance of the switches and hence their own noise current as well as the propagation of the voltage noise of $R_{b,LO}$ to the output. The other factors — the improvement in the pole and the reduced conversion efficiency of the switches — are less important.

In this design, the $V_{GS} - V_T$ of the switches is set to 0.15V in order to maximally reduce the contribution of the $1/f$ noise.

Until now, we have only talked about the $V_{GS} - V_T$ of the switches, but of course also their DC current must be determined. It should be clear that there is no point in increasing the current beyond the minimum current in (9.59); A larger current degrades the noise figure, increases the power consumption and makes the $V_{GS} - V_T$ window even smaller (see (9.61)). I_4 should therefore be set according to (9.59) and is 0.6 mA (for $s = \sqrt{2}$).

Now is the right time to check whether the DC offset is still reasonable. For the 160 μm × 0.25 μm (240 fF) transistor implemented on the IC, $\sigma(\Delta V_T)$ and $\sigma\left(\frac{\Delta\beta}{\beta}\right)$ are 3.3 mV and 0.6%, respectively. Later on, R_f is calculated to be 33 kΩ, so that the spread on the worst case DC offset (i.e. at maximum VGA gain) amounts to 0.6V. This number can be improved upon by increasing the area of the mixing transistors. Yet, in this particular design, the externally imposed constraint on the maximum load capacitance for the LO section has kept us from doing just that. Nevertheless, by adding an extra DC feedback at the output — as is done in zero-IF receivers — this problem can be alleviated.

current sources M_{5a} and M_{5b} Owing to the 10 times larger KF of nMOS transistors compared to pMOS transistors, and due to the subtle difference in the slope of their $1/f$ noise, $N(KF_n, AF_n, 200\text{ kHz}, 1\text{ kHz})$ is about 30 times larger than $N(KF_p, AF_p, 200\text{ kHz}, 1\text{ kHz})$. Hence, if no special measures are taken, nMOS current sources M_{5a} and M_{5b} inject a lot of $1/f$ noise into the VGA. Therefore, it is extremely important to maximize both the $V_{GS} - V_T$ and the area of M_{5a} and M_{5b}. In addition, this action reduces the white noise current ($\propto g_{m,5}$).

The $V_{GS} - V_T$ is constrained by the common-mode reference voltage:

$$s(V_{GS} - V_T)_5 \leq V_{CM,ref}. \tag{9.62}$$

In this design, the $V_{GS} - V_T$ is set to 0.75V, 33% lower than the common-mode reference voltage of 1V. The area of the transistors has been set to 1440 μm^2, which is sufficient to make the $1/f$ noise of M_{5a} and M_{5b} a few times lower than the $1/f$ noise of the pMOS switches. In addition, the large area makes that M_{5a} and M_{5b} do not play any role in the DC offset. It is worth noting that stability considerations put a maximum on the allowable area because the gate capacitance of M_{5a} and M_{5b} determines the non-dominant pole of the common-mode feedback.

Since the $1/f$ noise is ultimately dominated by the noise of the pMOS switches, and the noise figure contribution of the $1/f$ noise is eventually comparable to the contribution of the white noise, it is clear that pMOS devices were indeed the best choice for the switches. It would be virtually impossible to get the same result if nMOS switches are used: the factor 30 is simply too much to compensate.

transimpedance elements R_f and C_f The value of the resistors and the capacitors in the feedback of the VGA can be readily deduced from the required gain levels and the filter bandwidth.

The value of the largest resistor — the one which is always in the feedback — is given by

$$R_f = \frac{CVG_{max}}{g_{m,mixer}}, \tag{9.63}$$

where CVG_{max} is the maximum VGA gain level and $g_{m,mixer}$ is the conversion transconductance of the mixer core, defined in (9.6). In this design, $g_{m,mixer}$ and CVG_{max} are 4.26 mS and 43 dB, respectively, leading to an R_f of 33 kΩ. Since the ratio between the different gain levels of the VGA is 6 dB, the remaining resistors in the feedback are simply R_f, $R_f/2$, $R_f/4$ and $R_f/8$. When all resistor elements are put in parallel (i.e. at the lowest VGA gain), the total feedback resistance amounts to 2.1 kΩ. At that time, the contribution to the mixer noise figure is around 3 dB. At the highest gain, the noise of the resistors becomes negligible.

All the resistors are implemented in high-ohmic polysilicon using $R_f/8$ unit resistors. The resistors in the array are enclosed by dummy polysilicon strips to prevent edge effects. The area of the unit resistor must be chosen large enough in order no to cause too much gain mismatch between the I and Q paths. The worst case mismatch occurs at the lowest gain; At that time, half the signal current flows through $R_f/8$ — the resistor with the smallest area — while the other half flows through the resistors that had already been activated before. It can be shown that the gain mismatch due to mismatch in $R_f/8$ is given by

$$\sigma^2\left(\frac{\Delta A}{A}\right) \cong \frac{1}{2}\sigma^2\left(\frac{\Delta R}{R}\right), \tag{9.64}$$

where $\sigma^2\left(\frac{\Delta R}{R}\right)$ is the matching between two $R_f/8$ resistors. The factor 2 is caused by the damping effect of the rest of the array. The above equation provides an optimistic estimate for the total mismatch at the lowest gain level, yet, the effect of the other resistors in the array is less pronounced anyway because they carry less signal current and because their area is larger. In our design, the worst case gain mismatch due to the resistors can be made less than 0.1 dB.

The value of the smallest capacitor — the default capacitance in the feedback — is given by

$$C_f = \frac{1}{2\pi R_f BW}. \tag{9.65}$$

For the 300 kHz bandwidth and the 33 kΩ R_f, C_f is 16 pF. The remaining capacitors obviously need to have a value of C_f, $2C_f$, $4C_f$ and $8C_f$.

transimpedance switches The nMOS switches that connect the different RC elements in parallel need to have a low on-resistance in order not to disturb the gain levels:

$$R_{on} \ll \frac{R_f}{8}. \tag{9.66}$$

In addition, in order not to degrade the linearity of the current-to-voltage conversion, the switches always need to stay in the linear region: This requires that

$$\frac{1}{2} g_{m,mixer} v_{in,max} R_{on} \ll V_{GS} - V_T, \tag{9.67}$$

or equivalently, that

$$\frac{1}{2} g_{m,mixer} v_{in,max} \ll I_{DSsat}. \tag{9.68}$$

Since the over-drive voltage is small — at 2V V_{DD}, the $V_{GS} - V_T$ is about 0.2V—, this requires a large W/L.

OTA design The first OTA variable which is set is the OTA's open loop gain-bandwidth product. The minimum gain-bandwidth is related to the maximum allowable VGA input impedance. In this respect, two constraints must be met. First of all, a low input impedance is required at low frequencies to efficiently process the down-converted current. This is especially important at high VGA gains since at that time the feedback resistance is maximum. The maximum input impedance being given by (9.42), this requirement becomes

$$\frac{1}{2\pi GBW_{OL} C_{fb}} \ll Z_{out,mixer}. \tag{9.69}$$

Secondly, the total signal swing at the VGA input must be low enough to be compatible with the limited $V_{GS} - V_T$ of the first OTA (< 0.2V) and to make sure that the switch transistors M_{4a} and M_{4b} always remain in saturation. This translates into the requirement that

$$\frac{1}{2}\frac{g_{m,mixer} v_{in,max}}{2\pi GBW_{OL} C_{fb}} \ll \min\left((V_{GS} - V_T)_A, (V_{DS} - V_{DSsat})_{4a|4b}\right). \tag{9.70}$$

For a maximum signal swing of 50 mV at the highest gain level, the required GBW_{OL} is about 40 MHz. The input impedance then amounts to 250 Ω, compared to a few kΩs for the output impedance of the mixer core.

Of course, in order for the above expressions to be relevant, also the open loop gain of the OTA must be large enough, however, this is generally no problem in two-stage OTA's. The Miller OTA on the IC — OTA A and OTA B in Fig. 9.2 — features a gain of 68 dB, so that the input impedance seen by the wanted signal changes from 13 Ω at f_d (16 kHz) to 250 Ω at the 300 kHz cut-off frequency.

Analogously, the local oscillator signal which is injected onto the input node of the VGA must be deviated to ground without generating too much voltage swing. This leads to the requirement that

$$\frac{g_{m,4}V_{LO}}{2\pi f_{LO}C_v} \ll \min\left((V_{GS}-V_T)_A, (V_{DS}-V_{DSsat})_{4a|4b}\right). \tag{9.71}$$

For a swing of 50 mV, it turns out that only a C_v of 3 pF is required. Nevertheless, in the IC a 100 pF C_v is implemented, limiting the swing to below 2 mV. A large C_v eases the stability of the common-mode feedback of the mixer core by killing the gain-bandwidth of the elementary OTA M_5/C_v — we don't need it anyway — and relaxing the frequency of the non-dominant pole. This is highly welcome since M_5 features a large gate capacitance to reduce its $1/f$ noise.

The rest of the OTA design is focussed on achieving sufficient stability at a minimum power consumption while generating the required GBW_{OL}. The expressions for the poles and zeroes have been stated in 9.3.1.6. The design flow is as follows.

First, a good starting point for the design must be found. If we assume that $C_B \ll C_{L,eff}$ — which is always the case given the huge load capacitance — and if we assume for the moment that $C_c \gg C_B$ — which we have under control —, (9.48) becomes

$$f_{nd} \cong \frac{g_{m,B}}{2\pi C_{L,eff}}, \tag{9.72}$$

where $C_{L,eff}$ is given by (9.49):

$$C_{L,eff} = C_L + \frac{C_{fb}C_v}{C_v + C_{fb}}, \tag{9.73}$$

Based on this optimistic expression for f_{nd}, and assuming that $f_z \gg f_{nd}$ (i.e. $C_c \ll C_{L,eff}$), an absolute minimum for $g_{m,B}$ can be calculated to get 72 degrees of phase margin:

$$g_{m,B} \geq 3 \cdot 2\pi GBW_{CL}C_{L,eff}. \tag{9.74}$$

For a C_L of 50 pF, corresponding to a 25 cm BNC cable and 25 pF measurement equipment, $C_{L,eff}$ in (9.73) amounts to 122 pF, yielding

$$g_{m,B} \geq 66 \text{ mS}. \tag{9.75}$$

Even at a $V_{GS} - V_T$ of 0.15V, the realization of this transconductance value already requires a current of 6.2 mA [6] in each phase of the differential output stage. Note that we are allowed to choose such a low $V_{GS} - V_T$ for the output stage because of the dominance of $C_{L,eff}$ in the expression for the pole. In addition, the low $V_{GS} - V_T$ is very useful to guarantee a large output capability despite the low V_{DD}.

[6] The fact that $\Theta = 4.15$ must be taken into account!

Now the time has come to look for a suitable starting value for C_c. Until now, we have assumed that C_c is large enough in order not to affect the non-dominant pole — the condition that $C_c \gg C_B$ — and low enough in order not to endanger stability due to the associated zero — the requirement that $C_c \ll C_{L,eff}$. Because $C_{L,eff}$ is so large (122 pF) and C_B will most likely be only a few pF, there is no problem whatsoever to find a good starting value for C_c. 25 pF would for instance be a good compromise; In that case, $f_z \cong 5 f_{nd}$, degrading the phase margin by only 3.8 degrees.

In the final stage, the design is refined to minimize the total power consumption of the OTA. This is done as follows. First, it is examined how much power flows through the first stage (OTA A in Fig. 9.2). The required $g_{m,A}$ can be calculated from (9.43):

$$g_{m,A} = 2\pi GBW_{OL} C_c. \tag{9.76}$$

for a 25 pF C_c, this results in a $g_{m,A}$ of 6.3 mS, or equivalently in a current of 0.6 mA at a $V_{GS} - V_T$ of 0.15V. The total current consumption of the starting point is thus 2(0.6+6.2) mA = 13.6 mA.

By further decreasing C_c, the power in the first stage (OTA A) can be reduced linearly. Initially, this happens even without affecting the f_{nd} given in (9.48), and hence without requiring additional power in the output stage (OTA B). At the same time, f_z is pushed to higher frequencies, improving the stability. Eventually, decreasing C_c no longer pays off since the additional improvement in the current consumption of the first stage is counteracted by a correspondingly larger current in the output stage. As a result, the optimum C_c is obtained when

$$\frac{\partial I_A}{\partial C_c} = -\frac{\partial I_B}{\partial C_c}. \tag{9.77}$$

In this particular design, contour plots in Matlab revealed an optimum C_c of 8 pF, resulting in a $g_{m,A}$ of 2 mS (0.2 mA at 0.15V), a $g_{m,B}$ of 70 mS (6.5 mA at 0.15V) and a phase margin of 70 degrees. Obviously, the starting point already was an excellent design point in terms of power consumption; Little can be optimized because the optimum is very flat.

The contribution of the OTA to the noise figure of the mixer is only 2.5 dB. The internal and the external slew rates are 50 $V/\mu s$ and 53 $V/\mu s$, respectively.

summary Table 9.1 summarizes the transistor sizes, component values and biasing information of the mixer and the transimpedance VGA/filter. Essentially, everything in the table has been covered in detail in the preceding paragraphs. The resulting performance of the in-phase (I) signal path is listed in Table 9.2.

As can be derived from the table, each channel of the quadrature receiver — containing a mixer and a differential VGA-filter — draws 22.5 mA from a 2V supply, or equivalently, consumes 45 mW. However, a vast amount of the power in the VGA is burned to drive the 50 pF of the off-chip measurement equipment whereas in reality the VGA only needs to drive the on-chip data converters — merely a few pF. In that case, $C_{L,eff}$ is no longer 122 pF but 72 pF, which is 41% lower. As a result, instead of the 13 mA which now flows in the output stage, only 7.7 mA would be required in the fully integrated solution, resulting in a total VGA current of 9.7 mA instead of 15 mA. In a fully integrated receiver, each channel needs to draw only 17.2 mA, corresponding to a power consumption of 34.4 mW per channel.

$(I_{DS})_1$	5.1 mA	$(I_{DS})_2$	2.5 mA
$(V_{GS}-V_T)_1$	0.6 V	$(V_{GS}-V_T)_2$	0.19 V
$(W/L)_1$	75/0.5	$(W/L)_2$	110/0.25
$(g_m)_1$	13 mS	$(g_m)_2$	20 mS
$(I_{DS})_3$	3.7 mA	$(I_{DS})_4$	0.6 mA
$(V_{GS}-V_T)_3$	0.32 V	$(V_{GS}-V_T)_4$	0.15 V
$(W/L)_3$	320/0.25	$(W/L)_4$	160/0.25
$(g_m)_3$	18 mS	$(g_m)_4$	5.5 mS
$(I_{DS})_5$	0.6 mA	$(I_{DS})_6$	2.6 mA
$(V_{GS}-V_T)_5$	0.75 V	$(V_{GS}-V_T)_6$	0.75 V
$(W/L)_5$	120/12	$(W/L)_6$	50/0.25
$(g_m)_5$	1 mS	$(g_m)_6$	5.5 mS
$R_{b,LO}$	5 kΩ	$V_{ref,LO}$	0.4 V
$C_{b,LO}$	1.5 pF	$V_{CM,ref}$	1 V
R_{CM}	20 kΩ	R_1	333 Ω
C_{CM}	1.5 pF	R_2	2 kΩ
R_f	33 kΩ	C_v	100 pF
C_f	16 pF	C_c	8 pF
I_A	0.2 mA	I_B	6.5 mA
$g_{m,A}$	2 mS	$g_{m,B}$	70 mS

Table 9.1: *Transistor sizes, component values and biasing information of the mixer and the variable-gain amplifier.*

Mixer core	Noise figure (*NF*)	17.7 dB
	due to white noise	15.2 dB
	due to $1/f$ noise	14.1 dB
	Conversion transconductance ($g_{m,mixer}$)	4.26 mS
	Third order intercept point (IIP_3)	+17 dBm_V
	Maximum I/Q mismatch	0.22 dB
	Input capability ($V_{in,max}$)	-10 dBm_V
	LO signal amplitude (V_{LO})	0 dBm_V
	Input capacitance at LO side	240 fF/phase
	Minimum operating voltage	1.8 V
	Nominal supply voltage	2 V
	Current consumption (including biasing and CMFB)	7.5 mA
VGA/ Filter	Cut-off frequency	300 kHz
	Gain levels (mixer together with VGA)	19–43 dB 6 dB/step
	Additional contribution to mixer noise figure	< 2.5 dB
	Additional contribution to I/Q Mismatch	< 0.1 dB
	DC output level	1 V
	Output capability	> 11 dBm_V
	Maximum output load	50 pF/phase
	Minimum operating voltage	1.9 V
	Nominal supply voltage	2 V
	Current consumption (including biasing and CMFB)	15 mA

Table 9.2: *Projected spec-sheet of the down-conversion mixer and the VGA/filter.*

By decreasing C_v and at the same time reducing the gain of the common-mode feedback of the mixer core, the power consumption can be lowered as well: if C_v is decreased from 100 pF to 50 pF, $C_{L,eff}$ moves from 72 pF to 42 pF, leading to a total power consumption of only 28 mW per channel!

9.4 The Low Noise Amplifier

In this section, the design trajectory of the cascode low noise amplifier is covered. The trade-offs involved when designing this class of LNAs have already been discussed in detail in Chapter 6. In fact, with a single exception, the design of *this specific DCS-1800 LNA* has been the subject of the case study in Section 6.9, Section 6.10 and Section 6.13. In this section, we will cover the design trajectory of the LNA which is actually implemented on the IC. It will soon become clear what the differences are with the LNA designed in the case study.

9.4.1 Selection of the LNA Input Impedance

In the case study of Section 6.9, the fixed part of the input capacitance, $C_{p,fixed}$, is assumed to be 300 fF. At this $C_{p,fixed}$ value, there is no problem in matching the device to 50 Ω using the classical (direct) matching technique; All it requires is that the amplifying device is positioned in the region in Fig. 6.13(a) where $R_{S,max} \geq 50\ \Omega$. Yet, when $C_{p,fixed}$ increases, the forbidden region extends very fast (see Section 6.5). Precisely this happens in the LNA implemented on the IC, which features a $C_{p,fixed}$ of 370 fF— 270 fF from the bondpad and 100 fF from the protection — instead of the 300 fF used in the case study. In the presence of this capacitance, a direct 50 Ω match is no longer possible at 1.84 GHz [7]: $R_{S,max}$ is always smaller than 50 Ω. We are therefore *forced* to use a lower input impedance.

On the other hand, we *ask* for nothing better than lowering the input impedance; As explained in Section 6.10, a lower input impedance enables us to achieve the required performance at a lower current consumption. Whereas the optimum impedance in the case study was 32 Ω— the impedance value at which the linearity spec became binding —, here the optimum value must necessarily be lower than 32 Ω. After all, because of the larger $C_{p,fixed}$ compared to the one used in the case study, the signal at the gate of the amplifying device is smaller, so that linearity can simply not be binding. However, there are good reasons not to go lower than 32 Ω; The reason is that we will eventually choose to omit the extra matching network which transforms the 50 Ω source down to the LNA input impedance. To still comply to the filter termination spec, the input impedance may not deviate too much from 50 Ω. If one wants to guarantee a $\Gamma \leq -12$ dB, the minimum input impedance that can be allowed turns out to be 32 Ω.

9.4.2 Optimization of the Coupling Capacitor

Fig. 9.8 shows the LNA output section (forget R_{poly} for now). To enable the DC levels of the LNA and the mixer to be different, the LNA output node is capacitively coupled to the mixer. The

[7] In fact, this is the reason why we selected 300 fF in the case study; Otherwise the difference between a 50 Ω match and a 32 Ω match couldn't have been demonstrated.

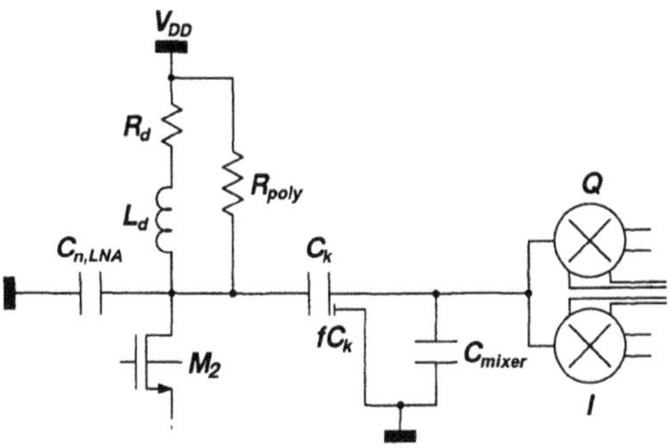

Figure 9.8: *The LNA output section.*

coupling capacitor C_k introduces a loss (due to capacitive division) which must be compensated for.

In principle, any loss introduced by the capacitive division can be counteracted by increasing the load resistance of the LNA core. However, this comes at the cost of an increased quality factor since

$$Q_{out} \propto C_{out} R_{out}. \tag{9.78}$$

Consequently, C_k can not be made too low. On the other hand, the bottom plate of C_k has a stray capacitance to the substrate which is a fraction of C_k (e.g. 8%). If C_k is too large, the stray capacitance dominates the capacitance on the output node of the LNA, which also increases the output quality factor.

It should be clear from the above that there exists an optimum C_k value that minimizes the quality factor of the output section. It can be calculated that the optimum C_k is given by

$$C_{k,opt} = \sqrt{\frac{C_{n,LNA} C_{mixer}}{f}}, \tag{9.79}$$

where $C_{n,LNA}$ is the capacitance on the output node owing to the LNA core (the parasitic capacitance of the cascode device and the inductor), C_{mixer} denotes the input capacitance of the mixer, and f is the ratio between the stray capacitance and the coupling capacitance (e.g. 0.08).

In this design, the influence of C_k on the net LNA gain (via the capacitive division) and on the total LNA quality factor (via its contribution to the total output capacitance) is taken into account in advance by making a first order estimate of $C_{n,LNA}$, calculating the optimum C_k using (9.79), and using this C_k value throughout the whole design procedure. Later on, we will be able to decide whether or not an extra iteration is needed. In the following design procedure, C_k is taken 1.5 pF— corresponding to the optimal C_k for a $C_{n,LNA}$ of 400 fF and a C_{mixer} of 450 fF [8]. The capacitive loss of 2.8 dB is compensated by increasing the gain of the LNA by the same amount.

[8]two 225 fF mixers (see page 199).

R'_S Ω	R_{in} Ω	NF dB	IP_3 dBm	Q	L_g nH	I_{DS} mA	$V_{GS} - V_T$ V	W/L	g_m mS	C_{gs} fF
32	32	✓ 1.15	✓-0.5	✓ 2.4	✓ 8	4.5	0.15	260/0.25	42	330
50	32	✓ 1.25	✓-0.2	✓ 2.3	✓ 8	4.5	0.15	260/0.25	42	330
50	32	✓ 1.5	✓ 0.3	✓ 2.4	✓ 7	4.5	0.12	350/0.25	48	440

Table 9.3: *Performance summary of different candidate LNA designs.*

9.4.3 Design Trajectory

Fig. 9.9(a)–(i) shows the contour lines of the LNA in the design space of the amplifying device. The device is considered to be conjugately matched to 32 Ω i.e. the matching network has not yet been omitted. It is readily derived from Fig. 9.9(i) that the minimum current for this design is 4.5 mA, and that the design is again fully determined by the constraints on Q and L_g. The properties of this LNA are summarized in the first row of Table 9.3. Note that the noise figure is slightly lower than the noise figure of the corresponding 32 Ω LNA with a $C_{p,fixed}$ of 300 fF in Table 6.6. This is explained by the fact that the current $R_{S,eq}$ value of 125 Ω is closer to the optimum source resistance (a noise match) than the 97 Ω in Subsection 6.13.1.

When omitting the matching network, essentially the identical thing happens as in Subsection 6.13.2; The $R_{S,eq}$ moves from 125 Ω to 185 Ω, slightly improving the total Q and the IP_3 and causing a small degradation in the noise figure. The performance of the unmatched LNA is summarized in the second row of Table 9.3. The noise figure in the table has been calculated assuming that the non-quasi static noise term and the classical noise term are approximately equal. This approximation holds near the minimum noise figure.

In a last step, the margin on the Q-factor in the previous solution is used up to decrease L_g. This is done by lowering the $V_{GS} - V_T$ (while keeping I_{DS} constant) until one meets the new '$Q = 2.4$' contour (the '$Q = 2.5$' contour in Fig. 9.9(i)). This leads to the solution in the last row of Table 9.3, the design which is implemented on the IC.

9.4.4 Practical Implementation

The above design has extensively been simulated in HSpice using the BSIM3v3 model. The non-quasi static gate resistance and the non-quasi static noise have been added manually according to the method described in Subsection 5.4.2 and Subsection 5.6.4. The design has resulted in the LNA spec-sheet in Table 9.4. The simulated noise figure includes the noise contribution of the bondpad structure, the ESD diodes and all the parasitic resistances in the circuit. Although the 2.2 dB noise figure is larger than the 1.8 dB target we have used in the design procedure, 2.2 dB is still much lower than the noise figure required by the DCS-1800 standard. As indicated in the table, the LNA is capable of operating perfectly at a supply voltage as low as 1.4V. However, since the front-end is to operate from a single supply, the supply voltage is ultimately set to 2V by the VGA.

The transistor sizes, component values and biasing information of the LNA are summarized in Table 9.5. The required load resistance of 165 Ω (see Fig. 9.9(e)) is implemented as a 250 Ω polysilicon resistor in parallel with the 490 Ω R_p of load inductor L_d (see Fig. 9.8). In this way, the resonance frequency of the load and the LNA gain are disconnected from each other so

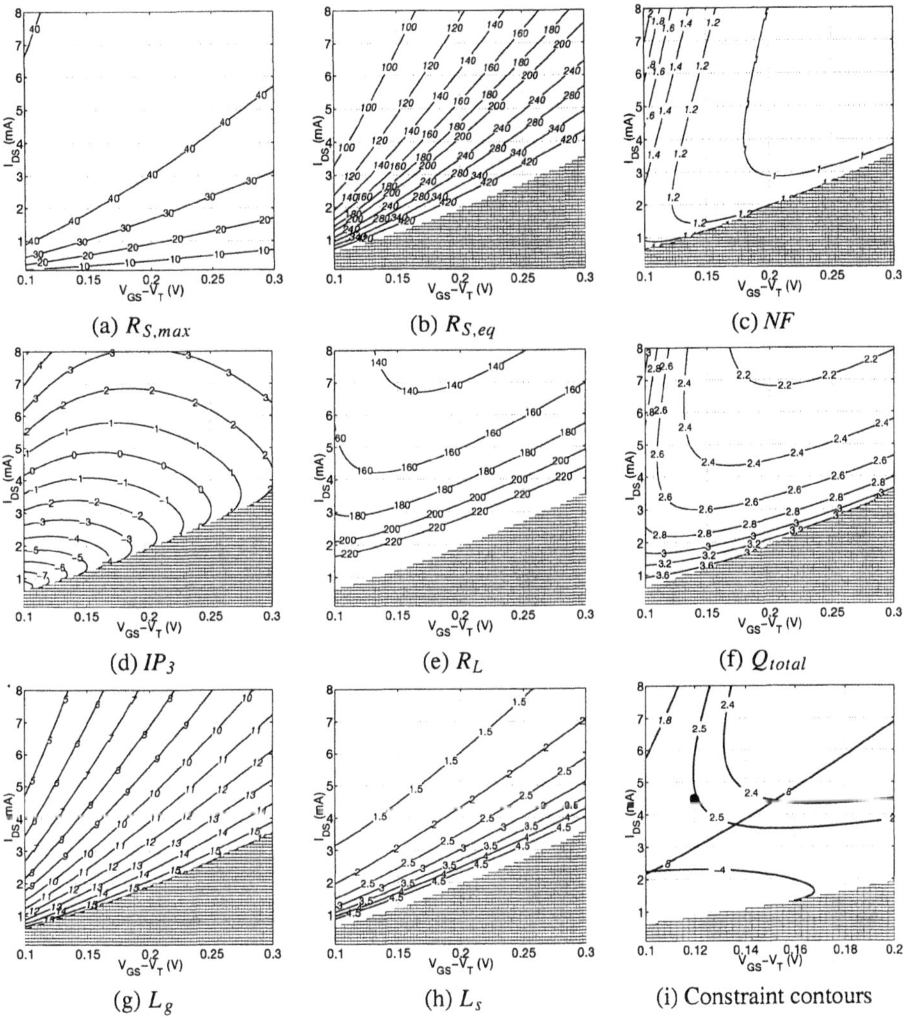

(a) $R_{S,max}$ (b) $R_{S,eq}$ (c) NF

(d) IP_3 (e) R_L (f) Q_{total}

(g) L_g (h) L_s (i) Constraint contours

Figure 9.9: *Contour lines of the LNA for $R'_S = R_{in} = 32\ \Omega$.*

Center frequency	1.84 GHz
Gain (PVC)	18.5 dB
Noise figure (NF)	2.2 dB
Third order intercept point (IIP_3)	2 dBm
Input reflection coefficient (S_{11})	-13 dB
Minimum operating voltage	1.4 V
Nominal supply voltage	2 V
Current consumption (including biasing)	5 mA

Table 9.4: *Projected spec-sheet of the realized LNA.*

$(I_{DS})_{amp}$	4.5 mA	$(I_{DS})_{casc}$	4.5 mA
$(V_{GS} - V_T)_{amp}$	0.12 V	$(V_{GS} - V_T)_{casc}$	0.17 V
$(W/L)_{amp}$	350/0.25	$(W/L)_{casc}$	200/0.25
$(g_m)_{amp}$	48 mS	$(g_m)_{casc}$	37 mS
$(C_{gs})_{amp}$	440 fF	$(C_{gs})_{casc}$	250 fF
C_{bp}	270 fF	L_g	7.5 nH
$C_{protect}$	100 fF	L_s	1.6 nH
L_d	9 nH	R_{poly}	250 Ω
R_d	23 Ω	C_k	1.5 pF

Table 9.5: *Transistor sizes, component values and biasing information of the LNA.*

Inductor	L (nH)	R_S (Ω)	$C_{L,sub}$ (pF)	r_{out} (μm)	r_{in} (μm)	W (μm)	n (–)
L_d	9.0	23	153	85	38	5	8
$L_{s,int}$	1.6	3	470	130	25	30	3
$L_{s,ext}$	min.	$\ll 1$	5 short bondwires to GND				
L_g	7.5	$\ll 1$	a single input bondwire				

Table 9.6: *Inductor dimensions.*

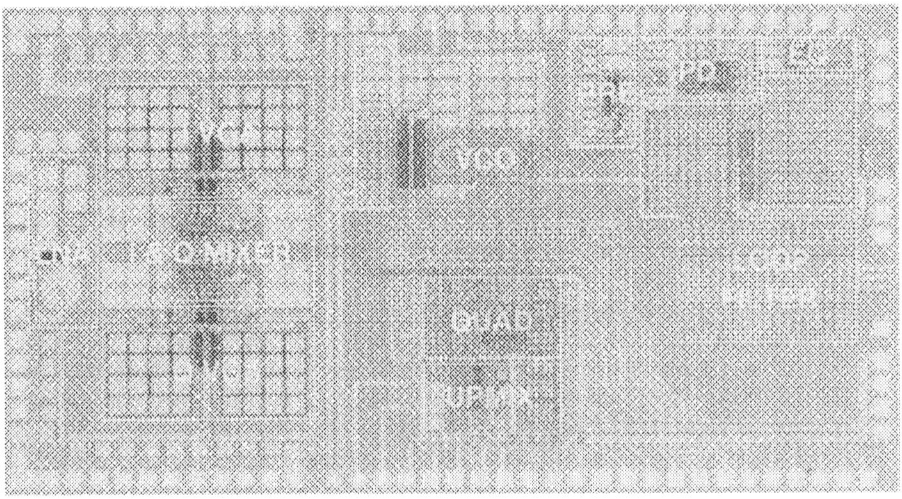

Figure 9.10: *Micro-photograph of the 2 Volt, 0.25 μm CMOS DCS-1800 transceiver.*

that future changes in the subsequent building blocks can easily be accounted for; It is no longer required to synthesize an inductor that features at the same time the right inductance value and the right R_p. Since there is practically no current flowing in the polysilicon resistor — nearly all the current flows through the inductor —, the drain voltage of the cascode device. almost equals the V_{DD}, so that the contribution of C_{db} on the output capacitance is minimized.

The load inductor L_d and the source inductor L_s are designed as on-chip spiral inductors and have been dimensioned in a two step process. The draft design is done using an analytical model [Crol96] which provides the starting values for the next design phase. The actual design of the inductors is done using the FastHenry finite element simulator [FastHen, Kamo94]. The properties of the inductors can be found in Table 9.6.

To conclude the design, it can be checked whether the 1.5 pF C_k is close enough to the optimum C_k. Based on the data in the two foregoing tables, $C_{n,LNA}$ can be calculated to be 360 fF, so that the optimum $C_{k,opt}$ according to (9.79) is 1.45 pF, which is within a few percent of the implemented C_k.

9.5 Layout

The receive path described in this chapter is embedded in a fully-integrated CMOS DCS-1800 transceiver[Stey00a, Stey00c]. The chip micro-photograph of the transceiver IC is shown in Fig. 9.10. The receive path — containing the LNA, the quadrature mixers and the VGA circuits — can be discerned at the left. The receiver is shielded from the rest of the IC by a vertical substrate strap. The only four signals that cross the substrate strap are the local oscillator signals coming from the quadrature VCO. The quadrature filter itself (QUAD) is placed close to the up-conversion mixers [Borr01] since the quality of the quadrature signals is most critical for this building block. Apart from the receive and transmit path one can distinguish the fully-integrated

PLL [Demu02], formed by the LC-tank VCO, the prescaler (PRE), the phase-frequency detector (PD) and the loop filter. The chip is implemented in a 0.25 μm 2M1P CMOS process and occupies a total area of 15.4 mm^2. A large part of the active area is consumed by the loop filter capacitors in the PLL and the feedback capacitors of the VGA.

Special measures have been taken to minimize the parasitic coupling between the different building blocks on the IC:

- The magnetic coupling between the VCO coil and the LNA load inductor has been reduced by placing the VCO far from the LNA.

- The parasitic coupling through the substrate has been minimized by actions at several levels [Lee99]. First of all, close to the injectors of the disturbances (mainly the VCO and the prescaler), substrate straps provide a low-ohmic path is provided for the injected noise. Due to the high-ohmic substrate, the noise is strongly attenuated before it reaches the sensitive building blocks. Finally, a lot of substrate straps are placed around all the high-frequency transistors and around the perimeter of the load coil in the LNA. In this way, the remaining noise is removed before it actually enters the sensitive part of the circuit.

- The symmetrical layout of the mixer reduces the LO feed-through and the mismatch between the I and Q path. Next to the transimpedance elements, dummies have been added to improve matching.

- All the on-chip grounds are connected to the off-chip ground using multiple bondwires. The same applies to the power supply lines. External interference is prevented from entering the IC by using a single external ground (star grounding).

- All reference voltages and supply voltages are decoupled towards the on-chip ground. Some extra resistance has been added in series with the power supply pin in order to kill the quality factor of the tank formed by the supply inductance and the decoupling capacitance.

9.6 Experimental Results

measurement setup To measure the transceiver in receive mode, the IC is glued and wire bonded onto a thick film ceramic substrate, as shown in Fig. 9.11. The RF input signal is provided through an SMA connector and a 50 Ω micro-strip line. The DC biasings are monitored and partially set by the external flat cable connections which are also carrying the VGA control bits, the low frequency quadrature output signals and the PLL reference frequency (the shielded line at the bottom-left). The transmit path (i.e. the up-converter) is disabled by shorting a set of its biasings to ground. Although the IC already contains a large amount of on-chip decoupling capacitance, especially at the power supply nodes of the high-frequency parts, additional SMD capacitors are used to short any external interference to ground. This is because the cables of the measurement setup easily pick up disturbances (e.g. AM radio signals) that might affect the performance of the transceiver. Eventually, the two inch by two inch substrate is mounted in a Copper-Beryllium box that serves as the common ground. A large amount of vias connects all ground planes and metal shielding to the reference ground.

Figure 9.11: *The transceiver die, mounted on a ceramic substrate inside of a CuBe box.*

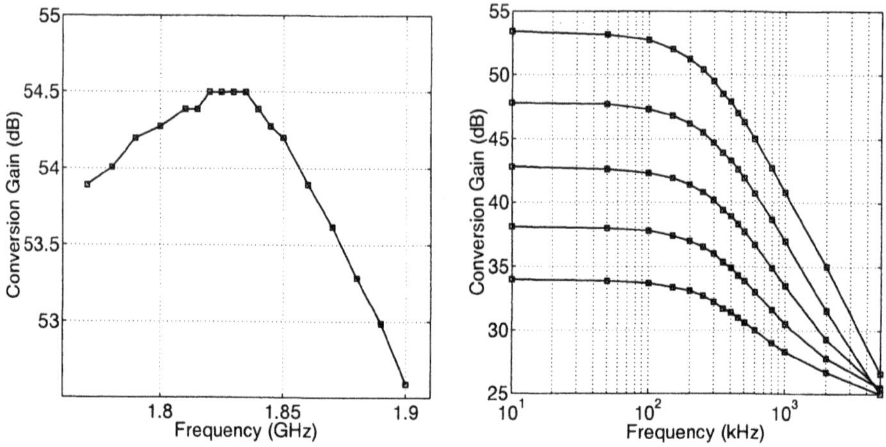

(a) Conversion gain in the DCS-1800 band.

(b) Conversion gain versus frequency offset at different VGA levels ($f_{center} = 1.88$ GHz)

Figure 9.12: *Conversion gain as a function of center frequency and frequency offset.*

conversion gain The conversion gain is measured by supplying the IC with a -50 dBm RF input signal which is swept over the DCS-1800 receive band. Each time, the local oscillator is adjusted to track the RF signal with a frequency offset of 10 kHz. The power of the 10 kHz output signal is subsequently measured in a spectrum analyzer. Since a spectrum analyzer only allows single-ended measurements, only a single phase of one of the two differential channels can be measured. The conversion gain of a differential channel is then obtained by adding 6 dB to the conversion gain value based on the measurement of a single output phase. The extracted conversion gain of the I-channel at maximum VGA gain is shown in Fig. 9.12(a). The center frequency of the receive path clearly lies in the middle of the DCS band; the gain increases from 54.3 dB at 1.8 GHz to 54.5 dB at 1.83 GHz, and decreases down to 53.3 dB at 1.88 GHz. The steep roll-off at higher frequencies and the somewhat flatter onset of the gain curve is a direct consequence of the pole on the switch node of the down-conversion mixer shaping the second-order roll-off of the LNA LC-tank. Note that the gain of the complete quadrature receiver, i.e. the gain from the RF signal towards a low-frequency $I + jQ$ vector signal, is another 3 dB higher and therefore equals 57.5 dB.

The gain of 54.5 dB is 6.5 dB lower than expected on the basis of simulations. The reason has been tracked down to the fact that, due to a processing flaw[9], the junction capacitance of the pMOS device in the batch of wafers is four times larger than normal. The gain degradation can largely be attributed to a single circuit node: the switch node of the down-conversion mixer. In fact, during the design phase, the optimization of the mixer core already resulted in a switch node pole around 1 GHz in order to satisfy all the boundary conditions like e.g. input capacitance. Since this pole lies below the DCS-1800 band, every additional node capacitance degrades the gain of the mixer by the same amount. Because the junction capacitance is responsible for more than 40 percent of the total capacitance on the switch node, the factor of four increase in the

[9]This can hardly be called a process 'fluctuation'.

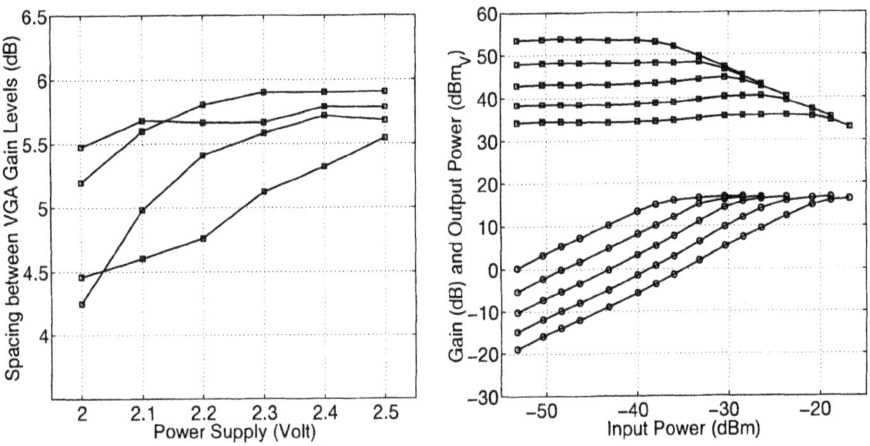

(a) VGA gain level spacing versus V_{DD}. (b) In-channel input-output capability

Figure 9.13: *VGA gain level spacing and input-output capability.*

junction capacitance leads to a uniform gain degradation of more than 6 dB. Note that a change in the position of the pole does not influence the center frequency value.

Fig. 9.12(b) and Table 9.7 show the conversion gain of the I-channel versus frequency offset from the carrier. The local oscillator frequency has been set to 1.88 GHz while the RF signal is swept from 1.88 GHz + 10 kHz to 1.88 GHz + 1 MHz. The first order channel selection annex blocking filtering can clearly be distinguished.

Obviously, only 20 dB of VGA is available whereas 24 dB (4 times 6 dB) had been expected. The reason can again be tracked down to a systematic process fluctuation; All the measured ICs feature a systematic increase of about 200 mV in the V_T of the nMOS transistors. Combined with the low supply voltage of 2 V and the 1 V common-mode level, this increase in V_T causes the switches in the VGA to operate in weak inversion instead of in strong inversion, so that the switches significantly increase the resistance of the RC tank. When the VGA operates at lower gain levels, the switch resistance contributes relatively more to the tank resistance because the impedance value of the RC elements that are incrementally switched in parallel exponentially decreases. It is precisely this effect that causes the non-uniform spacing of the gain levels. This is illustrated in Fig. 9.13(a), which shows the distance between the adjacent gain levels as a function of the power supply voltage. At the same time, all reference voltages are increased by

Parameter	++	+	~	-	- -
Conversion gain at 1.88 GHz (dB)	53.5 dB	48.0 dB	43.0 dB	38.3 dB	34.1 dB
In-channel input capability (dBm)	-38.0 dBm	-33.3 dBm	-29.2 dBm	-24.8 dBm	-20.4 dBm
DC offset (V)	1.1 V	0.58 V	0.32 V	0.19 V	0.12 V

Table 9.7: *Conversion gain, input capability and DC offset at different VGA gain levels. Apart from the reported processing flaw — causing the gain to be degraded by about 6 dB — the results are as expected.*

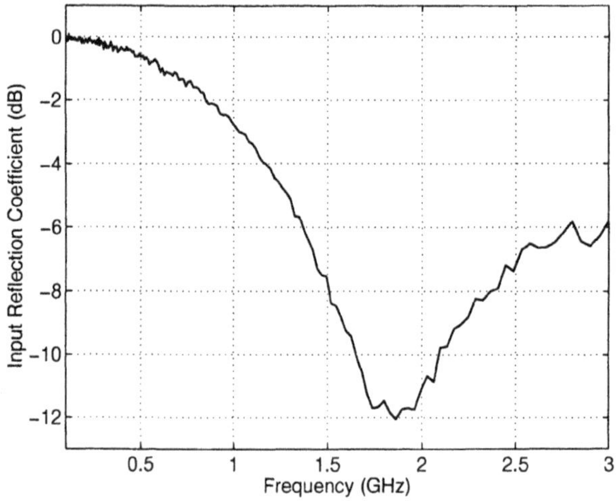

Figure 9.14: *Input reflection coefficient (S_{11}).*

an amount equal to $(V_{DD}-2)/2$, i.e. at a V_{DD} of 2.5 V, the common-mode output level is 1.25 V, so that the V_{GS} ($V_{GS}-V_T$) of the switches is 250 mV (200 mV) larger. Clearly, the VGA gain levels again become uniformly spaced (5.5–6 dB). Apart from the non-uniform spacing, one can also observe an increase in the channel bandwidth, along with a decrease in roll-off of the filter. This too is caused by the excessive resistance of the switches due to the higher threshold voltage; When two identical RC transimpedance elements are connected in parallel by a switch, the resulting transimpedance is given by

$$R_T(s) = \frac{R}{1+\tau s}\left(\frac{1+\dfrac{R_{SW}}{R}(1+\tau s)}{2+\dfrac{R_{SW}}{R}(1+\tau s)}\right), \tag{9.80}$$

where τ is the RC time constant and R_{SW} is the switch resistance. Obviously, apart from the single pole an extra zero-pole pair is generated, the position of which is determined by the ratio between R and R_{SW}: the lower the ratio (i.e. the lower the VGA gain), the lower the frequency of the zero-pole pair.

I/O capability To measure the input-output capability, a 1.88 GHz + 50 kHz tone is down-converted to 50 kHz while its power is swept from -55 dBm to -15 dBm. The fundamental of the output signal is monitored on the spectrum analyzer. Fig. 9.13(b) shows the measured output power versus input power. Obviously, the input capability is only limited by saturation and clipping at the output. Since the maximum power of the fundamental is +16.9 dBm_V, $4/\pi$ times larger than the real output capability, the output capability of a single channel is about +14.8 dBm_V, i.e. 3.4 V_ptp. The in-channel input capability ranges from -38 dBm at maximum gain to -20.4 dBm at minimum gain (Table 9.7).

input reflection The measured input reflection coefficient of the receive path is shown in Fig. 9.14. This number includes all SMA connector parasitics, strip-line non-idealities and the substrate to ground capacitance of the complete transceiver chip. The commonly accepted filter termination spec of -10 dB is met between 1.65 GHz and 2.1 GHz. The S_{11} is even better than -11.5 dB between 1.72 GHz and 1.97 GHz.

LO leakage Owing to the finite reverse isolation of the LNA and asymmetries in the differential mixer circuits, a small fraction of the local oscillator signal flows into the antenna. The LO leakage is measured by connecting the RF input terminal to the spectrum analyzer and subtracting 3 dB from the measured power to take into account the loss of the antenna filter. The net LO leakage is less than -63 dBm in the DCS-1800 band. The DCS-1800 specification ([ETSI, p. 19, 4.3.3] and Section 3.4.4) only requires the LO leakage to be lower than -47 dBm.

noise figure To measure the noise performance of the receive path, the local oscillator is adjusted to 1.8 GHz and a 50 Ω noise source is applied to the input. The differential output of a single channel is connected to a differential instrumentation amplifier to convert the output noise into a single-ended signal which is measured in a spectrum analyzer. Using the conversion gain of the complete measurement setup, the noise spectrum is referred back to the input of the receiver and is weighed against the reference noise (kTB). The resulting value is still not the noise figure of the complete quadrature receiver, but the noise figure of an image-rejectionless single-path IF receiver. Fortunately, the noise figure of the low-IF receiver can simply be obtained by subtracting 3 dB from the single channel measurement. The mathematical justification of this can be found in Subsection A.3.2, but the underlying reasons can also be explained as follows:

- The conversion gain of the complete low-IF receiver is two times larger than the conversion gain of a single path. On the other hand, it is shown in the next paragraphs that the integrated output noise of the quadrature receiver is identical to the measured output noise of a single path. This is true for the noise contribution of the mixers and the VGA circuits as well as for the noise coming from the source and from the LNA. Consequently, the quadrature output features a 3 dB improvement in *SNR* compared to the output of a single path. And, since the *SNR* at the input is the same in both cases, also the noise figure improves by 3 dB.

- The noise of the mixers and VGA circuits between -200 kHz and DC does not contribute to the noise figure of a low-IF receiver because the wanted $I + jQ$ vector is completely situated at positive frequencies. The noise power at negative frequencies is removed by the complex mixing operation in the digital zero-IF back-end. On the other hand, since there is no correlation between the noise of both paths, the noise density of the quadrature output is two times larger than the noise density of a single channel. In a single path measurement of a low-IF receiver, the measured noise includes the low-frequency mixer noise at both positive and negative frequencies — simply because a spectrum analyzer can not distinguish positive from negative frequencies. Therefore, both noise powers are the same.

- To understand what happens with the noise coming from the source or from the LNA, one must discriminate between the noise in the image band and the noise in the band of the

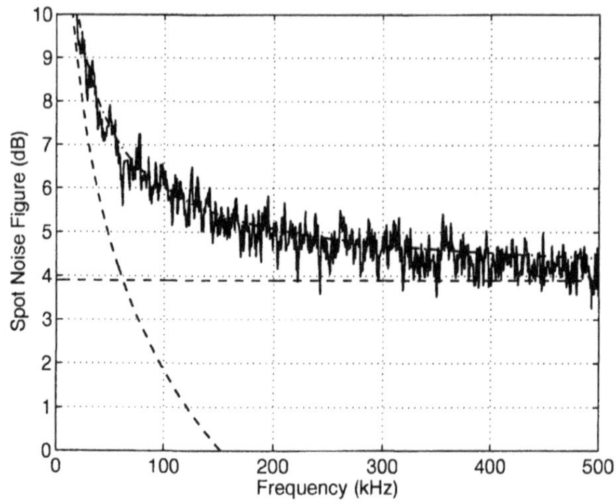

Figure 9.15: *Spot noise figure (NF_{I+jQ}) versus frequency offset.*

wanted signal. Noise at the image frequency is treated in exactly the same way as the image itself: this noise is suppressed at positive frequencies by combining the I and Q channels in the digital back-end. The noise in the band of the wanted signal on the other hand undergoes the same conversion gain as the wanted signal itself; Consequently, the output noise of the quadrature receiver is two times larger than the single channel output noise that is caused by noise at the frequency of the wanted signal. Since in a single path measurement, the noise that stems from the image frequency is still present, both noise powers are again the same.

Fig. 9.15 shows the spot noise figure of the receiver versus frequency offset from the carrier. The total noise figure of the complete low-IF receive path ($I + jQ$), i.e. resulting from integration of the output noise in the 200 kHz positive frequency band around the 100 kHz IF, is 6.2 dB.[10] Compared to the noise figure at 1.8 GHz, the noise figure at 1.88 GHz is hardly 1 dB larger. The largest part, 3.9 dB, is caused by white noise. The remaining 2.4 dB is caused by low-frequency $1/f$ noise. The white noise and 1/f noise contributions are indicated by the dashed lines in Fig. 9.15. With a noise figure of 6.2 dB, signals at the DCS-1800 reference sensitivity level of -102 dBm can be detected with an *SNR* of 9 dB. This is 2 dB better than the 7 dB *SNR* required by the DCS-1800 specification (see Subsection 3.4.1 and Subsection 3.4.2). The corresponding *BER* of 0.3% is considerably better than the 2% required by the standard.

The fact that the noise figure is still very low, in spite of the lower front-end gain, indicates that the noise is already superimposed onto the wanted signal *before* the attenuation takes place. This confirms that the conversion gain loss is mainly situated at the switch nodes in the mixers. As a consequence, only the $1/f$ corner frequency increases whereas the white noise floor remains largely unchanged.

[10] in [Stey00a], a higher noise figure value of 8.2 dB is stated because mistakenly only the high frequency image noise had been compensated for.

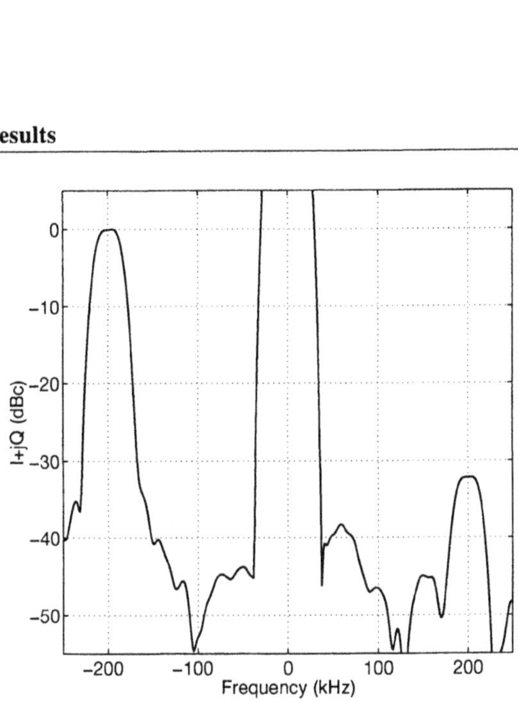

Figure 9.16: *Image rejection ratio at 200* kHz *offset (HP89410A).*

image rejection The image rejection ratio of a receiver is determined by the amplitude and phase accuracy of the quadrature local oscillator and by the circuit mismatch between the I and Q paths. Since circuit mismatch increases with decreasing area, the worst-case image rejection ratio occurs at maximum VGA gain, i.e. when the 'activated' area of the RC transimpedance bank is minimal. The image rejection ratio can accurately be measured using the HP89410A vector signal analyzer; In spectrum analyzer mode, this instrument decomposes the low-frequency $I + jQ$ signal into its positive and negative frequency components, directly exposing the amount of image suppression. As shown in Fig. 9.16, the worst-case image rejection ratio (*IMRR*) is 32.2 dB at 200 kHz offset from the carrier. The *IMRR* is measured at 200 kHz because it is precisely at this frequency that the tail of largest adjacent channel — the channel at 400 kHz offset from the wanted signal — interferes with the wanted signal. As derived in Subsection 3.4.3, an *IMRR* of 32 dB sufficiently suppresses this tail. The receiver therefore meets the DCS-1800 reference interference specification stated in [ETSI, p. 26, 6.3] and Subsection 3.4.3. The measured *IMRR* of 32.2 dB is very close to the measured image suppression in the transmit path (33 dB), which indicates that the limitation stems from the accuracy of the common quadrature VCO.

When dedicated measurement equipment is not available, the *IMRR* can still be extracted using a network analyzer or a digital scope. The *IMRR* can be calculated [Rude97] as

$$IMRR = 10\log_{10}\left(\frac{A_{IQ} + \dfrac{1}{A_{IQ}} + 2\cos(\Delta\Phi)}{A_{IQ} + \dfrac{1}{A_{IQ}} - 2\cos(\Delta\Phi)}\right), \tag{9.81}$$

where A_{IQ} and $\Delta\Phi$ are given by

$$A_{IQ} = \frac{A_I}{A_Q} \tag{9.82}$$

$$\Delta\Phi = \frac{\pi}{180}(\phi_I - \phi_Q), \tag{9.83}$$

with A_I and A_Q the amplitudes of the I and Q signals, respectively, and ϕ_I and ϕ_Q the phases of the I and Q signals, respectively. With a network analyzer, (9.82) and (9.83) can directly be measured by connecting the I and Q channel outputs to the 'reference' (R) channel and the 'reflect' (A) channel, respectively, and calculating A/R. Note that the network analyzer does not have a single clue about what frequency is to be measured unless it is told what frequency to listen to. The *IMRR* of 32.2 dB, measured by the HP89410A, corresponds to an amplitude accuracy of 0.3 dB and a phase accuracy of 2.1 degrees. Measurements with the network analyzer (scope) reveal roughly the same result: an amplitude accuracy of 0.31 dB (0.35 dB) and a phase accuracy of 2.4 degrees (2.1 degrees), resulting in an *IMRR* of 31.1 dB (31.3 dB).

interference performance The sensitivity to RF interferers is characterized by measuring the IP_3 and IP_2 values. Although IP_3 and IP_2 correctly describe the degradation of small signals by intermodulation of much larger interferers, they remain small-signal properties; In other words, they are answering the question 'How are μV signals degraded by the intermodulation of mV signals'.

If the spectrum analyzer were ideal and the output noise of the receiver would be extremely low, one would simply apply two low power interferers to the receiver input and observe the extremely low power intermodulation products at the output. In practice, the internal distortion of the spectrum analyzer — about −80 dB — limits the allowable ratio between the interferer and the intermodulation product to about 70 dB. The resolution bandwidth of the spectrum analyzer together with the output noise of the receiver forms an additional limit; At a practical resolution bandwidth of 300 Hz, the integrated noise power of the receiver is -88.5 dBm at 50 kHz offset, limiting the measurable intermodulation product to -80 dBm. These limitations, together with the severe linearity requirements of today's RF systems, make that the intermodulation must generally be measured at moderate (-40 dBm) to relatively high input powers. Yet, the pitfall is that at high input powers the receiver output saturates, resulting in clipping distortion. This is not what one is interested in; one is only interested in characterizing the internal non-linearities, not in hard clipping. Of course, this is only true when the interferers mentioned in the system specifications are low enough in order not to induce compression at the output. To allow measurements over a larger window of input powers, a way must be found to circumvent the clipping problem while still measuring the intermodulation products that are generated by all active intermodulation mechanisms.

The solution lies in the band-limited nature of the receive path, more specifically in the low-pass function associated with the current to voltage conversion in the VGA. If one positions the interferers in the DCS-1800 band, but out of the conversion channel, the interferers themselves are attenuated by the output filter so that there is no output clipping. Yet, because the filtering operation only takes place in the load element of the last building block, all intermodulation mechanisms remain active as before. Note that the intercept point values obtained by this procedure are relevant for all interferers that do not cause output clipping; The in-band, out-of-channel

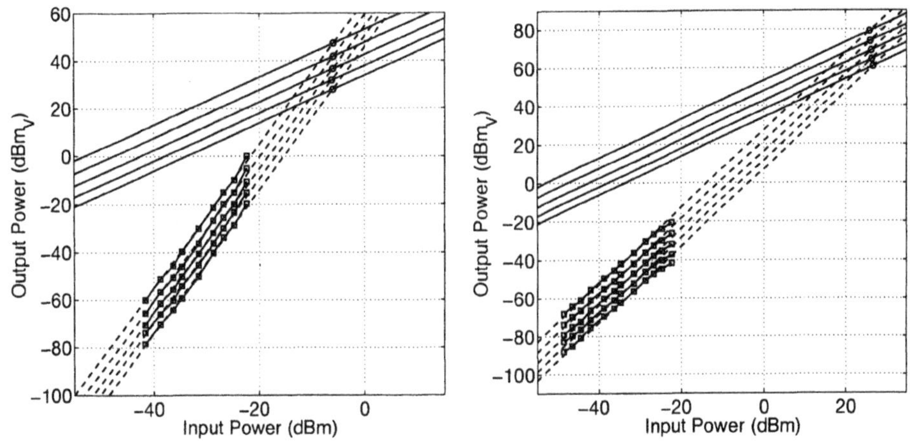

(a) Third-order intercept point (IP_3). (b) Second-order intercept point (IP_2).

Figure 9.17: *Intermodulation performance.*

IP_3 equals the in-band, in-channel IP_3 as long as the interferers are sufficiently low.

To measure the IP_3, two large out-of-channel interferers — positioned in the DCS-1800 band, but out of the conversion channel — are applied to the RF input of the receive path using two calibrated RF sources, some attenuators and a power combiner. The frequency of the interferers is 1.8 GHz + 50 kHz and 1.84 GHz, respectively. The two interferers generate an in-band, in-channel third-order intermodulation product at 1.87995 GHz that is being down-converted to 50 kHz by multiplication with a 1.88 GHz local oscillator. By sweeping the power of the interferers, IP_3 is extracted at all VGA gain levels. The outcome of the two-tone IP_3 test is shown in Fig. 9.17(a). Both the generated in-channel third-order intermodulation product (single-ended measurement with an additional 6 dB to do the single-ended to differential conversion) and the linear output (single-tone input power + in-band gain) are plotted versus the single-tone input power of the interferer. Each set of two curves extrapolates to a input-referred IP_3 of -6 dBm, regardless of the VGA gain level. This value is completely in agreement with the calculations. Consequently, the receiver complies to the DCS-1800 intermodulation specification stated in Subsection 3.4.5 with a margin of 11.5 dB! In addition, the third-order intermodulation products form a straight line, which indicates that no internal compression occurs until above -25 dBm.

The IP_2 of the receive path is measured in exactly the same way, but now the frequencies of the RF interferers are 1.84 GHz and 1.84 GHz + 50 kHz, respectively. These two signals generate a second-order intermodulation product at 50 kHz, which propagates towards the output as a result of mismatch. The two-tone IP_2 test is shown in Fig. 9.17(b). Each set of curves extrapolates to an IP_2 better than +26 dBm, again independent of the VGA gain level.

Note that the measured IP_3 and IP_2 values apply to the complete receive path. If the LNA would be bypassed, which is typically the case when the power of the wanted signal is high, the intercept points and the input capability immediately improve by 18 dB.

Finally, by combining Table 9.7 with Fig. 9.12(b), it can be shown that by meeting the specification on the IP_3 derived in Subsection 3.4.5, the receiver indeed complies to the DCS-1800 specification stated in [ETSI, p. 24, 5.3] . This can be done by proving that the power of the 800 kHz and 1.6 MHz interferers in [ETSI, p. 24, 5.3] lies in the window of input powers where the in-channel IP_3 equals the out-of-channel IP_3. The reasoning is as follows.

Since the attenuation of an 800 kHz interferer is 10 dB with respect to the in-channel gain, the -38 dBm in-channel input capability translates into an effective input capability at 800 kHz offset of -28 dBm. Because this power is still lower than the input power at which internal compression occurs (-25 dBm), the system is indeed limited by its output. Knowing that two tones of x dBm create voltage excursions as high as a single tone of $x + 6$dBm, the maximum single-tone input power during a two-tone test is -34 dBm. Since the intermodulation specification in [ETSI, p. 24, 5.3] and Subsection 3.4.5 is formulated with powers much less than -34 dBm— in fact, with two -48 dBm interferers —, the above-mentioned IP_3 value indeed proves compliance to the DCS-1800 specification in [ETSI, p. 24, 5.3].

blocking performance Compliance to the blocking performance requirements stated in [ETSI, p. 22, 5.1] could not be tested directly since testing this requires a digital back-end. However, to a high degree, the compliance to the blocking specification can be verified indirectly by proving that the PLL meets both the phase noise specifications and the spur levels, and by showing that the I/O capability of the receive path is much larger than the maximum blocking signal. Let us first look at the PLL.

As can be deduced from Table 9.8, the PLL meets the specifications in Subsection 3.4.6 and Subsection 3.4.7 with large margin. This is a first indication that the low-IF receive path meets the DCS-1800 blocking performance stated in [ETSI, p. 22, 5.2 and p. 25, 6.2] — at least concerning the PLL.

Due to the 3 dB attenuation in the antenna filter/duplexer, the -26 dBm blocking signal is reduced to -29 dBm at the input of the receiver. Since this blocking signal lies at an offset of 3 MHz from the wanted channel, it is attenuated by 21.5 dB with respect to the in-channel gain (see Fig. 9.12(b)), so that its output power is identical to the output power of an in-channel down-converted -50.5 dBm input signal.[11] This equivalent input power is much lower than the -38 dBm in-channel input capability corresponding to the onset of output compression (see Fig. 9.13(b)), so that the output is far enough from clipping. On the other hand, it can be seen in Fig. 9.17 that internal compression only becomes an issue when the receiver is steered with two -25 dBm interferers, corresponding to a total input power of -22 dBm. Again, -29 dBm is significantly lower than -22 dBm so that there is no internal compression in the presence of blocking signals.

DC offset At the the highest VGA gain, the DC offset due to LO self-mixing and static mismatch is 1.1 V. However, since a low-IF receiver does not employ the information at DC, this offset does not corrupt the received information: the GMSK signal is centered around 100 kHz and DC will be removed by the DSP anyway. As long as the A/D converters are not overloaded, the DC offset can be accounted for by putting half a bit extra in the A/D converter. In principle, there is no need for an offset compensation. Table 9.7 shows the DC offset as a function of the VGA gain level.

[11] For the -33 dBm blocking signal (at 1.6 MHz offset) and the -43 dBm blocking signal (at 600 kHz offset) the equivalent input powers are -52.5 dBm and 54.5 dBm, respectively.

performance summary For the sake of completeness, the measured performance of the complete transceiver is summarized in Table 9.8. Apart from the measurement results of the receive path — described in detail in this work — the table also contains the measured results of the PLL [Demu02] and the transmit path [Borr01] of the transceiver [Stey00a, Stey00c]. This transceiver clearly meets the DCS-1800 specifications that have been derived in Chapter 3 (summarized in Table 3.3 and Table 3.2).

illustrations with DCS-1800 GMSK signals To further illustrate the functionality of the receive path, the receiver is supplied with a 270.833 kbps PN23 bit-stream which is GMSK modulated onto a 1.8 GHz, -50 dBm carrier signal. The low-frequency I and Q outputs are fed to a HP89410A vector analyzer which acquires both channels simultaneously and displays the data on-the-fly in different formats. Each time, the instrument acquires 148 symbols — corresponding to the length of a normal GSM burst — by sampling 1471 data points in 542.8 μs (i.e. 10 samples per symbol). Nevertheless, the equipment imposes some important limitations on the measurements:

- Because the vector analyzer is not capable of doing a quadrature demodulation of a signal around a selectable center frequency but only allows demodulation of quadrature GMSK signals centered around DC, the low-IF receiver is necessarily measured in zero-IF mode. In the real system, the quadrature outputs are centered around a 100 kHz IF and a final zero-IF down-conversion is done in the digital domain.
- The measurement instrument only accepts single-ended signals, so that the differential nature of the I and Q channels can not be exploited.
- Differences between the common-mode DC level of the I and Q channels — which are not important when the I and Q channels are acquired differentially — along with the absence of an internal offset compensation mechanism in the receiver — which is in principle not necessary in a low-IF system — require AC coupling of both channels. The corresponding loss of DC information degrades the constellation in zero-IF mode.

Fig. 9.18(a) plots the normalized I and Q waveforms in the time domain. In order not to overload the graph, only the first 51 symbols out of a total of 148 symbols are depicted. The moment when each symbol is recovered is marked by the circles and the crosses. The I and Q signals can also be interpreted as the coordinates of a complex vector, leading to the constellation diagram in Fig. 9.18(b). All 148 symbols are represented by black dots. The symbol transitions are indicated by the dotted lines. Ideally, due to the inter-symbol interference, each cluster around $k\pi/4$ should consist of three separate dots. However, due to the above-mentioned non-idealities, the constellation is somewhat smeared out. But, as indicated in Subsection 2.3.2, only the detection of the quadrant is important.

Fig. 9.19 shows yet another representation of the received symbols. This eye diagram has been constructed by dividing the I channel waveform in 49 chunks of three consecutive symbols, and superimposing them onto each other. The timing jitter stems from the fact that at the moment when the vector analyzer equipment was available there were still some problems locking the PLL on the transceiver IC. In addition, in contrast to MSK — where there are only two possible

Receive path	Noise figure (*NF*)	6.2 dB
	White noise	3.9 dB
	$1/f$ noise	2.4 dB
	Image rejection ratio (*IMRR*)	32.2 dB
	Amplitude accuracy	0.3 dB
	Phase accuracy	2.1 deg
	Conversion gain	54.5 dB
	Reflection coefficient (S_{11})	-11.5 dB
	LO leakage	-63 dBm
	Third order intercept point (IP_3)	-6 dBm
	Second order intercept point (IP_2)	+26 dBm
	Output capability	+14.8 dBm
	Power consumption	113 mW
Transmit path	Image suppression	33 dBc
	LO feedthrough	-38 dBc
	Output power	-13 dBm
	Spectral mask	E T S I
	Power consumption	82 mW
Phase Locked Loop	Tuning range	10.5 %
	Settling time for a 104 MHz step	226 μs
	Phase noise	
	at 600 kHz offset	-125.8 dBc/Hz
	at 3 MHz offset	-143.7 dBc/Hz
	Spurious suppression	-84 dBc
	Power consumption	78 mW
	Supply voltage	2.0 V
	Total power consumption	
	Receive mode	191 mW
	Transmit mode	160 mW
	Area	15.4 mm^2

Table 9.8: *Summary of the transceiver measurements. Along with the experimental results of the receive path described in this work, the results of the PLL [Demu02] and the transmit path [Borr01] are shown as well. This transceiver clearly meets the DCS-1800 specifications.*

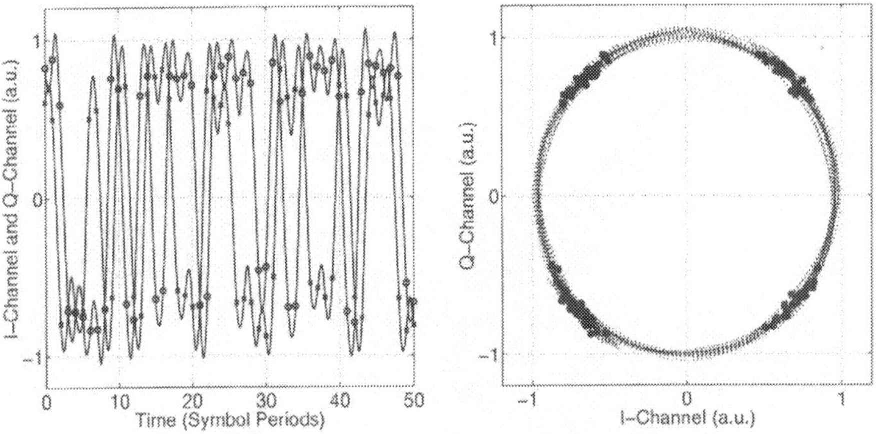

(a) I-channel and Q-channel outputs. (b) GMSK constellation diagram.

Figure 9.18: *GMSK constellation diagram in zero-IF mode.*

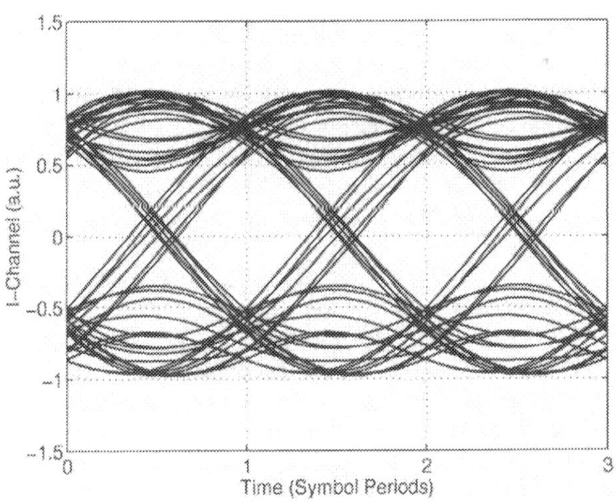

Figure 9.19: *Eye diagram in zero-IF mode.*

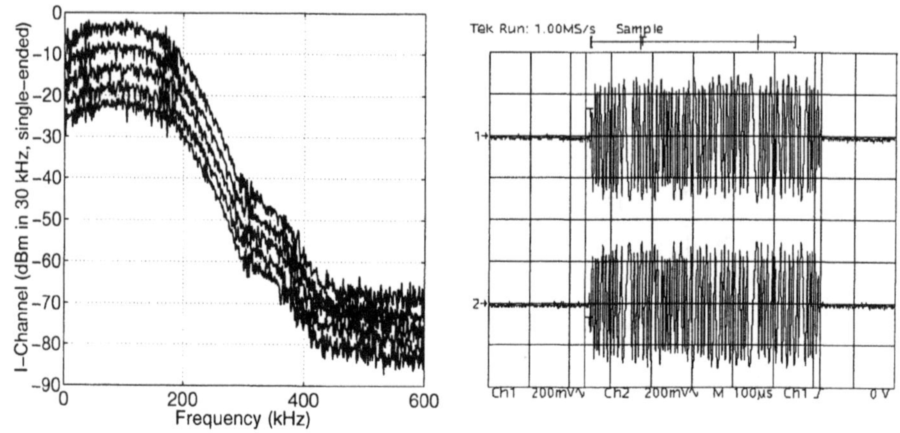

(a) Spectrum of a -50 dBm GMSK signal after down-conversion to the 100 kHz IF.

(b) A -50 dBm GSM burst, down-converted to the 100 kHz IF (I-channel).

Figure 9.20: *Receiving GMSK signals in low-IF mode.*

values at the sampling instant — GMSK features *six* possible values (Subsection 2.3.2). This gives rise to a family of transitions rather than a single clean transition.

To demonstrate how the low-IF receiver is used in practice, a 1.84 GHz, -50 dBm PN9 GMSK signal is applied to the input of the receiver and down-converted to the 100 kHz IF. The VGA control bits are swept in order to select all the gain levels. The resulting low-frequency GMSK signal is shown in Fig. 9.20(a). All measurements are performed with a resolution bandwidth of 30 kHz, the value that is generally used in the DCS-1800 specifications. In contrast to Fig. 9.20(a), where a continuous GMSK signal is applied to the input, Fig. 9.20(b) shows how the receiver reacts on a -50 dBm PN9 GMSK burst. The burst is a normal GSM burst at 1.84 GHz, carrying 156.25 bits in 577 μs. To illustrate the quadrature nature of the received signals, Fig. 9.21 shows the output of the receiver when receiving a custom GMSK burst. The burst consists of a repeated concatenation of 16 consecutive '1' bits and 16 consecutive '0' bits. One can clearly see that the bits are encoded differentially: only when there is a transition between a '1' and a '0' the instantaneous frequency is lower.

As a final illustration, a 1.8 GHz fractal antenna is connected to the receiver input. Knowing that the 3rd GSM operator in Belgium is in its start-up phase — the experiment was done on July 10th, 1999 — the ether is searched for GMSK signals in the DCS-1800 band. And indeed, there is some GMSK activity near 1.86449 GHz. The local oscillator is adjusted to down-convert these signals to the 100 kHz IF and one of the output phases of the receiver is connected to the spectrum analyzer. Of course, the finite sweep time of the spectrum analyzer along with the fact that the same frequency is reused by multiple users in different time slots (TDMA) makes it impossible to measure the GMSK spectrum of a single user. However, by putting the spectrum analyzer in 'max hold' mode, after a few sweeps the analyzer displays at each frequency the power of the most powerful frame that has come along. In this way, the envelope of the GMSK spectrum of the most powerful user around a certain frequency is obtained. The received spectrum is shown in Fig. 9.22. One can clearly see three GMSK spectra, separated by a multiple of 400 kHz. The

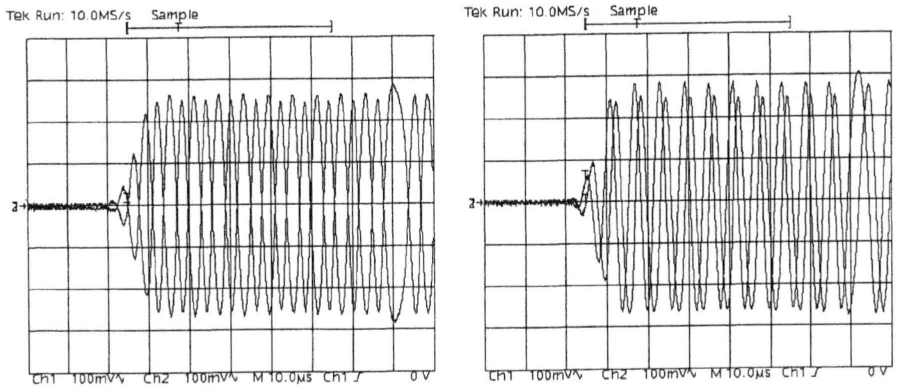

(a) I+ and I- output signals. (b) I+ and Q+ output signals.

Figure 9.21: *Receiving a custom GMSK burst in low-IF mode.*

200 kHz space in between is not occupied, as mentioned in Chapter 2. The dashed lines indicate the trajectories that are followed by the different GMSK spectra when the local oscillator is adjusted to select the respective signals.

9.7 Conclusion

In this chapter, the design of a 2V CMOS cellular receiver front-end has been presented. The front-end integrates a low noise amplifier together with two (I&Q) down-conversion mixers and two (I&Q) variable-gain amplifiers/filters. The input-matched LNA is designed for robustness to avoid tuning of the on-chip output section, and combines this with a power consumption as low as 10 mW. The LNA exploits a lower input impedance of 32 Ω to improve the gain compared to an exact 50 Ω match, while still providing an S_{11} of -12dB. The mixer core employs a highly linear current-folding switching mixer topology which is capable of operating at supply voltages as low as 2.0 V. The mixer is *seamlessly* integrated with the transimpedance VGA/filter. No extra interface circuits are required to connect the different building blocks or to do extra signal conversions: the mixer provides an output current and the VGA directly operates on this current. Even the single-ended to differential conversion of the LNA output signal is implemented implicitly in the mixer through multiplication with the differential LO.

All the blocks have been sized using a *systematic design procedure*; The design procedure is based on a bank of analytical equations describing the performance parameters (like e.g. noise figure, linearity, gain, etc.) in terms of the operating point of the involved transistors.

The receiver front-end has subsequently been embedded in a fully integrated CMOS DCS-1800 transceiver [Stey00a, Stey00c], the *first* CMOS transceiver with a receive path, a transmit path and a complete PLL integrated on a single die. Together with a power amplifier, an antenna filter/duplexer, low-frequency data converters and a reference crystal, this IC forms the complete analog part of a 1.8 GHz cellular system.

The IC operates from a 2V supply, which makes it — to the author's knowledge — the *lowest voltage CMOS RF front-end published to date.* All its building blocks feature circuit

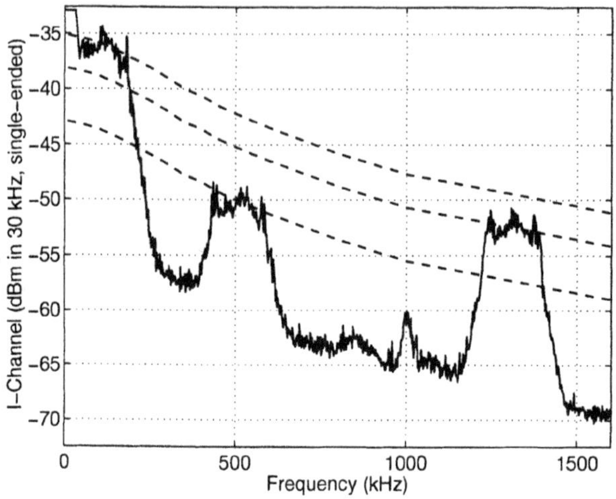

Figure 9.22: *Eavesdropping the start-up phase of KPN Orange.*

topologies that have been specially designed to be able to operate at 2.0V, despite the relatively large threshold voltages in the available 0.25 μm technology: an nMOS V_T of 0.65V and a pMOS V_T of 0.75V. In addition, all the reported specifications and performances have actually been measured and achieved at 2.0V.

The experimental results in Table 9.8 demonstrate the *feasibility* of achieving the analog performance required for the DCS-1800 cellular system in a standard 0.25μm 2M1P CMOS process without the use of thick metal layers, post-process etching, tuning or trimming.

Appendix A

Noise Figure of Receiver Systems

A.1 Sensitivity, Noise Factor and Noise Figure

The sensitivity of a receiver is defined as the minimum input signal level that is detectable with a given *SNR* value, the required *SNR* of course being dependent on the application. In the case of an ideal, noiseless system the sensitivity is only determined by the background noise that comes with the signal i.e. the noise of the signal source. However, since a practical system always adds noise, the effective input noise floor is always higher. The sensitivity thus critically depends on the noise contributions of the different building blocks in the receive path.

The relative degradation in *SNR* due to system noise, and hence also the degradation in sensitivity, is quantified by the noise factor F, which is defined as

$$F = \frac{SNR_{IN}}{SNR_{OUT}}. \tag{A.1}$$

System specifications generally use the noise figure *NF*, which is nothing else than the noise factor on a dB scale:

$$NF = 10\log_{10}(F)\ \mathrm{dB}. \tag{A.2}$$

In this work, most of the equations and expressions are written in terms of the noise factor F; The reason is that the noise factor lends itself perfectly for doing noise calculations since noise factors can simply be added taking into account the correct weight. In the graphical representations and when discussing noise performance, we systematically look at the noise figure *NF*, because this is the most widespread way of quantifying a circuit's noise performance. Keep in mind that the noise factor and the noise figure essentially represent the same quantity — be it on a linear scale and a log scale, respectively.

Note that in an ideal, noiseless linear system the *SNR* is invariant. This is not the case in non-linear systems, like mixers; It is only true provided that there is no background noise at the image frequencies of the wanted signal band.

A.2 Noise Figure of Receiver Building Blocks

A.2.1 Amplifiers

In the case of an amplifier, (A.1) is unambiguously converted into

$$F = \frac{\text{output noise power, including the noise contribution of the amplifier itself}}{\text{output noise power solely coming from amplified source noise}} \tag{A.3}$$

$$= \frac{\text{input-referred total output noise power (using the gain experienced by the signal)}}{\text{available noise power coming from the source}}. \tag{A.4}$$

The noise powers in the numerator and the denominator of (A.4) are the result of integrating a noise density over a bandwidth that is relevant to the system, e.g. the channel bandwidth. It is also possible to define a spot noise factor by leaving out the integration or by integrating over a very small bandwidth, e.g. 1 Hz; In this case the color of the noise can be distinguished.

The available (white) noise power coming from the source is given by

$$P_{N,av} = kT_nB, \tag{A.5}$$

where B is the bandwidth of interest and T_n is the noise temperature of the source. The noise temperature not necessarily equals the physical temperature of the source. This is for example the case in an antenna-driven system, where T_n is not the antenna temperature, but the noise temperature associated with the received background noise. Note that $P_{N,av}$ does *not* depend on the source impedance nor on the quality of the input match.

Though the noise factor should in principle be evaluated with a noise source featuring the noise temperature of the source actually driving the system, by convention T_n is taken to be 290 K [Lee98]; In this way, the noise powers of devices under test are always weighed against a common noise power so that their noise figures can be compared[1].

When calculating the *SNR* or the noise figure in CMOS circuits, it is often more convenient — and more natural — to reason in terms of the squared node voltages *at* a reference plane rather than in terms of the power *through* a reference plane. Both reasonings lead to exactly the same result for both the *SNR* and the *NF*; The only difference is that the former makes abstraction of the impedance level at the reference plane. In terms of the squared node voltages, the *SNR* is given by

$$SNR = \frac{\text{signal at the reference plane, in } \mathrm{V}^2_{rms}}{\text{noise at the reference plane, in } \mathrm{V}^2_{rms}}. \tag{A.7}$$

Fig. A.1 illustrates the two alternatives of specifying the noise factor of an amplifier. The amplifier is represented by its S-parameter equivalent. The original noise factor definition — graphically represented in Fig. A.1(a) — states that

$$F = \frac{P_{N,EQ} + kT_nB}{kT_nB}, \tag{A.8}$$

[1] If T_n significantly differs from 290 K, the real *SNR* degradation in the system can still be obtained as

$$F_{system} = \frac{290\text{ K}}{T_n}(F_{conventional} - 1) + 1. \tag{A.6}$$

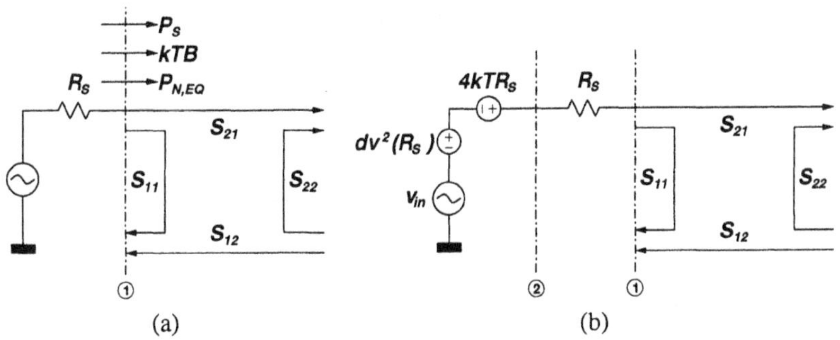

Figure A.1: *Alternative ways of specifying the noise figure of an amplifier.*

where $P_{N,EQ}$ is the equivalent noise power due to the amplifier itself through reference plane ①. An identical value is obtained by calculating the noise factor in terms of the squared noise voltages at a reference plane. This is illustrated in Fig. A.1(b). The expression for the noise factor at reference plane ② is then given by

$$F = \frac{dv^2(R_S) + 4kT_n R_S}{4kT_n R_S}, \tag{A.9}$$

where $dv^2(R_S)$ is nothing else than the equivalent noise density resulting from dereferencing the output noise of the amplifier itself to the voltage source, or in other words, the voltage excursions in R_S caused by the equivalent input noise sources of the amplifier. Note that (A.9) also holds at reference plane ①!

In the following sections we will always formulate the *SNR* and the noise factor in terms of the power flow. However, one must bear in mind that one can always replace the noise power values by the corresponding values in V_{rms}^2. Of course, then also the power gains must be replaced by the ratios of the squared input and output voltages. It is just a question of comparing apples to apples.

A.2.2 Mixers

The noise figure concept can also be employed to characterize the sensitivity of stand-alone mixers, yet the specific nature of a mixer gives rise to two different noise figure variants: the SSB and the DSB noise figure. Since these quantities are often being misinterpreted, the exact definition of the DSB and the SSB noise figure is reviewed in the following paragraphs. Note that the SSB and the DSB noise figure only describe the performance of a *single* mixer. How these numbers are related to the system noise figure or the noise figure of a quadrature mixer is covered in the next subsection.

To highlight the difference between the two noise figure definitions, it is assumed that the wanted signal is situated in only one of the down-converted side-bands, i.e. either at $LO + IF$ or at $LO - IF$. If the signal energy is equally spread out around the LO, the SSB and DSB noise figure amount to the same thing. In that case, the DSB noise figure provides the only 'logical' definition.

The definition of DSB noise figure is based on the observation that the mixer can't distinguish between input noise in the band of the wanted signal (e.g. in the upper side-band, residing at $LO + IF$) and input noise in the image band (i.e. in the lower side-band, residing at $LO - IF$). This means that, even though the signal itself only experiences the noise in its own band, the input *SNR* *seen* by the mixer is given by

$$SNR_{IN} = \frac{P_S}{P_{N,USB} + P_{N,LSB}}, \tag{A.10}$$

where P_S is the input signal power (completely situated in the USB), and $P_{N,USB}$ and $P_{N,LSB}$ denote the upper side-band noise power and the lower side-band noise power coming from the source, respectively. The *SNR* at the output of the mixer is given by

$$SNR_{OUT} = \frac{GP_S}{G(P_{N,USB} + P_{N,LSB}) + P_{N,MIXER}}, \tag{A.11}$$

where G is the mixer conversion gain and $P_{N,MIXER}$ is the output noise power that is caused by propagation towards the output of all the internal noise sources of the mixer itself. Thus, $P_{N,MIXER}$ includes internally generated noise that is being down-converted from the signal frequency and its image as well as low frequency noise. Application of (A.1) leads to the DSB noise factor:

$$F_{DSB} = \frac{G(P_{N,USB} + P_{N,LSB}) + P_{N,MIXER}}{G(P_{N,USB} + P_{N,LSB})} \tag{A.12}$$

$$= \frac{P_{N,USB} + P_{N,LSB} + \dfrac{P_{N,MIXER}}{G}}{P_{N,USB} + P_{N,LSB}}. \tag{A.13}$$

From the above it becomes clear that DSB noise figure is a questionable quantity when only one of the side-bands contains signal energy. Indeed, the definition states that noise at the image frequency was *already* undistinguishable from the noise at the signal frequency *before* the mixing operation even occurred. If the input noise at the image frequency is comparable to the noise at the signal frequency — which is the case in a low-IF receiver —, the input noise power is artificially increased by about 3 dB, lowering the numerical value of the noise figure by the same amount. Hence, stating the DSB noise figure of a mixer gives a false impression about the noise performance of a mixer down-converting SSB signals.

When only one of the side-bands contains energy, a much more representative way of quantizing mixer noise is found in the SSB noise figure. In contrast to the DSB noise figure, the SSB noise figure assumes no advance knowledge about what is going to happen with the signal when determining SNR_{IN}; The SNR_{IN} is just given by

$$SNR_{IN} = \frac{P_{S,USB}}{P_{N,USB}}, \tag{A.14}$$

where $P_{S,USB}$ is the input signal power in the upper side-band. This leads to the following

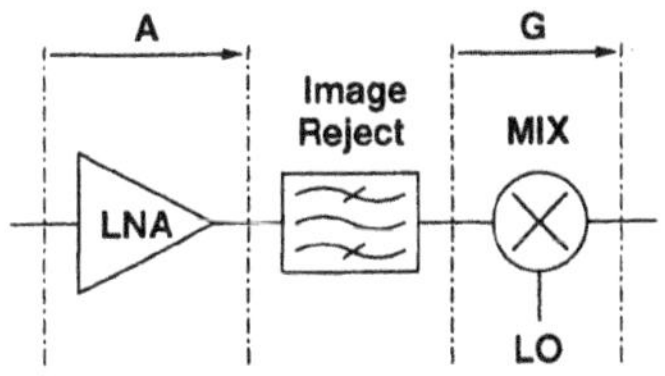

Figure A.2: *An elementary IF receiver.*

expression for the SSB noise factor:

$$F_{SSB} = \frac{G(P_{N,USB} + P_{N,LSB}) + P_{N,MIXER}}{GP_{N,USB}} \tag{A.15}$$

$$= \frac{P_{N,USB} + P_{N,LSB} + \dfrac{P_{N,MIXER}}{G}}{P_{N,USB}}. \tag{A.16}$$

Note that the SSB noise figure does take into account the ***output*** noise resulting from input noise at the signal frequency and at the image frequencies. Indeed, since the image noise is not canceled it *is* down-converted and contributes to the output noise. So, even if the mixer is noiseless, the SSB noise figure equals 3 dB. The two noise figure definitions only differ by the amount of noise that is used to normalize the total input-referred output noise. When only one side-band contains signal energy, the following relation applies:

$$NF_{SSB} = NF_{DSB} + 3 \text{ dB}. \tag{A.17}$$

For a single mixer operating with DSB signals — like e.g. in the I path of a zero-IF receiver —, there is only one relevant noise figure because there the SSB noise figure equals the DSB noise figure.

A.3 Noise Figure of Receiver Systems

A.3.1 Single-Path Receivers

Even though the above-mentioned figures correctly quantize the noise performance of stand-alone building blocks, a lot of care must be taken to correctly use these figures when calculating the noise performance of a receive system.

Consider for example the elementary IF receiver in Fig. A.2. The noise factor of this system is given by

$$F = \frac{AG(P_{N,USB} + SP_{N,LSB}) + G(P_{N,USB,LNA} + SP_{N,LSB,LNA}) + P_{N,MIXER}}{AGP_{N,USB}}, \tag{A.18}$$

where A is the gain of the LNA, G is the conversion gain of the mixer, S is the image suppression and $P_{N,USB,LNA}$ and $P_{N,LSB,LNA}$ are the output noise powers of the LNA in the respective bands. In practice, S can be considered zero, because the required image rejection in an IF receiver is extremely high to prevent large images from folding onto the wanted signal.

Reformulating (A.18) in terms of the noise factors of the building blocks, leads to

$$F \cong F_{LNA}(1+S) + \frac{P_{N,MIXER}}{AGP_{N,USB}} \tag{A.19}$$

$$\cong F_{LNA}(1+S) + \frac{F_{SSB,MIXER}-2}{A} \tag{A.20}$$

$$\cong F_{LNA}(1+S) + \frac{2F_{DSB,MIXER}-2}{A}. \tag{A.21}$$

Obviously, a few tricks are necessary to convert the noise figure of the building blocks to the system noise figure. The LNA forms no exception: the image rejection must still be accounted for. But, the clearest example remains the down-conversion mixer, where either a divide-by-two operation or a subtraction of two is required to remove the noise of the 50 Ω source — depending on whether the stand-alone mixer is characterized by its DSB or its SSB noise figure. Hence, since a mixer is almost never used as the first building block in a receive chain — except for low performance applications — it has little sense judging a mixer by its stand-alone SSB or DSB noise figure, because it never contributes to the system noise in that form.

(A.19)–(A.21) clearly state that only the noise of the mixer core contributes to the system noise figure. This leads to the definition of the so-called SSB eigen[2] noise factor, given by

$$F_{E,SSB} = \frac{P_{N,MIXER}}{GP_{N,USB}} = \frac{\frac{P_{N,MIXER}}{G}}{P_{N,USB}} \tag{A.22}$$

$$\cong F_{SSB,MIXER} - 2 \tag{A.23}$$

$$\cong 2F_{DSB,MIXER} - 2. \tag{A.24}$$

Thus, the eigen noise factor weighs the intrinsic noise of the mixer against the background noise power. In contrast to the DSB and SSB noise figure, this value can directly be used in the calculation of the system specification, without compensating for non-present noise sources. The 'eigen' noise figure can be measured by shorting the input of the mixer to ground, measuring the output noise of the mixer in a spectrum analyzer, referring it back to the input and comparing it against the reference noise power of a 'virtual' 50 Ω source. This shortcut can be made in CMOS, since the noise voltage that is generated by the equivalent input noise current of the mixer when it is flowing into the output impedance of the LNA, is generally negligible with respect to the equivalent input noise voltage of the mixer.

A.3.2 Quadrature Receivers

Noise figure computations become somewhat more complicated when the image rejection is done by quadrature mixing, i.e. two mixers performing a complex multiplication. In the following paragraphs, the noise performance of the zero-IF and the low-IF receiver is examined in detail.

Consider the following radio-frequency signal, representing a phase (frequency) modulated signal centered around a frequency ω_c:

$$RF(t) = \alpha \sin(\omega_c t + \phi(t)). \tag{A.25}$$

[2] As in 'eigen'value

In order not to obscure the calculations, we assume for now that there is no noise superimposed on the RF signal *before* it is being down-converted. The influence of this noise (LNA noise, etc.) will be determined later on.

In a zero-IF receiver, the RF signal is multiplied by a complex local oscillator of frequency $\omega_{LO} = \omega_c$ and amplitude Λ. After low-pass filtering, the phase information can directly be extracted:

$$I(t) + jQ(t) = LPF\Big(RF(t)\,(\Lambda\cos(\omega_{LO}(t)) + j\Lambda\sin(\omega_{LO}(t)))\Big) \tag{A.26}$$

$$= \frac{\Lambda\alpha}{2}(\cos(\phi(t)) + j\sin(\phi(t))). \tag{A.27}$$

Since the power of the original RF signal is $\alpha^2/2$ and the power of the resulting low-frequency vector is $(\Lambda\alpha/2)^2$, the power gain and the voltage gain of the complex mixing operation are $\Lambda^2/2$ and $\Lambda/\sqrt{2}$, respectively. Note that the signal in (A.25) has a bandwidth of BW — both at positive and negative frequencies —, while the I, Q and $I + jQ$ signals in (A.27) have a bandwidth of only $BW/2$ at positive and at negative frequencies.

The wanted signal vector of (A.27) is accompanied by a noise vector, $n_I(t) + jn_Q(t)$, resulting from down-conversion or propagation of internal mixer noise sources towards the output of the quadrature mixer. Because the signal vector itself is situated between $-BW/2$ and $BW/2$, only the noise power between $-BW/2$ and $BW/2$ contributes to the *SNR*. Since the noise of both channels is uncorrelated, the *SNR* of the quadrature output of the zero-IF receiver can be calculated as

$$SNR_{zero-IF} = \frac{\left(\frac{\Lambda\alpha}{2}\right)^2}{\int_{-BW/2}^{BW/2} p_{N,I-MIXER} + p_{N,Q-MIXER}\, df}. \tag{A.28}$$

where $p_{N,I-MIXER}$ is the PSD of the I-channel output noise and $p_{N,Q-MIXER}$ is the PSD of the Q-channel output noise, which is generally identical to $p_{N,Q-MIXER}$. Because measuring this *SNR* requires special equipment that is generally unavailable, it is useful to look at the relation between the noise of a complete quadrature system and the noise of a single channel. Knowing that the output power of a single channel is $(1/2)\cdot(\Lambda\alpha/2)^2$, the *SNR* of e.g. the I channel can be written as follows:

$$SNR_{zero-IF,I} = \frac{\frac{1}{2}\left(\frac{\Lambda\alpha}{2}\right)^2}{\int_{-BW/2}^{BW/2} p_{N,I-MIXER}\, df} \tag{A.29}$$

$$= SNR_{zero-IF}. \tag{A.30}$$

Obviously, calculating the *SNR* of just the I-channel or the Q-channel yields the same value as (A.28). At least, this is the case in a zero-IF receiver. The eigen noise factor of a zero-IF receiver thus becomes

$$F_{E,zero-IF} = \frac{\displaystyle\int_{-BW/2}^{BW/2} p_{N,I-MIXER}\, df}{\displaystyle\frac{1}{2}\left(\frac{\Lambda}{2}\right)^2 \int_{LO-BW/2}^{LO+BW/2} p_N\, df} \tag{A.31}$$

$$= F_{DSB} - 1 = F_{E,DSB}. \tag{A.32}$$

where p_N is the reference input noise floor and the $(1/2) \cdot (\Lambda/2)^2$ factor is the conversion gain of the I mixer.

The same calculations can be done for a low-IF receiver. Here, the output vector is given by

$$I(t) + jQ(t) = LPF\Big(RF(t)\,(\Lambda \cos(\omega_{LO}(t)) + j\Lambda \sin(\omega_{LO}(t)))\Big) \tag{A.33}$$

$$= \frac{\Lambda\alpha}{2}(\cos(\omega_{IF}(t) + \phi(t)) + j \sin(\omega_{IF}(t) + \phi(t))). \tag{A.34}$$

From now on it is assumed that the intermediate frequency, IF, equals $BW/2$. This makes that the output vector is situated entirely at positive frequencies: the $I + jQ$ signal lies completely between 0 and BW. Hence, only the vector noise between 0 and BW contributes to the overall *SNR*, given by

$$SNR_{low-IF} = \frac{\left(\dfrac{\Lambda\alpha}{2}\right)^2}{\displaystyle\int_0^{BW} p_{N,I-MIXER} + p_{N,Q-MIXER}\, df}. \tag{A.35}$$

Of course, in a complete low-IF receiver, there is still a quadrature down-conversion in the digital domain to mix the signal from the low IF towards a zero IF. However, this operation does not change the *SNR*, since it only shifts the vector along with its vector noise over $-BW/2$.

Again, because measuring the *SNR* of quadrature signals is quite impractical, one wants to extract the *SNR* based on measurements of a single channel. Unfortunately, almost directly a complication arises; In contrast to a zero-IF receiver where both the single channel and the vector signal reside between $-BW/2$ and $BW/2$, this is not the case in a low-IF receiver. In a low-IF receiver the I and Q signals lie between $-BW$ and BW, which means that they cover two times more bandwidth than the $I + jQ$ vector! This has an impact on the measured *SNR* of a single channel, which is given by

$$SNR_{low-IF,I} = \frac{\dfrac{1}{2}\left(\dfrac{\Lambda\alpha}{2}\right)^2}{\displaystyle\int_{-BW}^{BW} p_{N,I-MIXER}\, df} \tag{A.36}$$

$$= \frac{SNR_{low-IF}}{2}. \tag{A.37}$$

So, in contrast to a zero-IF receiver, the measured *SNR* of a single channel in a low-IF receiver is two times *worse* than the *SNR* of the quadrature output! As shown above, the reason is that a single channel measurement in a low-IF receiver (A.36) — by construction — integrates over a two times larger bandwidth than the same measurement in a zero-IF receiver (A.29). On the other hand, the quadrature measurements integrate over identical bandwidths ((A.35) and (A.28)) so that, as far as white noise is concerned, a low-IF and a zero-IF receiver feature exactly the same *SNR*.

Stating the eigen noise factor of a low-IF receiver in terms of the noise of a single channel leads to

$$F_{E,low-IF} = \frac{\frac{1}{2}\int_{-BW}^{BW} p_{N,I-MIXER}\,df}{\frac{1}{2}\left(\frac{A}{2}\right)^2 \int_{LO}^{LO+BW} p_N\,df} \tag{A.38}$$

$$= \frac{1}{2}(F_{SSB} - 2) = \frac{1}{2}F_{E,SSB}. \tag{A.39}$$

where $F_{E,SSB}$ is the SSB eigen noise factor of one of the mixer channels. Inspection of (A.32) and (A.37) clearly reveals that, as far as white noise is concerned, there is no difference between the noise figure values of a zero-IF and a low-IF receiver.

Incorporating the noise contribution of the LNA leads to the expressions for the total noise factor, which are given by

$$F_{zero-IF} = F_{LNA}(1 + S) + \frac{F_{DSB} - 1}{A} \tag{A.40}$$

$$= F_{LNA}(1 + S) + \frac{F_{E,DSB}}{A} \tag{A.41}$$

and

$$F_{low-IF} = F_{LNA}(1 + S) + \frac{1}{2}\frac{F_{SSB} - 2}{A} \tag{A.42}$$

$$= F_{LNA}(1 + S) + \frac{1}{2}\frac{F_{E,SSB}}{A}, \tag{A.43}$$

where F_{DSB} and F_{SSB} are the DSB and SSB noise factor of a single mixer. Comparing these formulae against (A.20) and (A.21) reveals that the noise figure contribution of a quadrature mixer in a I/Q receiver is two times less than the contribution of a single mixer in an IF receiver. Of course, the penalty is a two times larger power consumption!

Because in a low-IF receiver the rejection of the noise in the image band is done by a subtraction in the digital domain, some extra margin is necessary in the A/D converters to keep the system noise figure from degrading by kT/C noise or quantization noise. In practice, the margin that is required to implement the equalizer is generally large enough.

Appendix B

HD_x and IM_x Ratios based on Taylor Expansion of i_{DS}

Equation (4.12) in Section 4.4,

$$i(t) = \frac{c_0(c_1 + v(t))^2}{c_2 + c_3 v(t) + c_4 v^2(t)}, \tag{B.1}$$

can easily be expanded in a Taylor series around $v(t) = 0$,

$$i(t) = a_0 + a_1 v(t) + a_2 v^2(t) + a_3 v^3(t) + O(v^4(t)), \tag{B.2}$$

where the a_i coefficients are given by

$$a_0 = \frac{c_0 c_1^2}{c_2} \tag{B.3}$$

$$a_1 = -\frac{c_0 c_1(-2c_2 + c_1 c_3)}{c_2^2} \tag{B.4}$$

$$a_2 = \frac{c_0(c_2^2 - c_1^2 c_2 c_4 - 2c_1 c_2 c_3 + c_1^2 c_3^2)}{c_2^3} \tag{B.5}$$

$$a_3 = -\frac{c_0(2c_1 c_2^2 c_4 - 2c_1^2 c_2 c_3 c_4 + c_2^2 c_3 - 2c_1 c_2 c_3^2 + c_1^2 c_3^3)}{c_2^4}. \tag{B.6}$$

Applying the substitution

$$v(t) \mapsto U\cos(\omega_1 t) + U\cos(\omega_2 t), \tag{B.7}$$

reveals — after exploiting some trigonometric identities — the amplitudes of the different harmonics and intermodulation products. The distortion and intermodulation ratios can then be expressed in terms of the Taylor coefficients a_i and are in this case given by

$$HD_2[\mathrm{V/V}] = \frac{1}{2}\left|\frac{a_2}{a_1}\right| = \frac{1}{2}\left|\frac{(-c_2+c_1c_3)^2 - c_1^2c_2c_4}{c_1c_2(-2c_2+c_1c_3)}\right| \cdot U \tag{B.8}$$

$$HD_3[\mathrm{V/V}] = \frac{1}{4}\left|\frac{a_3}{a_1}\right| = \frac{1}{4}\left|\frac{(-c_2+c_1c_3)((-c_2+c_1c_3)c_3 - 2c_1c_2c_4)}{c_1c_2^2(-2c_2+c_1c_3)}\right| \cdot U^2 \tag{B.9}$$

$$IM_2[\mathrm{V/V}] = \left|\frac{a_2}{a_1}\right| = 2HD_2[\mathrm{V/V}] = \left|\frac{(-c_2+c_1c_3)^2 - c_1^2c_2c_4}{c_1c_2(-2c_2+c_1c_3)}\right| \cdot U \tag{B.10}$$

$$IM_3[\mathrm{V/V}] = \frac{3}{4}\left|\frac{a_3}{a_1}\right| = 3HD_3[\mathrm{V/V}] = \frac{3}{4}\left|\frac{(-c_2+c_1c_3)((-c_2+c_1c_3)c_3 - 2c_1c_2c_4)}{c_1c_2^2(-2c_2+c_1c_3)}\right| \cdot U^2. \tag{B.11}$$

Appendix C

Essentials of Two-port Noise Theory

Fig. C.1 shows the equivalent noise model of a two-port and its signal source. All internal noise sources have been externalized and are represented by the equivalent input noise sources $\overline{e_n^2}$ and $\overline{i_n^2}$. The noise current associated with the real part of the source admittance is represented by $\overline{i_s^2}$. Generally, the equivalent input noise current $\overline{i_n^2}$ is partly correlated with $\overline{e_n^2}$ and can be written as the sum of a fully correlated part, i_c, and an uncorrelated part, i_u:

$$i_n = i_c + i_u. \tag{C.1}$$

Because of the correlation, i_c and e_n are related by the correlation admittance Y_c:

$$i_c = Y_c e_n. \tag{C.2}$$

The noise factor in terms of $\overline{e_n^2}$, $\overline{i_s^2}$, $\overline{i_u^2}$, Y_c and Y_s is then given by

$$F = 1 + \frac{\overline{i_u^2} + |Y_c + Y_s|^2 \overline{e_n^2}}{\overline{i_s^2}}. \tag{C.3}$$

By defining R_n, G_u and G_s as

$$R_n \triangleq \frac{\overline{e_n^2}}{4kT\Delta f} \qquad G_u \triangleq \frac{\overline{i_u^2}}{4kT\Delta f} \qquad G_s \triangleq \frac{\overline{i_s^2}}{4kT\Delta f}, \tag{C.4}$$

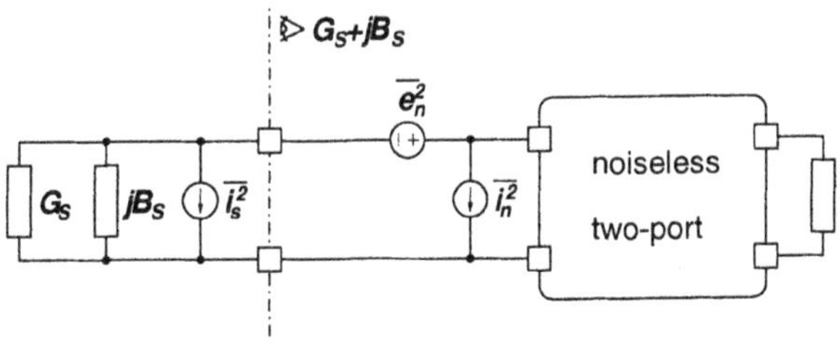

Figure C.1: *Equivalent noise model of a two-port and its signal source.*

(C.3) can also be written as

$$F = 1 + \frac{G_u + |Y_c + Y_s|^2 R_n}{G_s}. \tag{C.5}$$

The value of Y_s that optimizes the noise factor F is denoted by $Y_{opt} = G_{opt} + jB_{opt}$ and is given by

$$G_{opt} = \sqrt{\frac{G_u}{R_n} + G_c^2} \tag{C.6}$$

$$B_{opt} = -B_c. \tag{C.7}$$

The minimum noise factor is then given by

$$F_{min} = 1 + 2R_n(G_{opt} + G_c), \tag{C.8}$$

so that (C.5) can be rewritten as

$$F = F_{min} + \frac{R_n}{G_s}|Y_s - Y_{opt}|^2. \tag{C.9}$$

Bibliography

[Abid86] A.A.Abidi, "High-frequency noise measurements on FETs with small dimensions", *IEEE Transactions on Electron Devices*, Vol.ED-33, pp. 1801–1805, November 1986.

[Abid95] A.A.Abidi, "Direct Conversion Radio Transceivers for Digital Communications", *IEEE Journal of Solid-State Circuits*, Vol. 30, No. 12, pp. 1399–1410, July 1995.

[Aero01] Datasheet of the Aero GSM Transceiver Chipset, *http://www.silabs.com/products/aero.asp*, 2001.

[Anse01] Datasheet of a Third-Order Single-Loop Delta-Sigma ADC, *http://www.ansem.com/ dsResults.htm*, 2001.

[Behb00] F.Behbahani et al., "An Adaptive 2.4GHz Low-IF Receiver in 0.6μm CMOS for Wideband Wireless LAN", *IEEE International Solid-State Circuits Conference, Digest of Technical Papers*, San Francisco, USA, February 2000.

[Banu97] M.Banu et al., "A BiCMOS Double-Low-IF Receiver for GSM", *Proc. of the IEEE Custom Integrated Circuits Conference*, San Diego, USA, pp. 521–524, May 1997.

[Borr01] M.Borremans, "Highly-Integrated CMOS Transmitters for Wireless Communications", Ph.D. Thesis, K.U.Leuven, Leuven(Belgium), to be published in 2001.

[Bruc00] F.Bruccoleri, E.A.M.Klumperink and B.Nauta, "Generating All 2-Transistor Circuits Leads to New Wide-Band CMOS LNAs", *IEEE European Solid-State Circuits Conference, Digest of Technical papers*, Stockholm, Sweden, pp. 288–291, September 2000.

[BSIM3v3] P.K.Ko, C.Hu et al., *BSIM3v3 Manual*, Dept. Electrical Eng. Comp. Sci., Univ. California, Berkeley, 1995.

[Chan91] Z.Y.Chang and W.M.C.Sansen, *Low-Noise Wide-Band Amplifiers in Bipolar and CMOS Technologies*, Kluwer Academic Publishers, 1991.

[Cho99] T.Cho et al., "A Single-Chip CMOS Direct-Conversion Transceiver for 900MHz Spread Spectrum Digital Cordless Phones", *IEEE Solid-State Circuits Conference, Digest of Technical papers*, San Francisco, USA, February 1999.

[Colv99] J.T.Colvin, S.S.Bhatia and Kenneth K.O., "Effects of Substrate Resistances on LNA Performance and a Bondpad Structure for Reducing the Effects in a Silicon Bipolar Technology", *IEEE Journal of Solid-State Circuits*, Vol. 34, No. 9, pp. 1339–1344, September 1999.

[Cran95] J.Craninckx and M.Steyaert, "A CMOS 1.8 GHz Low-Phase-Noise Voltage Controlled Oscillator with Prescaler", *IEEE Solid-State Circuits Conference, Digest of Technical papers*, San Francisco, USA, pp. 266–267, February 1995.

[Cran97] J.Craninckx, "Low-Phase-Noise Fully Integrated CMOS Frequency Synthesizers", Ph.D. Thesis, K.U.Leuven, Leuven(Belgium), December 1997.

[Cran98] J.Craninckx and M.Steyaert, "A Fully-Integrated CMOS DCS-1800 Frequency Synthesizer", *IEEE Solid-State Circuits Conference, Digest of Technical papers*, San Francisco, USA, February 1998.

[Crol95a] J.Crols and M.Steyaert, "A 1.5GHz Highly Linear CMOS Downconversion Mixer", *IEEE Journal of Solid-State Circuits*, Vol. 30, No. 7, pp. 736–742, July 1995.

[Crol95b] J.Crols and M.Steyaert, "A Single-Chip 900MHz CMOS Receiver Front-End with a High Performance Low-IF Topology", *IEEE Journal of Solid-State Circuits*, Vol. 30, No. 12, pp. 1483–1492, December 1995.

[Crol96] J.Crols, P.Kinget, J.Craninckx and M.Steyaert, "An Analytical Model of Planar Inductors on Lowly Doped Silicon Substrates for High Frequency Analog Design up to 3 GHz", *IEEE Symposium on VLSI Circuits, Digest of Technical Papers*, Hawai, USA, pp. 28–29, June 1996.

[Crol98] J.Crols and M.Steyaert, "Low-IF Topologies for High-Performance Analog Front-Ends for Fully Integrated Receivers", *IEEE Transactions on Circuits and Systems II*, Vol. 44, No. 6, pp. 269–282, March 1998.

[Crol97] J.Crols, "Full Integration of Wireless Transceiver Systems", Ph.D. Thesis, K.U.Leuven, Leuven(Belgium), March 1997.

[Demu02] B.De Muer, "CMOS Fractional-n Frequency Synthesizers", Ph.D. Thesis, K.U.Leuven, Leuven(Belgium), to be published in 2002.

[ETSI] *ETS(I) 300 190 (GSM 05.05 version 5.4.1): Digital cellular communication system (Phase 2+); Radio transmission and reception*, European Telecommunications Standards Institute, August 1997.

[FastHen] ftp://rle-vlsi.mit.edu/pub/fasthenry/

[Floy99] B.A.Floyd, J.Mehta, C.Gamaro and Kenneth K.O., "A 900MHz, 0.8μm CMOS Low Noise Amplifier with 1.2dB Noise Figure", *Proc. of the IEEE Custom Integrated Circuits Conference*, Santa Clara, USA, pp. 661–664, May 1999.

[Fox93] R.M.Fox, "Comments on Circuit Models for MOSFET Thermal Noise", *IEEE Journal of Solid-State Circuits*, Vol. 28, No. 2, p. 184, February 1993.

[Fren01] L.E.Frenzel, "Transceiver Chip Set Wrings Out GSM Phone Costs", *Electronic Design*, Vol. 49, No. 5, March 5 2001.

[Geer98] Y.Geerts, A.Marques, M.Steyaert and W.Sansen, "A 3.3 V 15-bit Delta-Sigma ADC with a Signal Bandwidth of 1.1 MHz for ADSL-applications", *IEEE European Solid-State Circuits Conference, Digest of Technical papers*, The Hague, The Netherlands, pp. 168–171, September 1998.

[Geor01] N.George and D.Roberts, "A wake-up call for mobile investors: Market saturation is threatening telephone makers' plans to raise more cash for expansion, says Dan Roberts", article in *The Financial Times*, January 11 2001.

[Giel00] G.Gielen and R.Rutenbar, "Computer-aided design of analog and mixed-signal integrated circuits", *Proceedings of the IEEE*, Vol. 88, pp. 1825–1852, December 2000.

[Gier99] S.L.J.Gierkink, E.A.M.Klumperink, A.P.van der Wel, G.Hoogzaad, E.A.J.M. van Tuijl and B.Nauta, "Intrinsic 1/f Device Noise Reduction and Its Effect on Phase Noise in CMOS Ring Oscillators", *IEEE Journal of Solid-State Circuits*, Vol. 34, No. 7, p. 1022–1025, July 1999.

[GPS96] *Global Positioning System: Theory and Applications, Volume I (i.e. Progress in Aeronautics and Astronautics, Vol. 163)*, edited by B.W.Parkinson, J.J.Spilker Jr., P.Axelrad and P.Enge, American Institute of Aeronautics and Astronautics, Washington, 1996.

[Gram00a] G.Gramegna, A.Magazzu, C.Sclafani and M.Paparo, "Ultra-wide Dynamic Range 1.75dB Noise Figure, 900MHz CMOS LNA", *IEEE Solid-State Circuits Conference, Digest of Technical papers*, San Francisco, USA, pp. 380–381, February 2000.

[Gram00b] G.Gramegna, A.Magazzu, C.Sclafani, M.Paparo and P. Erratico, "A 9mW, 900MHz CMOS LNA with 1.05dB Noise Figure", *Proc. of the IEEE European Solid-State Circuits Conference*, Stockholm, Sweden, September 2000.

[GSMW] *http://www.gsmworld.com/membership/mem_stats.html*, web-site of the GSM Association.

[Haus60] H.A.Haus et al., "Representation of noise in linear two-ports", *Proc. IRE*, Vol. 48, pp. 69–74, January 1960.

[Haya98] G.Hayashi, H.Kimura, H.Simomura and A.Matsuzawa, "A 9mW 900MHz CMOS LNA with Mesh Arrayed MOSFETs", *IEEE Symposium on VLSI Circuits, Digest of Technical Papers*, Honolulu, Hawai, pp. 84–85, June 1998.

[Heij99] M.van Heijningen, J.Compiet, P.Wambacq, S.Donnay, M.Engels and I.Bolsens, "Modeling of Digital Substrate Noise Generation and Experimental Verification Using A Novel Substrate Noise Sensor", *Proc. of the IEEE European Solid-State Circuits Conference*, Duisburg, Germany, pp. 186–189, September 1999.

[HSpice] *Avant! Star-Hspice Manual*, Release 1998.2, July 1998.

[Huan98a] Q.Huang, P.Orsatti and F.Piazza, "Broadband, 0.25μm CMOS LNAs with sub-2dB NF for GSM Applications", *Proc. of the IEEE Custom Integrated Circuits Conference*, Santa Clara, USA, pp. 67–70, May 1998.

[Huan98b] Q.Huang, F.Piazza, P.Orsatti and T.Ohguro, "The Impact of Scaling Down to Deep Submicron on CMOS RF Circuits", *IEEE Journal of Solid-State Circuits*, Vol. 33, No. 7, pp. 1023–1036, July 1998.

[Huan98c] F.Piazza, P.Orsatti and Q.Huang, "A 0.25μm CMOS Transceiver Front-End for GSM", *Proc. of the IEEE Custom Integrated Circuits Conference*, Santa Clara, USA, pp. 413–416, May 1998.

[Hull93] C.D.Hull and R.G.Meyer, "A Systematic Approach to the Analysis of Noise in Mixers", *IEEE Transactions on Circuits and Systems*, Vol. 40, No. 12, pp. 909–919, December 1993.

[Hull96] C.D.Hull, J.L.Tham and R.C.Chu, "A Direct-Conversion Receiver for 900MHz (ISM Band) Spread-Spectrum Digital Cordless Telephone", *IEEE Journal of Solid-State Circuits*, Vol. 31, No. 12, pp. 1955–1963, December 1996.

[Jans97] J.Janssens, M.Steyaert and H.Miyakawa, "A 2.7 Volt CMOS Broadband Low Noise Amplifier", *IEEE Symposium on VLSI Circuits, Digest of Technical Papers*, Kyoto, Japan, pp. 87–88, June 1997.

[Jans98a] J.Janssens and M.Steyaert, "Design of Broadband Low-Noise Amplifiers in Deep-Submicron CMOS Technologies", in *Analog circuit design. 1 Volt electronics, mixed-mode systems, low-noise and RF power amplifiers for telecommunication*, edited by J.Huijsing, R.van de Plassche and W.Sansen, Kluwer Academic Publishers, Boston, 1999, pp. 317–335.

[Jans98b] J.Janssens, J.Crols and M.Steyaert, "A 10 mW Inductorless, Broadband Low Noise Amplifier for 900 MHz Wireless Communications", *Proc. of the IEEE Custom Integrated Circuits Conference*, Santa Clara, USA, pp. 75–78, May 1998.

[Jans98c] J.Janssens, M.Steyaert and T.Ohguro, "A 0.25 μm CMOS I/Q-channel Downconversion Mixer with Active Coil for DCS–1800 Applications", *Proc. of the IEEE European Solid-State Circuits Conference*, The Hague, The Netherlands, pp. 56–59, September 1998.

[Jans99a] J.Janssens and M.Steyaert, "Optimum MOS Power Matching by Exploiting Non-Quasi Static Effect", *IEE Electronics Letters*, Vol. 35, No. 8, pp. 672–673, April 15 1999.

[Jans99b] J.Janssens and M.Steyaert, "MOS Noise Performance under Impedance Matching Constraints", *IEE Electronics Letters*, Vol. 35, No. 15, pp. 1278–1280, July 22 1999.

[Jind84] R.P.Jindal, "Noise associated with distributed resistance of MOSFET gate structures in integrated circuits", *IEEE Transactions on Electron Devices*, Vol.ED-31, pp. 1505–1509, October 1984.

[Jind85] R.P.Jindal, "Noise associated with substrate current in fine line NMOS field effect transistors", *IEEE Transactions on Electron Devices*, Vol.ED-32, No. 6, pp. 1047–1052, June 1985.

[Jind86] R.P.Jindal, "Hot electron effects on channel thermal noise in fine line NMOS field effect transistors", *IEEE Transactions on Electron Devices*, Vol.ED-32, No. 9, pp. 1395–1397, September 1986.

[Kamo94] M.Kamon, M.J.Tsuk and J.K.White, "FASTHENRY: A Multipole-Acclerated 3-D Inductance Extraction Program", *Transactions on Microwave Theory and Techniques*, September 1994.

[Kara96] A.N.Karanicolas, "A 2.7V 900MHz CMOS LNA and Mixer", *IEEE Journal of Solid-State Circuits*, Vol. 31, No. 12, pp. 1939–1944, December 1996.

[Kish99] S.V.Kishore, G.Chang, G.Asmanis, C.Hull and F.Stubbe, "Substrate-Induced High-Frequency Noise in Deep Submicron MOSFETs for RF Applications", *Proc. of the IEEE Custom Integrated Circuits Conference*, San Diego, USA, pp. 365–368, May 1999.

[Koul93] I.A.Koullias, J.H.Havens, I.G.Post and P.E.Bronner, "A 900MHz Transceiver Chip Set for Dual-Mode Cellular Radio Mobile Terminals", *IEEE International Solid-State Circuits Conference, Digest of Technical Papers*, San Francisco, USA, pp. 140-141, February 1993.

[KS] Kawasaki Steel

[L&S94] K.R.Laker and W.M.C.Sansen, *Design of Analog Integrated Circuits and Systems*, MCGraw-Hill, New York, 1994.

[Lee98] T.H.Lee, *The Design of CMOS Radio-Frequency Integrated Circuits*, Cambridge University Press, 1998.

[Lee99] J.P.Z.Lee, F.Wang, A.Phanse and L.C.Smith, "Substrate Cross Talk Noise Characterization and Prevention in 0.35 μm CMOS Technology", *Proc. of the IEEE Custom Integrated Circuits Conference*, San Diego, USA, pp. 479–482, May 1999.

[Leen00] D.Leenaerts and P.de Vreede, "Influence of substrate noise on RF performance", *Proc. of the IEEE European Solid-State Circuits Conference*, Stockholm, Sweden, pp. 300–303, September 2000.

[LeJa01] P.Leroux, J.Janssens and M.Steyaert, "A 0.8 dB NF, ESD-protected, 9 mW CMOS LNA", *IEEE International Solid-State Circuits Conference, Digest of Technical Papers*, San Francisco, USA, February 2001.

[Lero99] P.Leroux, J.Janssens and M.Steyaert, "Design of a 0.25μm CMOS Low Noise Amplifier with a Local Match towards a Lower Impedance", *Master's Thesis KULeuven*, Dept.Electrical engineering, Leuven, May 2000.

[LiPu00] G.Li Puma et al., "A RF Transceiver for Digital Wireless Communication in a 25GHz Si Bipolar Technology", *IEEE International Solid-State Circuits Conference, Digest of Technical Papers*, San Francisco, USA, February 2000.

[Long95] J.R.Long and M.A.Copeland, "A 1.9GHz Low-Voltage Silicon Bipolar Receiver Front-End for Wireless Personal Communications Systems", *IEEE Journal of Solid-State Circuits*, Vol. 30, No. 12, pp. 1438–1448, December 1995.

[Mago00] R.Magoon et al., "A 900MHz/1.9GHz Integrated Transceiver and Synthesizer IC for GSM", *IEEE European Solid-State Circuits Conference, Digest of Technical papers*, Stockholm, Sweden, pp. 92–95, September 2000.

[Mank98] T.Manku, "Microwave CMOS - Devices and Circuits", *Proc. of the IEEE Custom Integrated Circuits Conference*, Santa Clara, USA, pp. 59–66, May 1998.

[Mank99] T.Manku, M.Obrecht and Y.Lin, "RF Simulations ans Physics of the Channel Noise Parameters within MOS Transistors", *Proc. of the IEEE Custom Integrated Circuits Conference*, San Diego, USA, pp. 369–372, May 1999.

[Marq98] A.Marques, V.Peluso, M.Steyaert and W.Sansen, "A 15-b Resolution 2-MHz Nyquist Rate Delta-Sigma ADC in a 1 μm CMOS Technology", *IEEE Journal of Solid-State Circuits*, Vol. 33, No. 7, pp. 1065–1075, July 1998.

[Mars95] C.Marshall et al., "2.7V GSM Transceiver ICs with On-Chip Filtering", *IEEE International Solid-State Circuits Conference, Digest of Technical Papers*, San Francisco, USA, pp. 148–149, February 1995.

[McDo93] M.D.McDonald, "A 2.5GHz BiCMOS Image-Reject Front-End", *IEEE International Solid-State Circuits Conference, Digest of Technical Papers*, San Francisco, USA, pp. 144-145, February 1993.

[Mert00] K.Mertens, M.Steyaert and B.Nauwelaers, "A 700 MHz, 1 W fully differential Class E power amplifier in CMOS", *IEEE European Solid-State Circuits Conference, Digest of Technical papers*, Stockholm, Sweden, pp. 104–107, September 2000.

[Mert03] K.Mertens, "CMOS Power Amplifiers for Wireless Communications", Ph.D. Thesis, K.U.Leuven, Leuven(Belgium), to be published in 2003.

[Meye94] R.G.Meyer and W.D.Mack, "A 1GHz BiCMOS RF Front-End IC", *IEEE Journal of Solid-State Circuits*, Vol. 29, No. 3, pp. 350–355, March 1994.

[Meye97] R.G.Meyer, W.D.Mack and J.J.E.M.Hageraats, "A 2.5GHz BiCMOS Transceiver for Wireless LAN's", *IEEE Journal of Solid-State Circuits*, Vol. 32, No. 12, pp. 2097–2104, December 1997.

[MOSCAL] P. Vancorenland, "MOSCAL V1.6 tutorial", *http://www.esat.kuleuven.ac.be/vanco/moscal/*.

[MRFIC98] Datasheet of the MRFIC1501 GPS GaAs Low Noise Amplifier, *Motorola Semiconductor Technical Data*, 1998.

[Muro81] K.Murota and H.Hirade, "GMSK Modulation for Digital Mobile Radio Telephony", *IEEE Transactions on Communications*, Vol. 29, No. 7, pp. 1044–1050, July 1981.

[Muro85] K.Murota and H.Hirade, "Spectrum Efficiency of GMSK Land Mobile Radio", *IEEE Transactions on Vehicular Technology*, Vol. 34, No. 2, pp. 69–75, May 1985.

[OptE00] F.Op 't Eynde, J.Craninckx and P.Goetschalckx, "A Fully-Integrated Zero-IF DECT Transceiver", *IEEE International Solid-State Circuits Conference, Digest of Technical Papers*, San Francisco, USA, pp. 140–141, February 2000.

[Orsa99] P.Orsatti, F.Piazza, Q.Huang and T.Morimoto, "A 20mA-Receive, 55mA-Transmit GSM Transceiver in 0.25μm CMOS", *IEEE Solid-State Circuits Conference, Digest of Technical papers*, San Francisco, USA, February 1999.

[Pelg89] M.J.M.Pelgrom, A.C.J.Duinmaijer and A.P.G.Welbers "Matching Properties of MOS Transistors", *IEEE Journal of Solid-State Circuits*, Vol. 24, No. 5, pp. 1433–1440, October 1989.

[Phil00] J.Phillips and K.Kundert, "Noise in Mixers, Oscillators, Samplers and Logic; An Introduction to Cyclostationary Noise", *Proc. of the IEEE Custom Integrated Circuits Conference*, San Diego, USA, pp. 431–438, May 2000.

[Piaz98] F.Piazza, P.Orsatti and Q.Huang, "A 0.25μm CMOS Transceiver Front-End for GSM", *Proc. of the IEEE Custom Integrated Circuits Conference*, Santa Clara, USA, pp. 413–416, May 1998.

[Raza94] B.Razavi, T.H. Yan and K.F.Lee, "Impact of distributed gate resistance on the performance of MOS devices", *IEEE Transactions on Circuits and Systems I*, Vol. 41, pp. 750–754, November 1994.

[Raza97] B.Razavi, "A 900MHz CMOS Direct Conversion Receiver", *IEEE Symposium on VLSI Circuits, Digest of Technical Papers*, Kyoto, Japan, June 1997.

[Raza97b] B.Razavi, "Design Considerations for Direct Conversion Receivers", *IEEE Transactions on Circuits and Systems II*, Vol. 44, No. 6, pp. 428–453, June 1997.

[Rofo96] A.Rofougaran, J.Y.C.Chang, M.Rofougaran and A.A.Abidi, "A 1GHz CMOS RF Front-End IC for a Direct Conversion Wireless Receiver", *IEEE Journal of Solid-State Circuits*, Vol. 31, No. 7, pp. 880–889, July 1996.

[Rofo98a] A.Rofougaran et al., "A Single-Chip 900MHz Spread-Spectrum Wireless Transceiver in 1μm CMOS — part I: Architecture and Transmitter Design", *IEEE Journal of Solid-State Circuits*, Vol. 33, No. 4, pp. 515–534, April 1998.

[Rofo98b] A.Rofougaran et al., "A Single-Chip 900MHz Spread-Spectrum Wireless Transceiver in 1μm CMOS — part II: Receiver Design", *IEEE Journal of Solid-State Circuits*, Vol. 33, No. 4, pp. 555–547, April 1998.

[Rude97] J.C.Rudell et al., "A 1.9GHz Wide-Band IF Double Conversion CMOS Receiver for Cordless Telephone Applications", *IEEE Journal of Solid-State Circuits*, Vol. 32, No. 12, pp. 2071–2088, December 1997.

[Sans99] W.Sansen, "Distortion in elementary transistor circuits", *IEEE Transactions on Circuits And Systems*, Vol. 46, pp. 315–325, March 1999.

[Seve94] J.Sevenhans et al., "An Analog Radio Front-End Chip Set for a 1.9GHz Mobile Radio Telephone Application", *IEEE International Solid-State Circuits Conference, Digest of Technical Papers*, San Francisco, USA, pp. 44–45, February 1994.

[Shae97] D.K.Shaeffer and T.H.Lee, "A 1.5V, 1.5GHz CMOS Low Noise Amplifier", *IEEE Journal of Solid-State Circuits*, Vol. 32, No. 5, pp. 745–759, May 1997.

[Shah97] A.R.Shahani, D.K.Shaeffer and T.H.Lee, "A 12-mW Wide Dynamic Range CMOS Front-End for a Portable GPS Receiver", *IEEE Journal of Solid-State Circuits*, Vol. 32, No. 12, pp. 2061–2070, December 1997.

[Stet95] T.D.Stetzler, I.G.Post, J.H.Havens and M.Koyama, "A 2.7–4.5V single chip GSM transceiver RF integrated circuit", *IEEE Journal of Solid-State Circuits*, Vol. 30, No. 12, pp. 1421–1429, December 1995.

[Stey93] M.Steyaert and W.Sansen "Opamp Design towards Maximum Gain-Bandwidth", in *Analog circuit design*, edited by W.Sansen, J.Huijsing and R.van de Plassche, Kluwer Academic Publishers, New York, Vol. 3, 1996, pp. 149–166.

[Stey96a] M.Steyaert, M.Borremans, J.Craninckx, J.Crols, J.Janssens and P.Kinget, "RF CMOS Design: Some Untold Pitfalls", in *Analog circuit design. MOST RF circuits, sigma-delta converters and translinear circuits*, edited by W.Sansen, J.Huijsing and R.van de Plassche, Kluwer Academic Publishers, Boston, 1996, pp. 63–88.

[Stey96b] M.Steyaert, M.Borremans, J.Craninckx, J.Crols, J.Janssens and P.Kinget, "RF Integrated Circuits in Standard CMOS Technologies", *Proc. of the IEEE European Solid-State Circuits Conference*, Neuchatel, Switzerland, pp. 11–18, September 1996.

[Stey98] M.Steyaert, M.Borremans, J.Janssens, B.De Muer, N.Itoh, J.Craninckx, J.Crols, E.Morifuji, H.S.Momose and W.Sansen, "A Single-chip CMOS Transceiver for DCS–1800 Wireless Communications", *IEEE Solid-State Circuits Conference, Digest of Technical papers*, San Francisco, USA, pp. 48–49, February 1998.

[Stey00a] M.Steyaert, J.Janssens, B.De Muer, M.Borremans and N.Itoh, "A 2 Volt CMOS Cellular Transceiver Front-End", *IEEE International Solid-State Circuits Conference, Digest of Technical Papers*, San Francisco, USA, pp. 142–143, February 2000.

[Stey00b] M.Steyaert, M.Borremans, J.Janssens, B.De Muer, N.Itoh, J.Craninckx, J.Crols, E.Morifuji, H.S.Momose and W.Sansen, "A Single-chip CMOS Transceiver Front-End for DCS–1800 Wireless Communications", *Analog Integrated Circuits and Signal Processing*, Vol. 24, pp. 83–99, September 2000.

[Stey00c] M.Steyaert, J.Janssens, B.De Muer, M.Borremans and N.Itoh, "A 2 Volt CMOS Cellular Transceiver Front-End", *IEEE Journal of Solid-State Circuits*, Vol. 35, No. 12, pp. 1895–1907, December 2000.

[Stub98] F.Stubbe, S.V.Kishore, C.Hull and V.Della Torre, "A CMOS RF-Receiver Front-End for 1 GHz Applications", *IEEE Symposium on VLSI Circuits, Digest of Technical Papers*, Honolulu, Hawai, June 1998.

[Taka95] C.Takahashi et al., "A 1.9GHz Si Direct Conversion Receiver IC for QPSK Modulation Systems", *IEEE International Solid-State Circuits Conference, Digest of Technical Papers*, San Francisco, USA, pp. 138–139, February 1995.

[Taki99] K.Takikawa et al., "RF Circuits Teqhnique of Dual-band Transceiver IC for GSM and DCS-1800 Applications", *Proc. of the IEEE European Solid-State Circuits Conference*, pp. 278–281, September 1999.

[Terr99] M.T.Terrovitis and R.G.Meyer, "Noise in Current-Commutating CMOS Mixers", *IEEE Journal of Solid-State Circuits*, Vol. 34, No. 6, pp. 772–783, June 1999.

[Tsiv87] Y.P.Tsividis, *Operation and Modeling of the MOS transistor*, New York: McGraw-Hill, 1987.

[vdZi86] A. van der Ziel, *Noise in Solid State Devices and Circuits*, New York: Wiley, 1986.

[Wang94] B.Wang, J.R.Hellums and C.G.Sodini, "MOSFET Thermal Noise Modeling for Analog Integrated Circuits", *IEEE Journal of Solid-State Circuits*, Vol. 29, No. 7, pp. 833–835, July 1994.

[Wu98] S.Wu and B.Razavi, "A 900MHz/1.8GHz CMOS receiver for Dual-Band Applications", *IEEE Journal of Solid-State Circuits*, Vol. 33, No. 12, pp. 2178–2185, December 1998.

[Xu00] M.Xu, D.K.Su, D.K.shaeffer, T.H.Lee and B.A.Wooley, "Measuring and Modeling the Effects of Substrate Noise on the LNA for a CMOS GPS eceiver", *Proc. of the IEEE Custom Integrated Circuits Conference*, San Diego, USA, pp. 353–356, May 2000.

[Yama97] T.Yamawaki et al., "A 2.7V GSM RF Transceiver IC", *IEEE Journal of Solid-State Circuits*, Vol. 32, No. 12, pp. 2089–2096, December 1997.

[Yama98] T.Yamawaki et al., "A Dual-band Transceiver for GSM and DCS-1800 Applications", *Proc. of the IEEE European Solid-State Circuits Conference*, The Hague, The Netherlands, pp. 84–87, September 1998.

[Yee00] D.Yee et al., "A 2GHz Low-Power Single-CHip CMOS Receiver for WCDMA Applications", *IEEE European Solid-State Circuits Conference, Digest of Technical papers*, Stockholm, Sweden, pp. 96–99, September 2000.

[Yoo00] C.Yoo and Q.Huang, "A Common-Gate switched, 0.9 W Class E power amplifier with 41% PAE in 0.25μm CMOS" *IEEE Symposium on VLSI Circuits, Digest of Technical Papers*, Honolulu, Hawai, pp. 56–57, June 2000.

[Yue98] C.P.Yue and S.S.Wong, "On-Chip Spiral Inductors with Patterned Ground Shields for Si-based RF IC's", *IEEE Journal of Solid-State Circuits*, Vol. 33, No. 5, pp. 743–752, May 1998.

[Zhou98] J.Zhou and D.J.Allstot, "Monolithic Transformers and Their Application in a Differential CMOS RF Low-Noise Amplifier", *IEEE Journal of Solid-State Circuits*, Vol. 33, No. 12, pp. 2020–2027, December 1998.